ROBERT
SHEETS

Intermediate Algebra

Second Edition

Marshall Fraser

Page-Ficklin Mathematics Series

Burgess Publishing Company
Minneapolis, Minnesota

Printed in the United States of America.

Library of Congress Catalog Card Number 79-83682

ISBN: 0-8087-0666-7

Marketed by Burgess Publishing Company

Preface

The objectives of this second edition of Intermediate Algebra are the same as those of the first edition: This book provides material for a course in intermediate algebra at the college level. The approach is intuitive rather than abstract or formal. Although the style is informal, the mathematics is accurate. There is a dual emphasis on expressing concepts verbally and in mathematical notation.

Some special features of this second edition are:

- Each section is organized to provide material for one class lesson.
- Exercises are graded according to difficulty.
- Applications or "word problems" are collected in the last section of each chapter.
- Each chapter contains a large set of review exercises.
- When appropriate, optional examples and exercises relating the material to use with an electronic calculator are included. These are set out in the text by a special typeface.
- The first chapter reviews topics from arithmetic and elementary algebra.
- Each mathematical concept is presented in a skill-oriented context.

The author was aided in the preparation of this edition by a great many teachers, users of the first edition, who responded to the publisher's questionnaire (many anonymously). Answers to some specific questions were forthcoming and appreciated.

A special word of thanks is due Professor Jule Hansen of the University of Texas, El Paso, who provided valuable suggestions based on her experience teaching from the first edition, and whose comments on the manuscript of this edition were valuable.

Marshall Fraser
San Francisco, California
December 1978

Contents

chapter 1

Review

1. Real Numbers

2. Operations

3. Order of Operations

4. Exponents

In this chapter we review some topics from elementary algebra. In Section 1.1 we review real numbers and graphing on the real number line. We review the operations of addition, subtraction, multiplication, and division in Section 1.2. We review the order used in performing operations and substituting numbers for a variable in Section 1.3. In Section 1.4 we review the definition of exponents and some properties of exponents.

section 1 • Real Numbers

In this section we review the system of real numbers. We define real numbers, construct the real number line, and graph inequalities on the real number line.

SYSTEM OF REAL NUMBERS

We use number systems to deal with situations that arise in everyday life and in the sciences. Many of these situations deal with counting or with measurement. The number system that is most useful for these purposes is the system of real numbers. We will use the following definition of a real number.

DEFINITION OF A REAL NUMBER. A *real number* is any number that possesses a decimal representation.

Some further remarks are necessary to clarify this definition.

DEFINITION OF A NATURAL NUMBER. A *natural number* is one of the numbers

$$1, 2, 3, 4, 5, \ldots .$$

(The three dots are read "dot, dot, dot" or "and so on." They imply that the reader understands how to continue the pattern.)

The natural numbers are the numbers we use for counting. Every natural number is a real number because it possesses a decimal representation. For example, 2.0 is a decimal representation for 2.

DEFINITION OF AN INTEGER. An *integer* is one of the numbers

$$\ldots, -4, -3, -2, -1, 0, 1, 2, 3, 4, \ldots .$$

Every integer is a real number. Thus the system of real numbers contains both positive and negative numbers. The set of positive integers is identical with the set of natural numbers.

DEFINITION OF A RATIONAL NUMBER. A *rational number* is any number that can be written in the form of a fraction a/b where a and b are integers and $b \neq 0$.

Every rational number is a real number. For example, $1/2$ is a rational number because it is a fraction in which the numerator and denominator are integers. A decimal representation for $1/2$ is 0.5.

DEFINITION OF AN IRRATIONAL NUMBER. An *irrational number* is any real number which is not a rational number.

Some examples of irrational numbers we will encounter are $\sqrt{2}$, $\sqrt{3}$, and π (the Greek letter pi), which is the ratio of the circumference of a circle to its diameter.

We mention some things which are not real numbers.

1. The symbol $a/0$ does *not* denote a real number, no matter what a denotes. That is, the denominator of a fraction cannot be 0.
2. Infinity, sometimes denoted by ∞, is not a real number.
3. Imaginary numbers, such as $\sqrt{-1}$, the square root of -1, are not real numbers. A number system that includes both real and imaginary numbers is the system of *complex numbers,* which we will study in Chapter 3.

Note: A real number can possess more than one decimal representation. For example, 2.0, 2.00, 2.000, and so on are all decimal representations for 2.

Some decimal representations are *terminating;* others are *nonterminating.*

Example 1. Terminating and nonterminating decimal representations.

a. A decimal representation for 1/2 is 0.5, which is terminating.
b. A decimal representation for 1/3 is $0.3333 \cdots$, which is nonterminating; the 3's in this decimal repeat indefinitely.
c. A decimal representation for 1/8 is 0.125, which is terminating.
d. A decimal representation for 2/11 is $0.181818 \cdots$, which is nonterminating; the 18's in this decimal repeat indefinitely.

If a decimal representation of a real number is nonterminating, then for practical applications, we round off the number to a desired degree of accuracy.

Note: Rounding off and accuracy are discussed in Appendix B.

Example 2. Every decimal representation of the number π is nonterminating. An approximation to π which is accurate to two decimal places is 3.14.

Exercise Set 1.1

In exercises 1–12, find a decimal representation for the given rational number and specify whether it is terminating or nonterminating.

1. 1/4
2. 1/5
3. 3/4
4. 3/5
5. 2/3
6. 3/11
7. 5/6
8. 2/9
9. 3/10
10. 3/20
11. 1/12
12. 1/7

THE REAL NUMBER LINE

The real number system is the number system we are most frequently concerned with in this book, so we want to have an intuitive way to think about real numbers. The *real number line* provides such a way.

To construct the real number line, we start with a straight line that extends indefinitely in both directions. The idea is to make real numbers *correspond* with points on the line making this correspondence in such a way that just one number corresponds with each point and vise versa. We start the correspondence by picking a point to correspond to the number 0 and a point to correspond to the number 1. We can pick any points to correspond to 0 and 1, so long as they're not the same point. It has become a convention that, if we draw the line horizontally, we choose 1 to the right of 0; and if we draw the line vertically, we choose 1 above 0. We shall always follow this convention in this book. Also, the distance between these two points is the unit distance; so we may wish to choose it conveniently, say an inch or centimeter. See Figure 1.1.

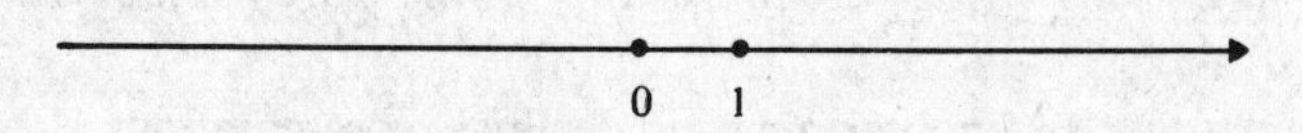

Figure 1.1

Once we've chosen points corresponding to 0 and 1, we think of marking off the unit distance on the line infinitely many times. Of course, in reality we can mark it off (say with a ruler) only a finite number of times. If the line is horizontal, we start with 1 and marking off the unit distance to the right, obtaining the points corresponding to 2, 3, 4, 5, . . . ; to the left we start with 0 and obtain the points corresponding to $-1, -2, -3, -4, \ldots$. In this way, we have chosen points corresponding to all the integers—positive integers to the right of 0, negative integers to the left. See Figure 1.2.

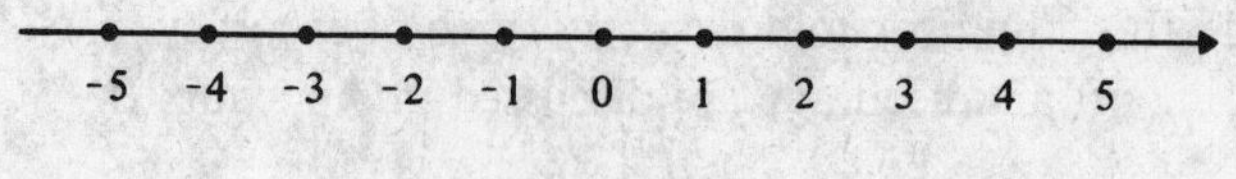

Figure 1.2

There is now a natural way to find points that correspond to fractions. For example, to find the point corresponding to the fraction $\frac{1}{2}$, we find the midpoint of the segment between 0 and 1. To find the point corresponding to $-2\frac{1}{2}$, we find the midpoint of the segment between -3 and -2. To find the point corresponding to $1\frac{5}{8}$, we divide the segment between 1 and 2 into eight equal parts and choose the fifth point; this is the same as taking one-eighth the unit distance and marking it off from 1 five times. These examples describe a method for finding points to correspond to all fractions of integers, that is, all rational numbers. See Figure 1.3.

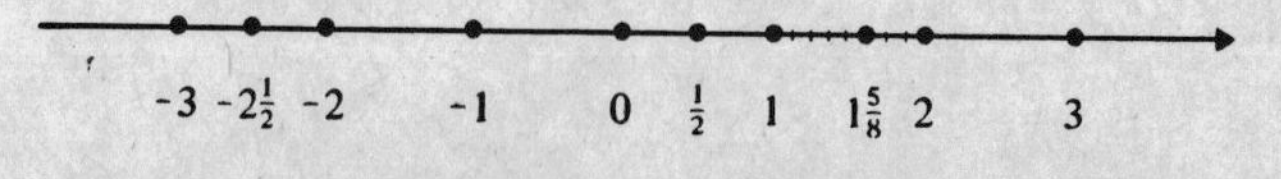

Figure 1.3

To find a point corresponding to *any* real number, we must have points corresponding to all decimals, terminating and nonterminating. For example, to find a point corresponding to π we can find points corresponding to each decimal in the following sequence:

$$3.1;\ 3.14;\ 3.141;\ 3.1415;\ 3.14159;\ \ldots .$$

Each decimal in this sequence is a rational number approximation to π. As the approximations get closer and closer to π, the corresponding points get closer and closer to a particular point which corresponds to π. (This description is theoretical; we would not carry out so many details in practice.)

Thus there is a correspondence between real numbers and points on a line. This is a one-to-one correspondence—for every real number, there corresponds just one point on the line; and conversely, for every point on the line, there corresponds just one real number.

The line together with its correspondence with real numbers is called the *real number line*. The point on the line corresponding to a real number is called the *graph* of that number. Conversely, the real number corresponding to a point on the line is called the *coordinate* of that point.

The *positive direction* for the line is from left to right. The graphs of positive real numbers are to the right of 0, and the graphs of negative real numbers are to the left of 0. The positive direction on the real number line is usually indicated by an arrow, as in Figure 1.4.

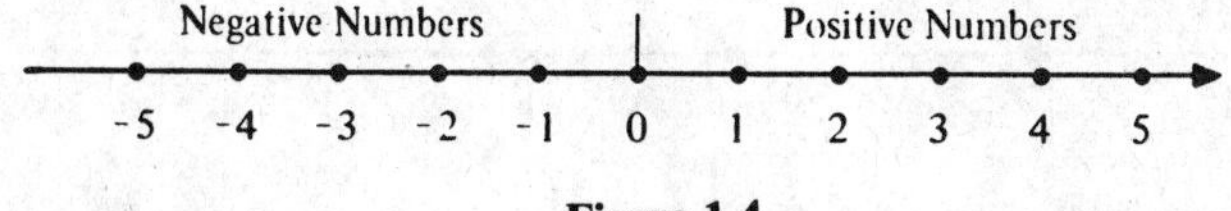

Figure 1.4

Exercise Set 1.1 (continued)

13. Sketch the graphs of the following real numbers on a real number line: $-6, \frac{3}{2}, 4\frac{1}{4}, -\frac{3}{2}, -\frac{1}{4}$.
14. Sketch the graphs of the following real numbers on a real number line: $3, -3, \frac{4}{3}, -\frac{4}{3}, 1\frac{3}{4}, -1\frac{3}{4}$.

GRAPHING INEQUALITIES

Recall the standard notation for equations and inequalities:

NOTATION FOR EQUATIONS AND INEQUALITIES

$a = b$ is read "a equals b"

$a \neq b$ is read "a is *not* equal to b."

$a > b$ is read "a is greater than b"

$a < b$ is read "a is less than b."

$a \geq b$ means that $a > b$ *or* $a = b$
and is read "a is greater than or equal to b."

$a \leq b$ means that $a < b$ *or* $a = b$
and is read "a is less than or equal to b."

On the real number line, the concepts greater than and less than express right and left. The number a is greater than b if a lies to the right of b, and a is less than b if a lies to the left of b (Figure 1.5).

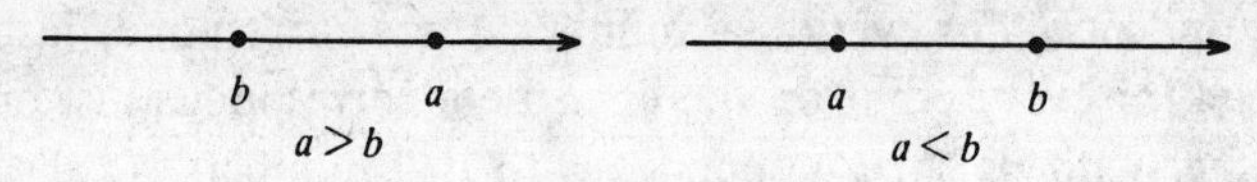

Figure 1.5

We can also express the fact that a real number lies between two numbers. For example, if a real number x is between 1 and 3, we can write $1 < x < 3$. This double inequality states that 1 is less than x *and* x is less than 3; in other words, x is between 1 and 3. We haven't specified what x actually is, except that x is a real number in a certain interval on the real line. See Figure 1.6.

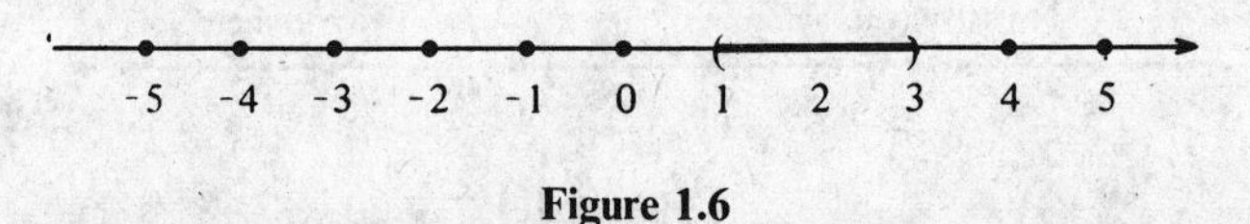

Figure 1.6

If we wish to draw the graph of an *interval* on the real line, say the points between a and b, we draw it as in Figure 1.7. The graph of a real number x is in this interval if $a < x < b$.

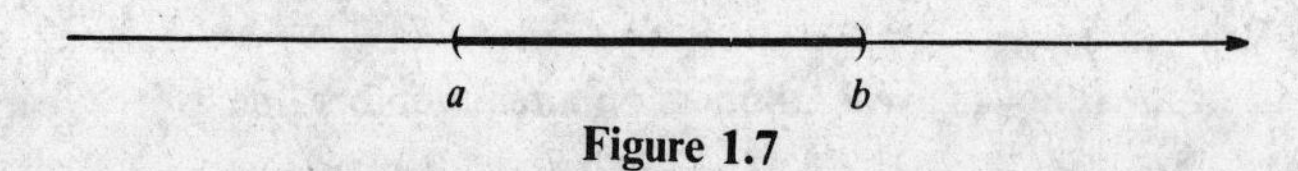

Figure 1.7

The interval corresponding to $a < x < b$ does not include its endpoints a and b. What if we wish to include the two endpoints? Then we draw the picture as in Figure 1.8, and we write $a \leq x \leq b$, which states that x is greater than or equal to a *and* less than or equal to b; that is, x is between a and b and may be either a or b.

Figure 1.8

In Figure 1.8, we examine the points outside the interval from a to b, not including the end points. We use two inequalities to express this: $x < a$ or $x > b$. Here we have represented the numbers that are either less than a or greater than b. Since we have two separate intervals on the real number line, it's necessary to use two separate inequalities; we cannot combine them.

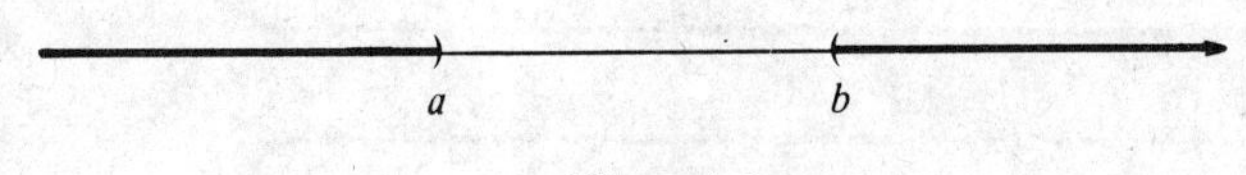

Figure 1.9

Figure 1.10 shows the points outside of a and b, but now we include the endpoints a and b. We express this by the two inequalities $x \leq a$ or $x \geq b$.

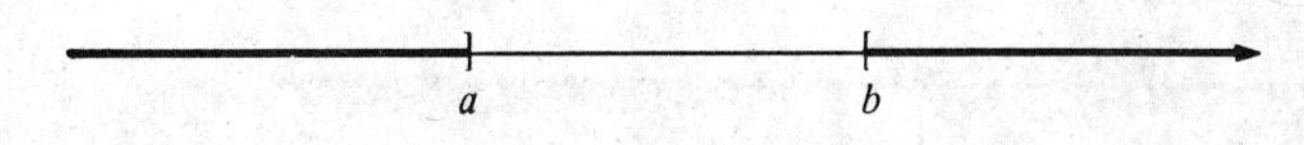

Figure 1.10

Example 3. Express the given interval in terms of inequalities.

a. -3 -2 -1 0 1 2 3 $\quad -1 < x \leq 2$

b. -3 -2 -1 0 1 2 3 $\quad x < 1$

c. -3 -2 -1 0 1 2 3 $\quad x \leq -1$ or $x > 3$

Example 4. Sketch the graph of the given inequality.

a. $x > 2$ $\quad$ -3 -2 -1 0 1 2 3

b. $-2 \leq x < 1$ $\quad$ -3 -2 -1 0 1 2 3

c. $x < -3$ or $x \geq 1$ $\quad$ -3 -2 -1 0 1 2 3

Exercise Set 1.1 (continued)

In exercises 15–30, express the given interval in terms of inequalities.

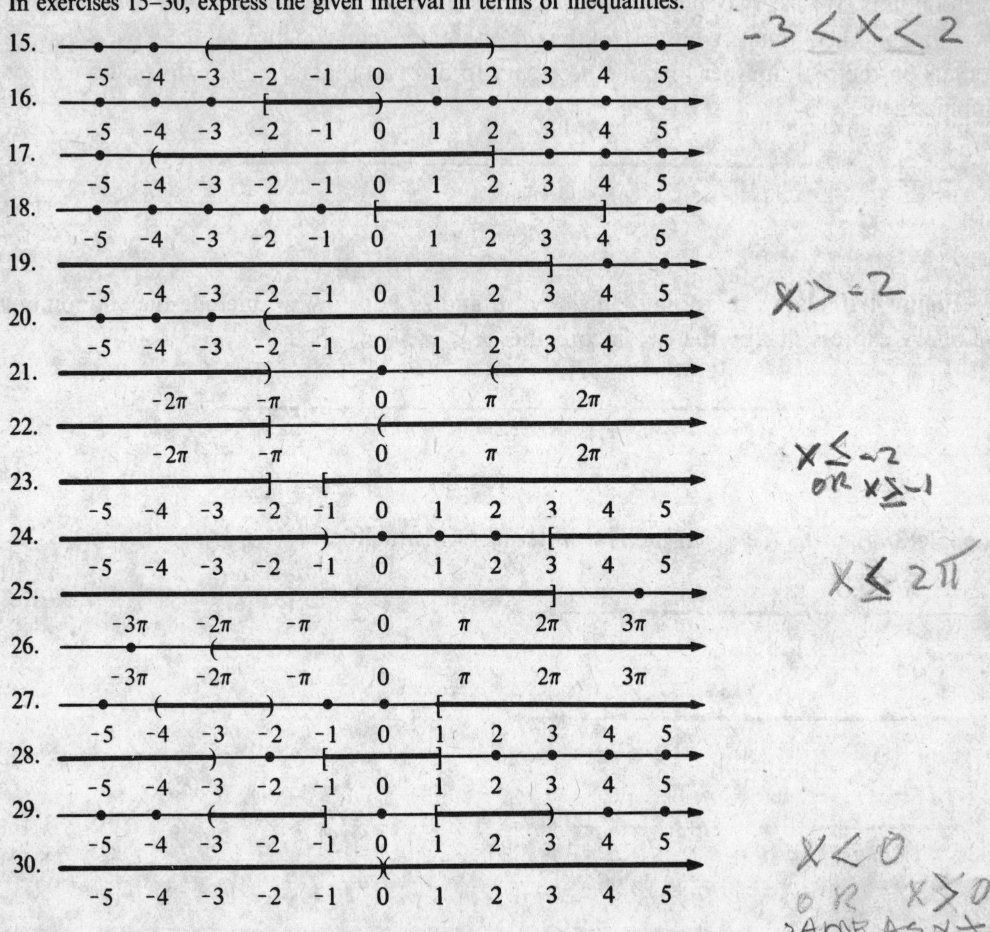

In exercises 31–46, sketch the graph of the given inequality.

31. $x > -3$
32. $x < 4$
33. $x \geq 2$
34. $x \leq -3$
35. $-1 < x < 5$
36. $-2 \leq x \leq 3$
37. $-3 < x \leq 0$
38. $-2 \leq x < -1$
39. $x < 2$ or $x > 5$
40. $x \leq -1$ or $x \geq 3$
41. $x < -2$ or $x \geq 1$
42. $x \leq 4$ or $x > 5$
43. $-1/2 \leq x \leq 3/2$
44. $x < 1/3$ or $x > 4/3$
45. $-\pi \leq x \leq \pi$
46. $x < \pi/2$ or $x > \pi$

section 2 • Operations

In this section we review the operations of addition, subtraction, multiplication, and division. We consider properties of these operations, apply the operations to algebraic expressions, and perform calculator computations.

PROPERTIES OF ADDITION AND SUBTRACTION

A familiar fact is that addition is denoted by a plus sign, $+$. So if a and b are numbers, then $a + b$ is read "a plus b" or "the sum of a and b." Subtraction is denoted by a minus sign, $-$. So if a and b are numbers, then $a - b$ is read "a minus b" or "the difference of a and b." The minus sign is also used for negative numbers. Thus, if a is a number, $-a$ is read "the negative of a" or "minus a." The minus sign is used for both subtraction and negative numbers because, as we shall see, subtraction and negative numbers are closely related.

We review some properties of addition; two important properties are commutativity and associativity.

COMMUTATIVE PROPERTY OF ADDITION. For any numbers a and b,

$$a + b = b + a.$$

The commutative property says that no matter in what *order* we perform the addition, the answer is the same.

ASSOCIATIVE PROPERTY OF ADDITION. For any numbers a, b, and c,

$$a + (b + c) = (a + b) + c.$$

The associative property says, when we add more than two numbers, no matter how we *group* the numbers together, the answer is the same.

The number 0 plays a special role in addition. If we add 0 to any number, we simply obtain that number. (It is sometimes said that 0 is the *additive identity*.)

PROPERTY OF ZERO UNDER ADDITION. For any number a,

$$a + 0 = a = 0 + a.$$

When we add a number and its negative, we obtain 0: for example, $3 + (-3) = 0$. The sum of two numbers is 0 if the numbers are negatives of each other. We use this relationship to define the concept of the negative of a number.

DEFINITION OF THE NEGATIVE OF A NUMBER. If a and b are numbers such that $a + b = 0$, then a is said to be the *negative of* b and b is said to be the *negative of* a. Also, a and b are said to be *negatives of each other*. We write $a = -b$ and $b = -a$.

Example 1. The number 3 is positive and the number -3 is negative. Also, the negative of the number 3 is -3. But this means that, also, the negative of the number -3 is 3. The numbers 3 and -3 are negatives of each other. We write this in symbols:

"The negative of 3 is -3" is written $-(3) = -3$.

"The negative of -3 is 3" is written $-(-3) = 3$.

PROPERTY OF NEGATIVES UNDER ADDITION. For any number a,

$$a + (-a) = 0 = (-a) + a \quad \text{and} \quad -(-a) = a.$$

Now we consider subtraction. Subtraction is the reverse of addition; it undoes what addition does. It is better to say that subtraction is defined in terms of addition and negatives. When we perform $a - b$, we add the negative of b to a.

SUBTRACTION DEFINED IN TERMS OF ADDITION AND NEGATIVES. If a and b are numbers, then

$$a - b = a + (-b).$$

Example 2. Subtraction.

a. $1 + (-2) = 1 - 2 = -1$

b. To perform $-2 + 1$ we can use the commutative law to write $-2 + 1 = 1 + (-2) = 1 - 2 = -1$.

Exercise Set 1.2

1. Check the commutative property for addition by finding $2 + 4$ and $4 + 2$; by finding $3 + 6$ and $6 + 3$; by finding $1 + 9$ and $9 + 1$.
2. Check the associative property for addition by finding $(2 + 3) + 6$ and $2 + (3 + 6)$; by finding $(1 + 5) + 4$ and $1 + (5 + 4)$; by finding $(3 + 5) + 7$ and $3 + (5 + 7)$.
3. What is the negative of 0 (i.e., what is -0)?

4. What are the properties of 0 under subtraction (i.e., what are $a - 0$ and $0 - a$)?
5. a. Find $5 - 2$ and $2 - 5$; find $3 - 7$ and $7 - 3$; find $9 - 4$ and $4 - 9$.
   b. Is subtraction commutative?—i.e., does $a - b = b - a$ for all numbers a and b?
6. a. Find $(5 - 2) - 3$ and $5 - (2 - 3)$; find $(3 - 7) - 6$ and $3 - (7 - 6)$.
   b. Is subtraction associative?—i.e., does $a - (b - c) = (a - b) - c$ for all numbers a, b, and c?

PROPERTIES OF MULTIPLICATION AND DIVISION

Now we study the operations of multiplication and division. There is an analogy between the pair consisting of multiplication and division and the pair consisting of addition and subtraction, and we want to watch for this analogy. In addition and subtraction, 0 plays a special role; in multiplication and division, 1 plays a special role. In addition and subtraction, we deal with the negative of a number; in multiplication and division, we deal with the *reciprocal* of a number.

There are four notations for multiplication. The first is the cross, $\times$. So if a and b are numbers, $a \times b$ means a multiplied by b, or the *product* of a and b. The second notation is a dot, $\cdot$; if a and b are numbers, $a \cdot b$ means the product of a and b. The third notation is *juxtaposition*, or simply writing the numbers together. If a and b are numbers, ab denotes the product of a and b. The fourth way uses parentheses. So $a(b)$, $(a)b$, and $(a)(b)$ denote the product of a and b. We have to be careful with multiplication notation. In particular, juxtaposition does not *always* mean multiplication. We are accustomed to using juxtaposition of digits to hold decimal places. For example, 23 means "twenty three" and *not* "two times three." We could write "two times three" using the dot notation: $2 \cdot 3$. If we want to write "2 times the square root of 2," we can use juxtaposition: $2\sqrt{2}$. Similarily, "5 times π" would be written 5π. On the other hand, if we want to write "three *times* one seventh," we would write $3 \cdot \frac{1}{7}$, because the juxtaposition $3\frac{1}{7}$ means the mixed fraction "three *and* one seventh."

In algebra we use letters to represent numbers, and we frequently use juxtaposition to denote multiplication. So xy means "x times y." We also combine letters and numbers, so we will write $3x$, which means "three times x." When we combine letters and numbers, we usually write the number to the left of the letter.

There are two notations for division. If a and b are numbers, we write "a divided by b" as either $a \div b$ or a/b. The second notation also means the fraction a/b; but since fractions are a way of expressing division, this notation should not cause confusion.

A natural interpretation of multiplication of two positive numbers is *area*. If a and b

are positive real numbers, then ab represents the area of the rectangle with sides a and b (Figure 1.11). This procedure works for two numbers; if we have three positive numbers a, b, c, then abc represents the volume of a three-dimensional rectangular solid with sides a, b, and c (Figure 1.11).

Figure 1.11

To see how negatives behave under multiplication, we examine the number patterns in Example 3.

Example 3. Multiplying negatives.

$3 \cdot 5 = 15$	$3(-5) = -15$	$4(3) = 12$	$(-4)(3) = -12$
$2 \cdot 5 = 10$	$2(-5) = -10$	$4(2) = 8$	$(-4)(2) = -8$
$1 \cdot 5 = 5$	$1(-5) = -5$	$4(1) = 4$	$(-4)(1) = -4$
$0 \cdot 5 = 0$	$0(-5) = 0$	$4(0) = 0$	$(-4)(0) = 0$
$-1 \cdot 5 = -5$	$(-1)(-5) = 5$	$4(-1) = -4$	$(-4)(-1) = 4$
$-2 \cdot 5 = -10$	$(-2)(-5) = 10$	$4(-2) = -8$	$(-4)(-2) = 8$
$-3 \cdot 5 = -15$	$(-3)(-5) = 15$	$4(-3) = -12$	$(-4)(-3) = 12$

We see that we have the following properties for multiplying negatives:

MULTIPLYING NEGATIVES. For any numbers a and b,

$$(-a)b = a(-b) = -ab \quad \text{and} \quad (-a)(-b) = ab.$$

We now consider some properties of multiplication.

COMMUTATIVE PROPERTY OF MULTIPLICATION. For any numbers a and b,

$$ab = ba.$$

ASSOCIATIVE PROPERTY OF MULTIPLICATION. For any numbers a, b, and c,

$$a(bc) = (ab)c.$$

The commutative property says that, no matter in what *order* we perform the multiplication, the answer is the same. The associative property says that when we multiply more than two numbers, no matter how we *group* the numbers together, the answer is the same.

A further property of multiplication is the role played by the number 1. The property is simply that, if we multiply any number by 1, we get that number. (It is sometimes said that 1 is the *multiplicative identity*.)

PROPERTY OF ONE UNDER MULTIPLICATION. For any number a,

$$a \cdot 1 = a = 1 \cdot a.$$

The number 0 also enjoys a property with respect to multiplication; namely, that any number multiplied by 0 is 0.

PROPERTY OF ZERO UNDER MULTIPLICATION. For any number a,

$$a \cdot 0 = 0 = 0 \cdot a.$$

Because the number 1 plays a special role with respect to multiplication, we want to look at numbers whose product is 1. Such numbers are called *reciprocals*.

DEFINITION OF THE RECIPROCAL OF A NUMBER. If a and b are numbers such that $ab = 1$, then a is said to be the *reciprocal of* b and b is said to be the *reciprocal of* a. Also, a and b are said to be *reciprocals of each other*. We write $a = 1/b$ and $b = 1/a$.

Example 4. Reciprocals.

a. Since $2(1/2) = 1$, 2 and 1/2 are reciprocals of each other. So 1/2 is the reciprocal of 2 and 2 is the reciprocal of 1/2.
b. Since $(-3)(-1/3) = 1$, -3 and $-1/3$ are reciprocals of each other. So $-1/3$ is the reciprocal of -3 and -3 is the reciprocal of $-1/3$.

An important fact about reciprocals is that the number 0 does not have any reciprocal. Because 0 times anything is 0, we cannot possibly multiply 0 times something to get 1. But 0 is the only number which does not have a reciprocal; every other number has a reciprocal

We define division in terms of multiplication and reciprocals. The definition is: when we divide a by b, we multiply a by the reciprocal of b.

DIVISION DEFINED IN TERMS OF MULTIPLICATION AND RECIPROCALS. If a and b are numbers and $b \neq 0$, then

$$a \div b = \frac{a}{b} = a \cdot \frac{1}{b}.$$

The rules for dividing negatives are analogous to those for multiplying negatives.

DIVIDING NEGATIVES. For any numbers a and b,

$$\frac{-a}{b} = \frac{a}{-b} = -\frac{a}{b} \quad \text{and} \quad \frac{-a}{-b} = \frac{a}{b}.$$

Example 5. Dividing negatives.

a. $9/3 = 3$ b. $(-9)/3 = -3$ c. $9/(-3) = -3$ d. $(-9)/(-3) = 3$

One of the most important and most frequently misunderstood facts about division is that we do *not* divide by 0. There is nothing mysterious about not dividing by 0; the reason follows from the definition of division. We have defined a divided by b to be a times the reciprocal of b. So in order to divide a by b, the number b must have a reciprocal. Every real number except 0 has a reciprocal, but 0 does *not* have a reciprocal. So we cannot divide by 0. We say that division by 0 is *undefined.*

Exercise Set 1.2 (continued)

7. Check the associative property for multiplication by finding $(3 \cdot 2)4$ and $3(2 \cdot 4)$; by finding $(2 \cdot 5)6$ and $2(5 \cdot 6)$; by finding $(4 \cdot 7)2$ and $4(7 \cdot 2)$.
8. Check the commutative property for multiplication by finding $3 \cdot 5$ and $5 \cdot 3$; by finding $6 \cdot 8$ and $8 \cdot 6$; by finding $2(10)$ and $(10)2$.

In exercises 9–20, perform the given multiplication and division.

9. a. $5 \cdot 7$ b. $(-5)7$ c. $5(-7)$ d. $(-5)(-7)$
10. a. $4 \cdot 3$ b. $4(-3)$ c. $(-4)3$ d. $(-4)(-3)$
11. a. $\frac{16}{2}$ b. $\frac{16}{-2}$ c. $\frac{-16}{2}$ d. $\frac{-16}{-2}$
12. a. $10/2$ b. $(-10)/2$ c. $10/(-2)$ d. $(-10)/(-2)$
13. $3(-2)$
14. $(-4)3$
15. $(-7)(2)(3)$
16. $(-6)(-6)(-1)$
17. $\left(\frac{4}{2}\right)(-3)$
18. $7\left(\frac{-6}{3}\right)$
19. $\left(\frac{-5}{5}\right)\left(\frac{4}{-2}\right)\left(-\frac{9}{3}\right)$
20. $\left(-\frac{12}{3}\right)\left(-\frac{14}{7}\right)(5)$

OPERATIONS ON ALGEBRAIC EXPRESSIONS

Algebraic expressions involve letters as well as numbers. These letters are thought to stand for numbers; however, the strength of algebra derives from the fact that we can perform the operations of arithmetic on the algebraic expressions without substituting numbers for the letters.

Frequently algebraic expressions of the *form* nx occur, where n denotes a positive integer; for example, $2x$, $3x$, $4x$, ... We can think of nx either as n multiplied by x or as the sum of x taken n times. Thus

$$1x = x$$
$$2x = x + x$$
$$3x = x + x + x$$
$$4x = x + x + x + x$$

Example 6. Addition and subtraction.

a. $2x + 3x = 5x$
b. $7x - 3x = 4x$
c. $(x + y) + (2x - 4y) = (x + 2x) + (y - 4y) = 3x - 3y$

Example 7. Multiplication and division.

a. $(3x)(4y) = 12xy$
b. $(-6ab)(3c) = -18abc$
c. $(-4ax)(-7by) = 28abxy$
d. $(2cx)(-12yz) = -24cxyz$

Exercise Set 1.2 (continued)

In exercises 21–36, perform the indicated operations on the algebraic expressions.

21. $7x + 5x$
22. $9x + x$
23. $7x - 3x$
24. $9x - 2x$
25. $(3x + y) + (4x - 2y)$
26. $(2x + 5y) + (2x - 5y)$
27. $(3x + y) - (4x - 2y)$
28. $(2x + 5y) - (2x - 5y)$
29. $(a + 4b - 2c) + (2a - b - 6c)$
30. $(3a - b - 3c) + (a - b + 2c)$
31. $(a + 4b - 2c) - (2a - b + 6c)$
32. $(3a - b - 3c) - (a - b + 2c)$
33. $(9x)(3y)$
34. $(4a)(7bc)$
35. $(2a)(6b)(-3c)$
36. $(-2a)(4b)(-3c)$

CALCULATOR COMPUTATIONS

Some subsections of this book relate algebra and the electronic calculator. These calculator subsections are optional and can be omitted, but they should improve your understanding of both algebra and the calculator.

Various calculators are programmed differently; you should see how to enter numbers and perform operations on your calculator.

An important feature of the electronic calculator is that it displays decimals; that is, the calculator displays the decimal 0.5 rather than the fraction 1/2. As a result, the calculator is ideal for finding a decimal representation of a fraction. We find a decimal representation for a/b by entering $\boldsymbol{a \div b =}$.

Every calculator has a *capacity*; that is, a certain maximum number of digits are displayed. Your calculator may have a capacity of six digits, eight digits, or possibly more. Sometimes the result of an operation may exceed the capacity of your calculator. For example, see what happens on your calculator when you try to find the product $345{,}100 \times 206{,}101$. In Section 3.2 we will consider how to perform products and quotients that exceed the capacity of your calculator.

If you perform the division **2 ÷ 3 =** on your calculator, the display will probably read 0.666666 (six-digit capacity). Of course, the decimal representation of 2/3 is nonterminating; the calculator displays digits to its capacity and drops the rest. This process, called *truncation*, is not the same process as *rounding off*. Some calculators do round off and display 0.666667.

Both truncation and rounding off can lead to inaccuracies. For example, see what happens on your calculator when you find $2 \div 3 \times 3$. Usually these inaccuracies are not serious, but sometimes, when you are performing repeated operations, they can accumulate and affect the result significantly.

You must be careful with negatives when using your calculator. Various calculators are programmed differently with respect to negatives. You should check how negatives are entered on your calculator. What do you get, for example, if you enter **−3** and **−3 =**? Can you enter double negatives on your calculator? What do you get if you enter **− −3=**?

Be careful when subtracting negatives on the calculator. For example, your calculator may not be programmed to perform $3 - (-2)$ because of the double negative. Try entering **3− −2=** on your calculator and see if you get 5. It will probably be necessary to perform subtraction of negatives mentally; that is, you may have to enter $3 - (-2)$ as $3 + 2$.

Finally, be careful when using the calculator to multiply or divide negatives. You should check out simple products such as $(-2)(3)$, $(2)(-3)$, and $(-2)(-3)$. It is always a good idea to keep track of the negative signs mentally when finding products and quotients.

Don't expect the calculator to do all the work—always check your answer with your common sense.

Exercise Set 1.2 (continued)

37. a. Find 6(1/6) on the calculator; that is, $6 \times 1 \div 6 =$
    b. Find (1/6)6 on the calculator; that is, $1 \div 6 \times 6 =$
38. a. Find 11(24/6) on the calculator; that is, $11 \times 24 \div 6 =$
    b. Find (11/6)24 on the calculator; that is, $11 \div 6 \times 24 =$
39. Find the product (1.00001)(1.00001) both on the calculator and "by hand;" compare the results.

40. Find the following quotients on the calculator: 1/(0.999999); 1/(0.333333); 1/(0.666666); 1/(0.666667).

section 3 • Order of Operations

In this section we review the order in which we perform operations and the use of parentheses as symbols of inclusion. We apply this to substituting numbers for a variable in an algebraic expression. We also relate these topics to using an electronic calculator.

ORDER OF PERFORMING OPERATIONS

Whenever we have an algebraic expression involving two or more operations, we must know which operation to perform first, second, and so on to the last operation. Sometimes the order doesn't matter. For example, $2 + 6 + 3$ equals 11 whether we first add $2 + 6$ to get 8 and then add $8 + 3$ to get 11, or first add $6 + 3$ to get 9 and then add $2 + 9$ to get 11 (associative property). But usually it does matter which operation is done first. Consider $2 + 6 \cdot 3$. If we do the addition first and then the multiplication, we get $2 + 6 = 8$ and $8 \cdot 3 = 24$. On the other hand, if we do the multiplication first and then the addition, we get $6 \cdot 3 = 18$ and $2 + 18 = 20$. Now 20 is the correct answer—we do multiplication before addition.

The standard mathematical convention is to do multiplications and divisions first from left to right and then to do additions and subtractions from left to right.

Example 1. Order of Operations.

a. $3 + 5(7) = 3 + 35 = 38$ (multiplication first)

b. $3 - \frac{8}{2} = 3 - 4 = -1$ (division first)

c. $2 \cdot 6 - 4 = 12 - 4 = 8$ (multiplication first)

d. $\frac{9}{3} + 2 \cdot 7 = 3 + 14 = 17$ (division and multiplication first)

We build more complicated algebraic expressions by the use of parentheses and brackets. These are symbols of *inclusion*—they tell us that certain other symbols are included or grouped together. Sometimes a fraction bar serves as a symbol of inclusion. The general rule is that operations are performed in the following order:

1. An expression within a symbol of inclusion is simplified first, starting with the innermost inclusion symbol.
2. Multiplications and divisions are performed from left to right.
3. Additions and subtractions are performed from left to right.

Example 2. Symbols of inclusion.

a. $4(2 + 6) = 4(8) = 32$

b. $\frac{7 + 5}{4} = \frac{12}{4} = 3$

c. $-2 + (6 - 3)(2 + 5) = -2 + (3)(7) = -2 + 21 = 19$

d. $\left[\frac{3 + 3}{2 - 4} \cdot (2 + 7)\right][4 + 3(-1)] = \left[\frac{6}{-2} \cdot 9\right][4 - 3]$

$= [(-3)9][1] = (-27)1 = -27$

Exercise Set 1.3

In exercises 1–24, perform the operations to obtain a real number.

1. $2 + 4(6)$
2. $3 - 3(3)$
3. $-5 + \frac{10}{2}$
4. $\frac{12}{3} + \frac{6}{-3}$
5. $(6 + 2)7$
6. $(-4 + 3)(-12)$
7. $2 - 3(6 + 5)$
8. $(7 - 9)(-1 + 5)$
9. $\frac{6 + 5}{2 - 3}$
10. $\frac{6 - 8}{-4 + 3 \cdot 2}$
11. $\frac{3(13 - 5)}{2}$
12. $\frac{(-3 + 7)6}{12}$
13. $\left(2 + \frac{12}{4}\right)\left(2 - \frac{12}{4}\right)$
14. $\left(\frac{16 - 4}{-3} + 2\right)\left(7 - \frac{24}{5 - 3}\right)$
15. $\frac{\frac{4 + 6}{2}}{-5}$
16. $\frac{\frac{6}{3 - 2}}{-1}$
17. $\frac{\frac{4}{2 + 6}}{4} \cdot \frac{\frac{18}{3}}{-3 + 0}$
18. $\left(\frac{4 + 7}{13 - 2}\right)\left(\frac{6 - 15}{9 - 6}\right)$
19. $\left[\frac{5(2 - 8)}{3}\right]\left[\frac{6 + 4 \cdot 3}{6}\right]$
20. $\left[\frac{(-6 + 3)3}{-9}\right]\left[\frac{13 - 4}{-3}\right]$

21. $\frac{1}{2} \cdot \frac{3+7}{5} - \frac{17-1}{4}$

22. $\frac{6}{2+1} \cdot \frac{2+5}{7}$

23. $\frac{-1+5}{-1-1}(3-1)$

24. $\left(6 + \frac{9-3}{2}\right)\left(7 - \frac{8}{2+2}\right)$

OPERATIONS ON ALGEBRAIC EXPRESSIONS

We follow the same rules for order of operations and use of parentheses for algebraic expressions as for real numbers.

Example 3. Operations on algebraic expressions.

a. $(2a)(3b) + (4x)(2y) = 6ab + 8xy$

b. $(4x)(3y) - (6x)y = 12xy - 6xy = 6xy$

c. $(5ab)(2c) + (3a)(4bc) = 10abc + 12abc = 22abc$

d. $(4a)(4x) - (5a)(3y) = 16ax - 15ay$

Exercise Set 1.3 (continued)

In exercises 25–38, perform the operations on the given algebraic expression.

25. $(2x)(4y) + (5x)(2z)$

26. $(6x)(3y) - (2a)(7b)$

27. $(4a)(7b) + a(3b)$

28. $(9a)(2b) - (4a)b$

29. $(-3a)b - (4a)(-2b)$

30. $(7a)(-b) + (2a)(2b)$

31. $4(4ab) + (2a)(2b) - (11a)(2b)$

32. $3(3ab) - (2a)(4b) + (6a)b$

33. $(2a + 3a)(6x - x)$

34. $(3x + x)(2y + 2y)$

35. $(4a - 6a)(2b + b)$

36. $(3a - 7a)(-b - 2b)$

37. $(2a - 3a)(4b + b)(7c - 5c)$

38. $(4a - 7a)(b + 3b)(3c - 8c)$

SUBSTITUTING NUMBERS FOR A VARIABLE

Usually an algebraic expression contains both variables and constants.

DEFINITIONS OF VARIABLE AND CONSTANT. A *variable* in an algebraic expression is a symbol that can assume at least two different number values; a *constant* is a symbol that can assume only one number value.

A constant is a symbol for a number, whereas we substitute various numbers for a variable.

Example 4. In the algebraic expression $3x + 8$, x is a variable and 3 and 8 are constants.

a. If we substitute 2 for x: $3x + 8 = 3(2) + 8 = 14$
b. If we substitute 5 for x: $3x + 8 = 3(5) + 8 = 23$
c. If we substitute -1 for x: $3x + 8 = 3(-1) + 8 = 5$

Example 5. The equation $A = wl$ expresses the area A of a rectangle in terms of its width w and length l. Since rectangles of different sizes have different widths, lengths and areas, w, l, and A are variables.

a. If we substitute 2 for w and 6 for l: $A = (2)(6) = 12$; that is, the rectangle with width 2 and length 6 has area 12.
b. If we substitute 0.5 for w and 14 for l: $A = (0.5)(14) = 7$; that is, the rectangle with width 0.5 and length 14 has area 7.
c. If we substitute 4 for w and 5.2 for l: $A = (4)(5.2) = 20.8$; that is, the rectangle with width 4 and length 5.2 has area 20.8.

Example 6. The equation $c = 2\pi r$ expresses the circumference c of a circle in terms of its radius r. Since circles of different sizes have different radii and circumferences, r and c are variables. However, 2 and π are constants. Using 3.14 as an approximation to π, we can write

$$c = 2(3.14)r = 6.28r.$$

a. If we substitute 3 for r: $c = (6.28)(3) = 18.84$; that is, the circumference of a circle with radius 3 is 18.84.
b. If we substitute 200 for r: $c = (6.28)(200) = 1256$; that is, the circumference of a circle with radius 200 is 1256.
c. If we substitute 1.1 for r: $c = (6.28)(1.1) = 6.908$; that is, the circumference of a circle with radius 1.1 is 6.908.

Exercise Set 1.3 (continued)

In exercises 39–48, substitute the numbers 2, 5, 0, and -1 for the variable x in the given algebraic expression.

39. $4x + 15$
40. $3x + 20$
41. $2x + 2(x + 8)$
42. $2x + 2(x + 6)$
43. $x(x + 1)$
44. $x(x - 1)$
45. $(x + 4)(x - 2)$
46. $(x + 5)(x - 4)$
47. $1/x$
48. $36/x$

In exercises 49–52, substitute the specified numbers for the variables in the given equation.

49. $A = wl$; find A when $w = 9$ and $l = 8$; when $w = 0.5$ and $l = 16$; when $w = 5$ and $l = 3.1$.
50. $c = 2\pi r$; find c when $r = 0.1$; $r = 0.01$; $r = 10$; $r = 100$; $r = 50$. Use 3.14 as an approximation to π.
51. $C = 5(F - 32)/9$; find C when $F = 0$; $F = -20$; $F = 32$; $F = 65$; $F = 100$; $F = 98.6$; $F = 212$. This equation converts temperature from degrees Fahrenheit to degrees Centigrade.
52. $V = wlh$; find V when $w = 3$, $l = 4$, $h = 6$; when $w = 0.2$, $l = 0.8$, $h = 1$; when $w = 10$, $l = 12$, $h = 4$. This equation expresses the volume V of a rectangular box in terms of its width w, length l, and height h.

CALCULATOR COMPUTATIONS

When performing computations on the calculator, we must be careful to do the operations in an appropriate order. Notice that the calculator performs an operation on whatever number occurs in the display as we key the operation and whatever number is keyed immediately after the operation. Consider how to key some types of expressions on the calculator. (In what follows, we assume that the calculator has algebraic notation.)

To compute a number written in the form $ab + c$, we can enter $\mathbf{a \times b + c =}$. To compute a number written in the form $a + bc$, we can write the number as $bc + a$ (commutative property) and then enter $\mathbf{b \times c + a =}$: In both cases we perform the multiplication first.

Similarly, to compute a number written in the form $(a/b) + c$ or $a + (b/c)$, we enter the division first.

To compute a number written in the form $ax + by$, we can first compute by and write down the answer. Then we can compute ax and add the answer we have written down. If your calculator has a memory or a reciprocal key, you can find ways to compute a number having this form without writing anything down. In general, however, you may find that writing down steps as you do computations will help to prevent errors.

To compute a number written in the form $(a + b)c$, we can enter $\mathbf{a + b \times c =}$, since the multiplication acts on the sum of a and b. To compute a number written in the form $a(b + c)$, we can write the number as $(b + c)a$ (commutative property) and then enter $\mathbf{b + c \times a =}$. (Some calculators require an extra = in these computations.)

To compute a number written in the form $(a + b)/c$, we can enter $\mathbf{a + b \div c =}$, since the division acts on the sum of $a + b$. However, to compute a number written in the form $a/(b + c)$, we compute $b + c$, write down the answer, and then find a divided by that answer. If your calculator has a memory or a reciprocal key, you can find ways to compute a number having this form without writing anything down.

Return to exercises 1–24 and perform the computation using a calculator.

Exercise Set 1.3 (continued)

In exercises 53–56, substitute the specified numbers for the variables in the given equation.

53. $A = wl$; find A when $w = 9.12$ and $l = 8.03$; when $w = 0.497$ and $l = 16.113$; when $w = 4.8981$ and $l = 3.0932$.

54. $c = 2\pi r$; find c when $r = 0.121$; $r = 0.01302$; $r = 8.686$; $r = 48.565$; $r = 154.271$. Use 3.14159 as an approximation to π.

55. $C = 5(F - 32)/9$; find C when $F = 1.03$; $F = 30.06$; $F = 76.072$; $F = 121.566$; $F = -12.021$; $F = -29.336$.

56. $V = wlh$; find V when $w = 3.6$, $l = 4.1$, $h = 5.8$; when $w = 0.212$, $l = 0.811$, $h = 0.987$; when $w = 10.16$, $l = 11.49$, $h = 3.98$; when $w = 5.5555$, $l = 6.6666$, $h = 3.3333$.

In exercises 57–58, write the algebraic expression represented by the calculator entry.

57. a. $a \times b - c =$ b. $a - b \times c =$
c. $a + b - c =$ d. $a - b + c =$

58. a. $a \div b \times c =$ b. $a \times b \div c =$
c. $a \div b + c =$ d. $a + b \div c =$

section 4 • Exponents

We review the definition of exponents and the laws for exponents. Also, we substitute numbers for a variable and consider calculator computations.

DEFINITION OF EXPONENTS

Recall that the area of a square whose sides are x is $x \cdot x$. It becomes cumbersome to keep writing $x \cdot x$, and so we introduce a shorthand notation, namely x^2. This notation is read "x squared" in honor of the geometric example. (Figure 1.12).

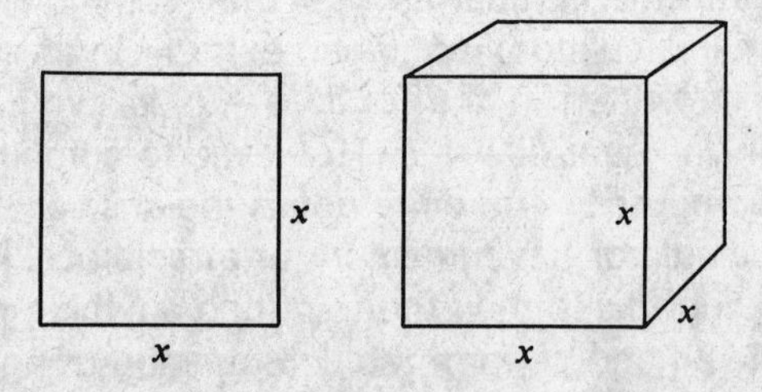

Figure 1.12

The volume of a cube whose sides are x is $x \cdot x \cdot x$. Here it is even more convenient to use a shorthand notation, and we define $x^3 = x \cdot x \cdot x$. Again in honor of the geometric example, this notation is read "x cubed."

The figures from dimensions higher than three don't have common geometric names, but expressions such as $x \cdot x \cdot x \cdot x$ and $x \cdot x \cdot x \cdot x \cdot x$ occur frequently in algebra. So we define $x^4 = x \cdot x \cdot x \cdot x$ and $x^5 = x \cdot x \cdot x \cdot x \cdot x$ and read these as "x fourth" and "x fifth." We define x^n in a natural way for any positive integer n.

DEFINITION OF EXPONENTS. If n is a positive integer,

$$x^n = x \cdot x \cdot \cdot \cdot x, n \text{ factors of } x.$$

This is read "x n^{th}" or "the nth power of x."

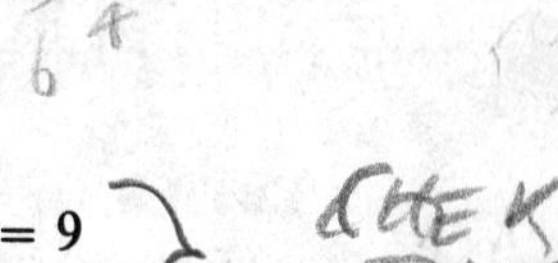

Example 1. Finding powers of numbers.

a. $3^2 = 3 \cdot 3 = 9$
b. $3^3 = 3 \cdot 3 \cdot 3 = 27$
c. $5^4 = 5 \cdot 5 \cdot 5 \cdot 5 = 625$
d. $(-3)^2 = (-3)(-3) = 9$
e. $-3^2 = -(3 \cdot 3) = -9$
f. $(-3)^3 = (-3)(-3)(-3) = -27$

Exercise Set 1.4

In exercises 1–20, find the indicated power.

1. 2^3	2. 2^4	3. 3^4	4. 5^2
5. 4^3	6. 4^4	7. 6^3	8. 6^5
9. $(-1)^2$	10. $(-1)^3$	11. $(-1)^4$	12. $(-1)^5$
13. $(-2)^3$	14. $(-2)^4$	15. -2^3	16. -2^4
17. $(-5)^2$	18. -5^2	19. $(-5)^4$	20. -5^4

EXPONENTIAL NOTATION

We practice writing an expression in exponential notation.

Example 2. Writing an expression in exponential notation.

a. $x \cdot x \cdot x \cdot y = x^3y$
b. $x \cdot y \cdot x \cdot y \cdot y = x^2y^3$
c. $(ab)(ab)(ab) = (ab)^3$

Exercise Set 1.4 (continued)

In exercises 21–30, write the given expression in exponential notation.

21. $x \cdot x \cdot x \cdot y \cdot y$	22. $x \cdot x \cdot y \cdot y \cdot y \cdot z \cdot z \cdot z \cdot z$
23. $a \cdot x \cdot y \cdot x \cdot x \cdot x \cdot y$	24. $a \cdot a \cdot b \cdot a \cdot a$
25. $(abc)(abc)$	26. $(xy)(xy)(xy)$

27. $(ab)(ab)(ab)(ab)c \cdot c$

28. $a \cdot a \cdot a(bc)(bc)$

29. $\dfrac{x \cdot x}{y \cdot y \cdot y \cdot y}$

30. $\dfrac{(ab)(ab)(ab)}{c \cdot c \cdot c}$

PROPERTIES OF EXPONENTS

To simplify algebraic expressions involving exponents we use properties of exponents. We now review these properties. To appreciate them better, we compare these properties with corresponding properties of multiples.

PROPERTIES OF EXPONENTS AND MULTIPLES. If m and n are positive integers, then for all x and y:

	multiples	*exponents*
1.	$mx + nx = (m + n)x$	$x^m x^n = x^{m+n}$
2.	$n(mx) = (mn)x$	$(x^m)^n = x^{mn}$
3.	$n(x + y) = nx + ny$	$(xy)^n = x^n y^n$

We illustrate the properties of exponents in Example 3.

Example 3. Properties of exponents.

a. $x^2 \cdot x^3 = (x \cdot x)(x \cdot x \cdot x) = x^5$

b. $(a^3)^2 = a^3 \cdot a^3 = (a \cdot a \cdot a)(a \cdot a \cdot a) = a^6$

c. $x^2(x^2)^3 = x^2 \cdot x^2 \cdot x^2 \cdot x^2 = (x \cdot x)(x \cdot x)(x \cdot x)(x \cdot x) = x^8$

Exercise Set 1.4 (continued)

In exercises 31–50, multiply the algebraic expressions and write the answer in terms of exponents.

31. $x \cdot x^3$

32. $a^2 \cdot a \cdot a^2$

33. $y^2 \cdot y^4$

34. $b^2 \cdot b \cdot b \cdot b^2$

35. $a \cdot a^2 \cdot b^2 \cdot b^2$

36. $x^2 \cdot x \cdot x^3 \cdot b^2 \cdot b^2$

37. $x^3 \cdot x^2 \cdot y \cdot y^4$

38. $x^2 \cdot y^2 \cdot y^3 \cdot z^3 \cdot z$

39. $(a^2)^3$

40. $(a^2)^4$

41. $(a^3)^3$

42. $(a^5)^2$

43. $a^2 \cdot (a^2)^2 \cdot a$

44. $b \cdot (b^3)^3$

45. $t^4(t^3)^2$

46. $s^2 \cdot s^3 \cdot s^4$

47. $\dfrac{a^2 \cdot a}{b \cdot b^3}$

48. $\dfrac{(a^2)^4}{b^2 \cdot b^3}$

49. $\dfrac{(a^3)^2 \cdot a}{b^4}$

50. $\dfrac{a \cdot a^4 \cdot a}{b \cdot (b^2)^2 \cdot b}$

SUBSTITUTING NUMBERS FOR A VARIABLE

We consider examples of substituting numbers for variables in an equation.

Example 4. The equation $A = \pi r^2$ expresses the area A of a circle in terms of its radius r. We will use 3.14 as an approximation to π.

a. When we substitute 3 for r we obtain $A = (3.14)3^2 = 28.26$; that is, the area of a circle with radius 3 is 28.26.
b. When we substitute 6 for r we obtain $A = (3.14)6^2 = 113.04$; that is, the area of a circle with radius 6 is 113.04.
c. When we substitute 100 for r we obtain $A = (3.14)100^2 = 31{,}400$; that is, the area of a circle with radius 100 is 31,400.

Example 5. The equation $E = I/r^2$ expresses the illumination E on a surface which is a distance r from a light source of luminous intensity I.

a. When we substitute 1.5 for I and 5 for r, we obtain $E = 1.5/5^2 = 1.5/25 = 0.06$
b. When we substitute 200 for I and 10 for r, we obtain $E = 200/10^2 = 200/100 = 2$
c. When we substitute 120 for I and 20 for r, we obtain $E = 120/20^2 = 120/400 = 0.3$.

Exercise Set 1.4 (continued)

In exercises 51–58, substitute the specified numbers for the variables in the given equation.

51. $A = \pi r^2$; find A when $r = 1$; when $r = 8$; when $r = 50$; when $r = 200$.
52. $V = 4\pi r^3/3$; find V when $r = 1$; when $r = 8$; when $r = 50$; when $r = 200$.
53. $s = \frac{1}{2}gt^2$; find s when $t = 2$; when $t = 10$; when $t = 20$; when $t = 200$. This equation expresses the distance s that a freely falling object falls in terms of the time t; the constant g is the acceleration due to gravity; 9.8 is an approximation to g.
54. $h = v^2/2g$; find h when $v = 1$; when $v = 2$; when $v = 10$; when $v = 20$. This expresses Torricelli's theorem of fluid dynamics; the constant g is the acceleration due to gravity; 9.8 is an approximation to g.
55. $E = I/r^2$; find E when $I = 20$ and $r = 1$; when $I = 16$ and $r = 2$; when $I = 100$ and $r = 10$; when $I = 80$ and $r = 40$.
56. $F = mv^2/r$; find F when $m = 1$, $v = 6$, and $r = 2$; when $m = 4$, $v = 3$, and $r = 8$; when $m = 5$, $v = 4$, and $r = 12$. This equation expresses the centripetal force F of a mass m in circular motion with velocity v and radius r.
57. $V = \pi r^2 h$; find V when $r = 1$ and $h = 3$; when $r = 3$ and $h = 6$; when $r = 5$ and $h = 5$; when $r = 4$ and $h = 2$. This equation expresses the volume V of a right circular cylinder in terms of the radius r of its base and its height h.
58. $P = VI - I^2R$; find P when $V = 30$, $I = 5$, and $R = 50$; when $V = 100$, $I = 20$, and $R = 400$; when $V = 10{,}000$, $I = 200$, and $R = 800$. This equation expresses the power P in an electrical transmission line in terms of the voltage V, current I, and resistance R.

CALCULATOR COMPUTATIONS

Example 6. In the equation $A = \pi r^2$, find A when $r = 7.6382$.

Solution. Using 3.14159 as an approximation to π,

$$A = (3.14159)(7.6382)(7.6382) = 183.287.$$

We have rounded off the answer to three decimal places.

Exercise Set 1.4 (continued)

In exercises 59–62, substitute the specified numbers for the variables in the given equation. Use 3.14159 as an approximation to π and use 9.806 as an approximation to g. Round off your answers to three decimal places.

59. $A = \pi r^2$; find A when $r = 1.362$; when $r = 7.926$; when $r = 48.763$; when $r = 206.21$.

60. $V = 4\pi r^3/3$; find V when $r = 0.965$; when $r = 8.027$; when $r = 52.138$; when $r = 206.21$.

61. $s = \frac{1}{2}gt^2$; find s when $t = 1.898$; when $t = 10.737$; when $t = 18.913$; when $t = 204.16$.

62. $h = v^2/2g$; find h when $v = 0.873$; when $v = 2.613$; when $v = 12.194$; when $v = 18.715$.

63. By the third law for exponents, $(xy)^2 = x^2y^2$ for all x and y. Let $x = 1.00001$ and $y = 0.999999$ and find both $(xy)^2$ and x^2y^2 using your calculator. You may also wish to do these computations "by hand." Also try $x = 1.00002$ and $y = 0.999998$.

Review Exercises

In exercises 1–4, find a decimal representation for the given rational number and specify whether it is terminating or nonterminating.

1. 3/4
2. 5/8
3. 1/6
4. 1/9

In exercises 5–12, sketch the graph of the given inequality on the real number line.

5. $-3 < x < 1$
6. $-2 \leq x \leq 4$
7. $x < -1$ or $x > 2$
8. $x < -3$ or $x > 0$
9. $x \leq 2$ or $x \geq 3$
10. $x \leq 3$ or $x \geq 2$
11. $-2 \leq x \leq 2$
12. $1 < x < 5$

In exercises 13–32, perform the indicated operations on the given algebraic expression.

13. $(2x + 5y) + (4x - 2y)$
14. $(3x - 7y) + (x + 2y)$
15. $(5x - 3y) - (x - 4y)$
16. $(6x + 2y) - (3x - 5y)$
17. $(4x)(6y)$
18. $(3x)(8y)$

19. $(-2x)(-3y)$

20. $(-5x)(-4y)$

21. $(5x)(-6y)$

22. $(7x)(-3y)$

23. $(3x)(3y) + (2x)(5y)$

24. $(2x)y + (4x)(4y)$

25. $3(4xy) - (6x)(2y)$

26. $5(6xy) - (10x)(3y)$

27. $(2x)(3y) - (-3x)(6y)$

28. $(4x)(-7y) - (6x)(-3y)$

29. $(-5x)(7y) - (-3x)(-4y)$

30. $x(-7y) - (-5x)(-10y)$

31. $(3x)(-5y) + (-2x)(-7y)$

32. $(5x)(-4y) + (-4x)(-5y)$

In exercises 33–40, find the indicated power.

33. 10^2

34. 10^3

35. $(-5)^2$

36. $(-5)^3$

37. 1^4

38. $(-1)^4$

39. 2^5

40. $(-2)^5$

In exercises 41–48, write the given expression in exponential notation.

41. $x \cdot x$

42. $a \cdot a \cdot a$

43. $r \cdot r \cdot r \cdot r$

44. $y \cdot y \cdot y$

45. $a \cdot a \cdot a \cdot b$

46. $a \cdot b \cdot b \cdot b$

47. $s \cdot s \cdot r \cdot r$

48. $r \cdot r \cdot s \cdot s \cdot s \cdot s$

In exercises 49–56, find the product and write the answer in exponential notation.

49. $x \cdot x^4$

50. $x^3 \cdot x^2$

51. $a^4b \cdot ab^2$

52. $a^2b^2 \cdot ab$

53. $a^3(ab)^2b^3$

54. $a^2(ab)^2b^2$

55. $(a^2)^2$

56. $(a^3)^2$

In exercises 57–60, substitute the specified numbers for the variables in the given equation. Use 3.14 as an approximation for π.

57. $A = wl$; find A when $w = 2$ and $l = 13$; when $w = 18$ and $l = 20$; when $w = 5$ and $l = 100$; when $w = 20$ and $l = 80$.

58. $C = 5(F - 32)/9$; find C when $F = -100$; when $F = -50$; when $F = 10$; when $F = 40$.

59. $c = 2\pi r$; find c when $r = 12$; when $r = 30$; when $r = 90$; when $r = 1000$.

60. $A = \pi r^2$; find A when $r = 0.3$; when $r = 3$; when $r = 30$; when $r = 300$.

chapter 2

Algebraic Expressions

In this chapter we consider the operations of addition, subtraction, multiplication, and division of algebraic expressions. Basically, we develop two themes—the distributive property and fractions. The distributive property relates the operations of addition and multiplication. It provides a method of forming products of algebraic expressions and of factoring algebraic expressions. Fractions arise naturally as quotients of algebraic expressions.

In Section 2.1 we introduce the distributive property, and we form special products arising from the distributive property in Section 2.2. We examine various types of factoring in Sections 2.3, 2.4, and 2.5.

We consider fractions in Sections 2.6 to 2.9. In Section 2.6 we examine equality of fractions and proportions. We add, subtract, multiply, and divide fractions in Sections 2.7 and 2.8. In Section 2.9 we simplify complex fractions.

In Section 2.10 we apply these topics to geometry (right circular cylinders), business (markups and discounts), population estimates, and medical dosages.

section 1 • The Distributive Property

The purpose of this section is to gain some practice in multiplication when it is combined with addition and subtraction—for example, multiplying out algebraic expressions of the form

$$a(b + c) \quad \text{or} \quad (a + b)(c + d) \quad \text{or} \quad (a + b)(c + d + e).$$

To do multiplications such as these, we use the distributive property and extensions of it.

BASIC FORM OF THE DISTRIBUTIVE PROPERTY

An algebraic expression involving no sums or differences—that is, one involving only products and quotients—is called a *monomial.* Examples of monomials are

$$3x, \quad 7t^2, \quad xy^2z, \quad 16ab.$$

If there is just one sum or difference, the expression is called a *binomial.* Examples of binomials are

$$3 + 6, \quad x - y^2, \quad 6x + 17, \quad a^2b - c.$$

If the algebraic expression involves just two sums or differences, it is called a *trinomial.* Examples of trinomials are

$$6x - y + z, \quad -x^2 - 3x + 2, \quad ab^2 + d - x.$$

What happens if we multiply a monomial by a binomial? That is, we wish to multiply an expression of the form $a(b + c)$. The rule is to multiply the monomial by each term in the binomial and then add. In symbols,

$$a(b + c) = ab + ac.$$

If we think of multiplication of two numbers as representing an area (Section 1.2), then Figure 2.1 shows the validity of this rule. The quantity $a(b + c)$ is the area of the large rectangle; this area is the sum of the areas of the small rectangles ab and ac. This rule for multiplying a monomial by a binomial is very important in algebra; it's called the *distributive property.*

DISTRIBUTIVE PROPERTY. For any numbers a, b, and c,

$$a(b + c) = ab + ac.$$

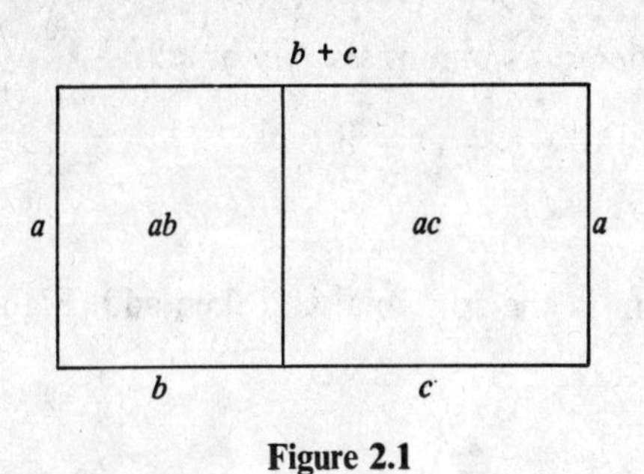

Figure 2.1

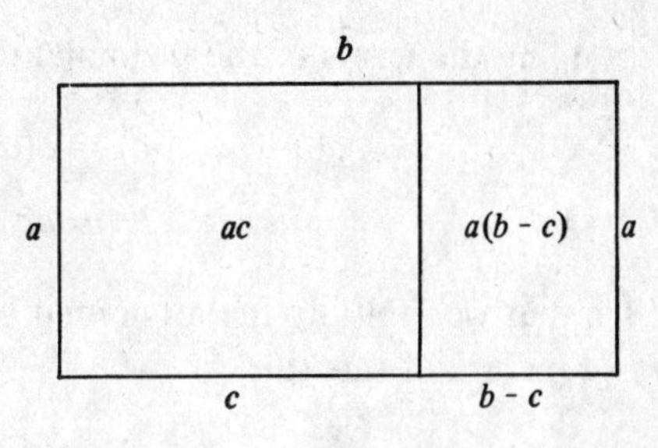

Figure 2.2

Suppose in the distributive property we substitute the negative of c for c. We obtain

$$a(b + (-c)) = ab + a(-c) = ab + (-ac).$$

Using the fact that subtraction is the addition of a negative (Section 1.2), we have

$$a(b - c) = ab - ac.$$

We see that we can use the distributive property for subtraction when we pay proper attention to multiplying negatives. Figure 2.2 illustrates this subtraction.

Example 1. Multiplication of monomials by binomials.

a. $ab(x + y) = abx + aby$
b. $3xy(6x + 2y) = 3xy(6x) + 3xy(2y) = 18x^2y + 6xy^2$
c. $6ac(b - 12) = 6ac(b) + 6ac(-12) = 6abc - 72ac$
d. $\frac{1}{2}ab(10x - 4y) = \frac{1}{2}ab(10x) + \frac{1}{2}ab(-4y) = 5abx - 2aby$
e. $-x(x^2 - y^2) = -x(x^2) + (-x)(-y^2) = -x^3 + xy^2$

Exercise Set 2.1

In exercises 1–10, multiply the algebraic expressions.

1. $6(x + 12)$	2. $x^2(y - z)$	3. $5x(xy + 3z)$	4. $4a(a - 3b)$
5. $-12(x^2 - 3y)$	6. $-xy(ax + 1)$	7. $(4x + 3)x^2$	8. $(a - 2b)ab$
9. $(x - y)xy$	10. $(a + b)a^2b^2$		

5. $-12x^2+36y$

EXTENSION OF THE DISTRIBUTIVE PROPERTY

Multiplication of a monomial by a trinomial proceeds in the same way, because we can consider this case as an extension of the distributive property. What happens if we wish to multiply something of the form $a(b + c + d)$? We can use the distributive property here by grouping the terms of the trinomial to be considered as binomials. We obtain

$$a(b + c + d) = a((b + c) + d) = a(b + c) + ad$$
$$= ab + ac + ad.$$

We see that we multiply the monomial by each term in the trinomial and then add. Figure 2.3 is a picture of this rule.

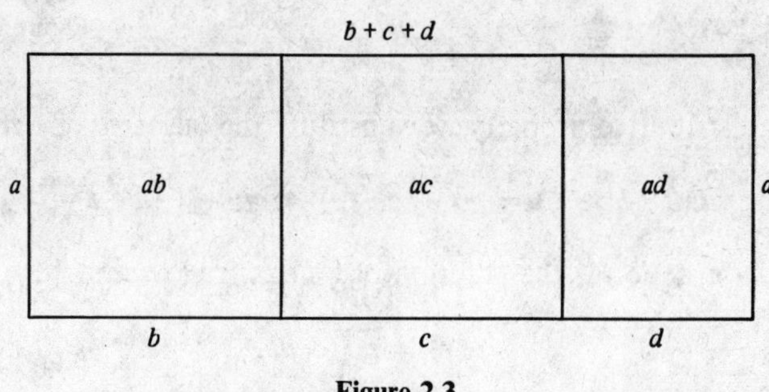

Figure 2.3

Example 2. Multiplication of a monomial by a trinomial.

a. $x^2(xy + yz + 6) = x^2(xy) + x^2(yz) + x^2(6) = x^3y + x^2yz + 6x^2$

b. $a(bc - ab + ac) = a(bc) + a(-ab) + a(ac) = abc - a^2b + a^2c$

c. $-a(ab - ac - bc) = -a(ab) + (-a)(-ac) + (-a)(-bc) = -a^2b + a^2c + abc$

Exercise Set 2.1 (continued)

In exercises 11–18, multiply the algebraic expressions.

11. $3a(a + b + ab)$
12. $2xy(x + y - xy)$
13. $xyz(xy + xz + yz)$
14. $xy(x + y - 3)$
15. $4x^2y(-x + y + y^2)$
16. $-ab(2x - y + z)$
17. $(x^2 - xy + y^2)(-xy)$
18. $(r + s - rs)rs^3$

(15) $-4x^3y + 4x^2y^2 + 4x^2y^3$

FURTHER EXTENSION OF THE DISTRIBUTIVE PROPERTY

Suppose now we wish to find the product of two binomials—a product of the form $(a + b)(c + d)$. Using the distributive property twice and momentarily considering $a + b$ as one term, we obtain

$$(a + b)(c + d) = (a + b)c + (a + b)d = ac + bc + ad + bd.$$

We see that the rule is to form all the possible products we can make and add them together. Figure 2.4 illustrates the product of two binomials.

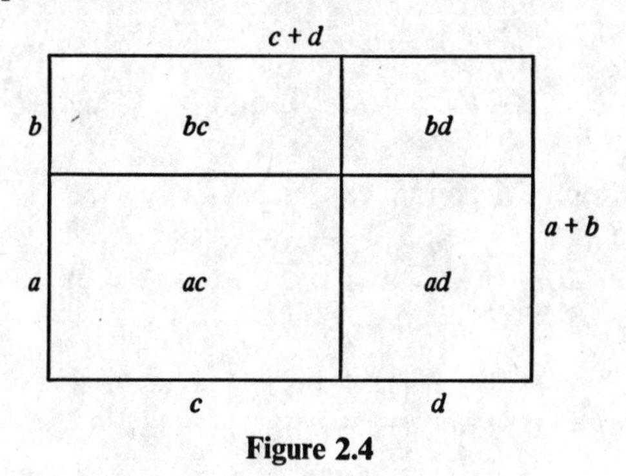

Figure 2.4

Example 3. Multiplication of two binomials.

a. $(x + 1)(y + 2) = xy + 1(y) + x(2) + (1)(2) = xy + y + 2x + 2$

b. $(a + b)(x - y) = ax + bx + a(-y) + b(-y) = ax + bx - ay - by$

c. $(3x + 2)(a - 4y) = 3x(a) + 2(a) + 3x(-4y) + 2(-4y)$
$= 3ax + 2a - 12xy - 8y$

d. $(3x - 2)(x + 6) = 3x(x) + (-2)x + 3x(6) + (-2)6$
$= 3x^2 - 2x + 18x - 12 = 3x^2 + 16x - 12$

Exercise Set 2.1 (continued)

In exercises 19–34, multiply the algebraic expressions.

19. $(x + 1)(y + 3)$
20. $(x + 2)(y - 2)$
21. $(x - 3)(y - 5)$
22. $(x + a)(y + b)$
23. $(a - 5)(b - 3)$
24. $(a + 2)(b - 4)$
25. $(a - 1)(b + 1)$
26. $(a + 7)(b + 5)$
27. $(6x + 1)(2y + 1)$
28. $(x - 3)(4y + 3)$
29. $(7x + 5)(2y + 1)$
30. $(3x - 2)(y - 11)$
31. $(a - b)(a^2 + b^2)$
32. $(a + b)(a + 2b)$
33. $(x^2 + 1)(x + 2)$
34. $(x^3 - 1)(x^2 + 1)$

THE GENERAL DISTRIBUTIVE PROPERTY

The multiplication of a binomial by a trinomial or the multiplication of two trinomials is more complicated in practice but not in theory. We extend the rule we have found in previous cases. Actually, we can extend the distributive property to cover the product of any algebraic expressions.

RULE FOR MULTIPLYING ALGEBRAIC EXPRESSIONS. To multiply algebraic expressions, make all possible monomial products and add them together. Follow the rules for multiplying negatives:

$$(-a)b = a(-b) = -ab \qquad \text{and} \qquad (-a)(-b) = ab.$$

Example 4. Multiplication of algebraic expressions.

a. $(a + b)(x - y + z) = ax + a(-y) + az + bx + b(-y) + bz$
$\qquad = ax - ay + az + bx - by + bz$

b. $(a + b + 1)(x + y) = ax + ay + bx + by + 1 \cdot x + 1 \cdot y$
$\qquad = ax + ay + bx + by + x + y$

c. $(a - b)(a^2 + ab + b^2) = a(a^2) + a(ab) + a(b^2) + (-b)(a^2) + (-b)(ab)$
$\qquad + (-b)(b^2)$
$\qquad = a^3 + a^2b + ab^2 - a^2b - ab^2 - b^3$
$\qquad = a^3 - b^3$

If we have a product of three or more algebraic expressions, we can use the same rule for multiplying them. We can also find the product by grouping the expressions in pairs.

Example 5. Multiplication of three algebraic expressions.

a. $a(a + b)(x - y) = a \cdot a \cdot x + a \cdot b \cdot x + a \cdot a(-y) + a \cdot b(-y)$
$\qquad = a^2x + abx - a^2y - aby$

b. $(a + b)(a - b)(a + b) = [(a + b)(a - b)]\,(a + b)$
$\qquad = [a \cdot a + a(-b) + b \cdot a + b(-b)]\,(a + b)$
$\qquad = [a^2 - ab + ab - b^2]\,(a + b)$
$\qquad = (a^2 - b^2)(a + b)$
$\qquad = a^2 \cdot a + (-b^2)a + a^2 \cdot b + (-b^2)b$
$\qquad = a^3 - ab^2 + a^2b - b^3$

Exercise Set 2.1 (continued)

In exercises 35–48, multiply the algebraic expressions.

35. $(a + b)(a + b - ab)$
36. $(a + b)(a^2 + ab + b^2)$
37. $(x + y + 1)(x + y - 1)$
38. $(x^2 + xy + y^2)(x^2 - xy + y^2)$
39. $(a - b)(a^2 - b^2)(a + b)$
40. $(a + 1)(a^2 + 1)(a^3 - 1)$
41. $(a - 1)(a^2 - 1)(a^3 - 1)$
42. $(a + b)(a + 2b)(a + 3b)$
43. $(a + b)(a - b)$
44. $(a + b)(a + b)$
45. $(x + a)(x + b)$
46. $(x - a)(x - b)$
47. $(a - b)(a^2 + ab + b^2)$
48. $(a + b)(a^2 + 2ab + b^2)$

section 2 • Some Special Products

The purpose of this section is to aquaint you with certain products that come up frequently enough to warrant special attention. Recognizing the form of these special products will permit you to multiply more quickly and easily.

SQUARE OF A BINOMIAL

Frequently we must find the square of a sum; that is, we must find $(a + b)^2$. We use the rule of multiplication of Section 2.1 to obtain

$$\begin{aligned}(a + b)^2 &= (a + b)(a + b) = a \cdot a + a \cdot b + b \cdot a + b \cdot b \\ &= a^2 + 2ab + b^2.\end{aligned}$$

SQUARE OF A BINOMIAL, $(a + b)^2 = a^2 + 2ab + b^2$.
Figure 2.5 is the picture for this special product. Notice that, if we substitute $-b$ for b in this equation, we obtain

$$(a + (-b))^2 = a^2 + 2a(-b) + (-b)^2,$$
$$(a - b)^2 = a^2 - 2ab + b^2.$$

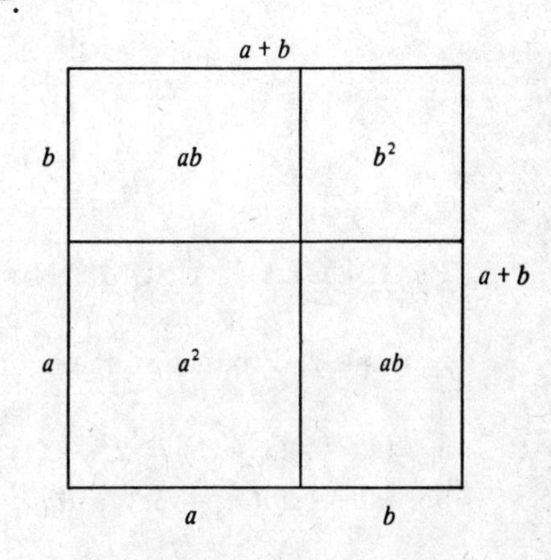

Figure 2.5

or

Example 1. Square of a binomial.

a. $(x + 3)^2 = x^2 + 2 \cdot x \cdot 3 + 3^2 = x^2 + 6x + 9$

b. $(xy - 1)^2 = (xy)^2 + 2(xy)(-1) + (-1)^2 = x^2y^2 - 2xy + 1$

A particularly common mistake in algebra is to say that $(a + b)^2 = a^2 + b^2$. This is a *mistake*. The square of a binomial must contain the term $2ab$.

Exercise Set 2.2

In exercises 1–8, find the given square.

1. $(x + 1)^2$ 2. $(x - 1)^2$ 3. $(x + 5)^2$ 4. $(y - 2)^2$
5. $(xy + 6)^2$ 6. $(xy - 3z)^2$ 7. $(x^2 + y^2)^2$ 8. $(x^3 - yz)^2$

PRODUCT OF A SUM AND DIFFERENCE

Here we have the product of two binomials, one of which is the sum of two terms and the other is the difference of the same two terms. When we find this product, we obtain

$$\begin{aligned}(a + b)(a - b) &= a(a) + a(-b) + b(a) + b(-b)\\ &= a^2 - ab + ab - b^2\\ &= a^2 - b^2.\end{aligned}$$

So we obtain a difference of squares. The picture for this special product is shown in Figure 2.6.

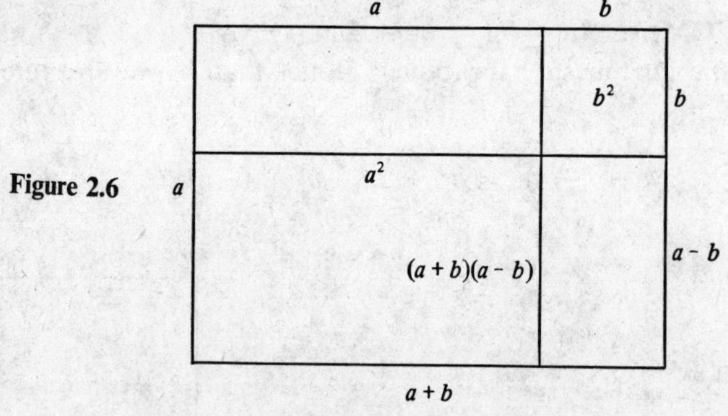

Figure 2.6

PRODUCT OF A SUM AND DIFFERENCE, $(a + b)(a - b) = a^2 - b^2$.

Example 2. Product of a sum and difference.

a. $(x + 1)(x - 1) = x^2 - (1)^2 = x^2 - 1$
b. $(ab + 3c)(ab - 3c) = (ab)^2 - (3c)^2 = a^2b^2 - 9c^2$

Exercise Set 2.2 (continued)

In exercises 9–16, find the product.

9. $(x + 2)(x - 2)$ 10. $(a + 6)(a - 6)$ 11. $(2x + 5y)(2x - 5y)$

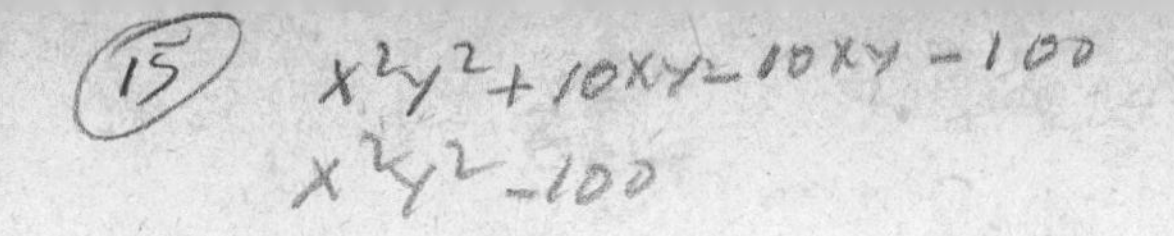

12. $(3x - 2y)(3x + 2y)$
13. $(x^2 + y^2)(x^2 - y^2)$
14. $(x^3 - 1)(x^3 + 1)$
15. $(xy + 10)(xy - 10)$
16. $(xyz + a)(xyz - a)$

BINOMIALS WITH A COMMON TERM

We wish to find a product of the form $(x + a)(x + b)$.

Here we have a product of two binomials, each of which has a common term, which we've called x. When we multiply these binomials, we obtain

$$\begin{aligned}(x + a)(x + b) &= xx + ax + xb + ab\\ &= x^2 + ax + bx + ab.\end{aligned}$$

Now the two terms $ax + bx$ can be combined, $(a + b)x$ (Distributive property). So the product is actually a trinomial,

$$(x + a)(x + b) = x^2 + (a + b)x + ab.$$

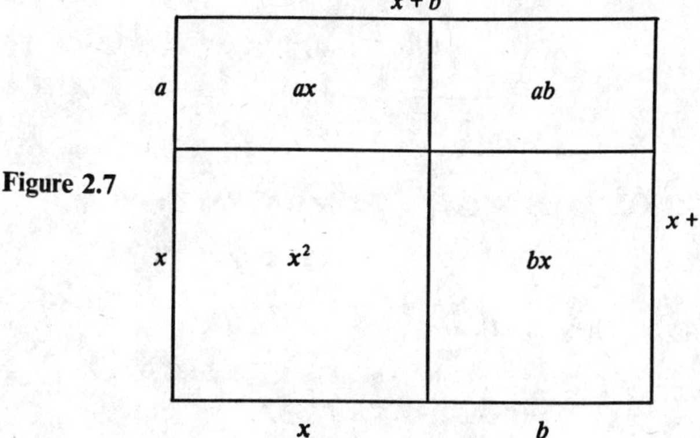

Figure 2.7

PRODUCT OF BINOMIALS WITH COMMON TERM,
$(x + a)(x + b) = x^2 + (a + b)x + ab.$

Example 3. The product $(x + a)(x + b)$.

a. $(x + 2)(x + 3) = x^2 + (2 + 3)x + 2 \cdot 3 = x^2 + 5x + 6$
b. $(x + 1)(x - 4) = x^2 + (1 - 4)x + (1)(-4) = x^2 - 3x - 4$
c. $(y - 1)(y - 3) = y^2 + (-1 - 3)y + (-1)(-3) = y^2 - 4y + 3$

Exercise Set 2.2 (continued)

In exercises 17–24, find the indicated product.

17. $(x + 3)(x + 5)$
18. $(x - 4)(x + 2)$
19. $(x - 6)(x + 7)$
20. $(x - 12)(x - 2)$
21. $(x + 12)(x - 4)$
22. $(x + 10)(x + 8)$

23. $(x + y)(x + 2y)$ 24. $(x - 3y)(x + 5y)$

BINOMIALS WITH A COMMON FACTOR

We wish to find a product of the form $(ax + b)(cx + d)$.

Here we have a product of two binomials; each has a term with a common factor we've called x. When we find the product, we obtain

$$\begin{aligned}(ax + b)(cx + d) &= ax(cx) + ax(d) + b(cx) + bd \\ &= acx^2 + adx + bcx + bd \\ &= acx^2 + (ad + bc)x + bd.\end{aligned}$$

So we have a trinomial for the product because we can combine two terms. The following scheme suggests how to remember the three terms in the product.

Last Term

First Term

$(ax + b)(cx + d)$

Middle Term

PRODUCT OF TWO BINOMIALS WITH COMMON FACTOR, $(ax + b)(cx + d) = acx^2 + (ad + bc)x + bd$.

Example 4. Product of $(ax + b)(cx + d)$.

a. $(2x + 3)(5x + 1) = 2 \cdot 5x^2 + (2 \cdot 1 + 3 \cdot 5)x + 3 \cdot 1$
$= 10x^2 + 17x + 3$

b. $(4x - 2)(3x - 7) = 4 \cdot 3x^2 + (4(-7) + (-2)3)x + (-2)(-7)$
$= 12x^2 - 34x + 14$

c. $(2y + 5)(y - 6) = 2 \cdot 1y^2 + (2(-6) + 5(1))y + (5)(-6)$
$= 2y^2 - 7y - 30$

Exercise Set 2.2 (continued)

In exercises 25–36, find the indicated product.

25. $(2x + 8)(3x + 4)$ 26. $(2x + 7)(x - 6)$ 27. $(2x - 15)(5x - 3)$

28. $(4x - 5)(2x - 1)$ 29. $(5x + 1)(3x + 1)$ 30. $(x - 2)(4x + 7)$

31. $(3x + 7)(3x + 4)$ 32. $(2x - 5)(x - 4)$ 33. $(2x + a)(3x - a)$

34. $(2x - y)(3x - y)$ 35. $(6x + ay)(x + ay)$ 36. $(4x + 3y)(5x - y)$

SOME CUBES

When we cube the binomial $a + b$, we obtain

$$\begin{aligned}(a + b)^3 = (a + b)(a + b)^2 &= (a + b)(a^2 + 2ab + b^2)\\ &= a(a^2) + a(2ab) + a(b^2) + b(a^2) + b(2ab) + b(b^2)\\ &= a^3 + 2a^2b + ab^2 + a^2b + 2ab^2 + b^3\\ &= a^3 + 3a^2b + 3ab^2 + b^3\end{aligned}$$

CUBE OF A BINOMIAL, $(a + b)^3 = a^3 + 3a^2b + 3ab^2 + b^3$.

Example 5. Cube of a binomial.

a. $(x + 2)^3 = x^3 + 3(x^2)2 + 3(x)2^2 + 2^3$
$\qquad = x^3 + 6x^2 + 12x + 8$

b. $(x^2 - 1)^3 = (x^2)^3 + 3(x^2)^2(-1) + 3(x^2)(-1)^2 + (-1)^3$
$\qquad = x^6 - 3x^4 + 3x^2 - 1$

SPECIAL BINOMIAL–TRINOMIAL, $(a - b)(a^2 + ab + b^2) = a^3 - b^3$.

Here we have a product of a difference and a special trinomial. We obtain

$$\begin{aligned}(a - b)(a^2 + ab + b^2) &= a \cdot a^2 + a(ab) + a(b^2) + (-b)a^2 + (-b)ab + (-b)b^2\\ &= a^3 + a^2b + ab^2 - a^2b - ab^2 - b^3\\ &= a^3 - b^3\end{aligned}$$

So we have a difference of cubes.

Example 6. Product $(a - b)(a^2 + ab + b^2)$.

$$(x - 1)(x^2 + x + 1) = x^3 - 1^3 = x^3 - 1$$

SPECIAL BINOMIAL–TRINOMIAL, $(a + b)(a^2 - ab + b^2) = a^3 + b^3$.

Here we have a product of a sum and a special trinomial. We obtain

$$\begin{aligned}(a + b)(a^2 - ab + b^2) &= a(a^2) + a(-ab) + a(b^2) + b(a^2) + b(-ab) + b(b^2)\\ &= a^3 - a^2b + ab^2 + a^2b - ab^2 + b^3\\ &= a^3 + b^3\end{aligned}$$

Example 7. Product $(a + b)(a^2 - ab + b^2)$.

$$(x + 1)(x^2 - x + 1) = x^3 + 1^3 = x^3 + 1$$

The purpose of this section is to help you recognize and perform special products quickly and efficiently. Many steps that have been included in the examples can be done mentally.

Exercise Set 2.2 (continued)

In exercises 37–48, find the product.

37. $(x + 1)^3$
38. $(x - 1)^3$
39. $(xy + 3)^3$
40. $(x^2 - 2)^3$
41. $(a - 2)(a^2 + 2a + 4)$
42. $(2a - 1)(4a^2 + 2a + 1)$
43. $(x - 3y)(x^2 + 3xy + 9y^2)$
44. $(x^2 - y^2)(x^4 + x^2y^2 + y^4)$
45. $(x + 5)(x^2 - 5x + 25)$
46. $(2x + 1)(4x^2 - 2x + 1)$
47. $(x^2 + 3)(x^4 - 3x^2 + 9)$
48. $(x^3 + y^3)(x^6 - x^3y^3 + y^6)$

section 3 • Factoring, I

We have learned how to multiply algebraic expressions. Sometimes when we deal with fractions or solve equations, we have to find the factors of an algebraic expression. That is, we wish to *factor* the expression. Factoring is generally trickier than multiplying because we have to look at the end product and try to figure out what the factors were. Doing this successfully is mostly a matter of practice.

COMMON MONOMIAL FACTORS

Suppose we have a factor that occurs in each term of a sum. For example, a occurs in each term of the binomial $ab + ac$ and the trinomial $ab + ac + ad$. We use the distributive property (Section 2.1) to write

$$ab + ac = a(b + c) \quad \text{and} \quad ab + ac + ad = a(b + c + d).$$

That is, we factor an a from each term of the sum and bring it outside the remaining sum. When you are factoring, a good rule is to look for monomial factors first.

Example 1. Common monomial factors.

a. $6xy + 3xz = 3x(2y) + 3x(z) = 3x(2y + z)$

b. $4ax^2 + 4a^2x = 4ax(x) + 4ax(a) = 4ax(x + a)$

c. $x^4y + x^2y^3 = x^2y(x^2) + x^2y(y^2) = x^2y(x^2 + y^2)$

d. $6a^3b^2 + 3a^4b^2 - 3a^3b^3 = 3a^3b^2(2) + 3a^3b^2(a) + 3a^3b^2(-b)$
$= 3a^3b^2(2 + a - b)$

Exercise Set 2.3

In exercises 1–8, factor the expression by finding a monomial factor.

1. $5xy + 25x$
2. $11ab + 4bc$
3. $3a - 9ax$
4. $4xy - 2x^2$
5. $a^3b + ab^2$
6. $a^3b^2 + abc$
7. $x^2y + xy^2 - xy$
8. $ax^2 - bxy - cx$

FACTORING THE SQUARE OF A BINOMIAL

Suppose we have a trinomial of the form $a^2 + 2ab + b^2$. From the special products of Section 2.2, we recognize this expression as the square of the binomial $a + b$. Thus, we have the factorization

$$a^2 + 2ab + b^2 = (a + b)^2.$$

Example 2. Factoring the square of a binomial.

a. $x^2 + 2x + 1 = (x + 1)^2$
b. $y^2 - 12y + 36 = y^2 - 2 \cdot y \cdot 6 + 6^2 = (y - 6)^2$
c. $x^2y^2 + 2xyz + z^2 = (xy)^2 + 2(xy)z + z^2 = (xy + z)^2$
d. $x^4 - 10x^2y + 25y^2 = (x^2)^2 - 2(x^2)(5y) + (5y)^2 = (x^2 - 5y)^2$

Exercise Set 2.3 (continued)

In exercises 9–16, factor each given expression as the square of a binomial.

9. $x^2 + 14x + 49$
10. $a^2 + 4ab + 4b^2$
11. $a^2 - 6a + 9$
12. $x^2 - 8xy + 16y^2$
13. $x^2y^2 + 10xy + 25$
14. $a^2 + b^2c^2$
15. $4x^2 + 12x + 9$
16. $9x^2 - 30x + 25$

FACTORING A DIFFERENCE OF SQUARES

If we have an algebraic expression of the form $a^2 - b^2$, which is a difference of two squares, we can use a special product of Section 2.2 in reverse to factor it:

$$a^2 - b^2 = (a + b)(a - b).$$

This factorization comes up frequently and is a good one to watch for.

Example 3. Factoring a difference of squares.

a. $x^2 - 1 = x^2 - 1^2 = (x + 1)(x - 1)$
b. $x^2y^2 - 9 = (xy)^2 - 3^2 = (xy + 3)(xy - 3)$

c. $x^6 - a^2 = (x^3)^2 - a^2 = (x^3 + a)(x^3 - a)$
d. $x^2 - 100y^2z^4 = x^2 - (10yz^2)^2 = (x + 10yz^2)(x - 10yz^2)$

Exercise Set 2.3 (continued)

In exercises 17–24, factor the difference of squares.

17. $x^2 - 64$	18. $a^2 - 81$	19. $4x^2 - 225y^2$	20. $25x^2 - 121z^2$
21. $36a^2b^2 - 49$	22. $a^2x^2 - b^2y^2$	23. $-49 + 4x^2$	24. $-100t^2 + 1$

IRREDUCIBLE FACTORS

A factorization frequently requires a combination of two or more types of factoring.

Example 4. Factor $3x^5 - 6x^3 + 3x$.

Solution. We start with a monomial factor of $3x$ and obtain

$$3x^5 - 6x^3 + 3x = 3x(x^4 - 2x^2 + 1)$$

We aren't finished yet, because

$$x^4 - 2x^2 + 1 = (x^2)^2 - 2(x^2) + 1 = (x^2 - 1)^2.$$

But $x^2 - 1$ is a difference of squares, so

$$x^2 - 1 = (x + 1)(x - 1).$$

Putting all this together, we obtain the factorization

$$3x^5 - 6x^3 + 3x = 3x(x^2 - 1)^2 = 3x[(x + 1)(x - 1)]^2.$$

This cannot be factored further. It is also correct to write

$$3x^5 - 6x^3 + 3x = 3x(x + 1)^2(x - 1)^2.$$

When factoring an algebraic expression, we keep factoring until we reach factors that cannot themselves be factored. An algebraic expression that cannot be factored is called *irreducible*. So we wish, then, to factor an algebraic expression into irreducible factors.

How do we recognize an irreducible algebraic expression? To obtain a precise answer to this question, we must specify what numbers we permit in a factorization. For example, $x^2 - 2$ may be considered irreducible but, if we permit square roots of numbers, then $x^2 - 2$ can be factored as

$$x^2 - 2 = (x + \sqrt{2})(x - \sqrt{2}).$$

In Chapter 5 we will consider factorizations of this type. However, if we confine ourselves to factorizations in which the terms only involve *integers*, then $x^2 - 2$ is irreducible. We can say that $x^2 - 2$ is *irreducible over the integers* but is not *irreducible over the real numbers*. In this chapter, we confine ourselves to factorizations over the integers.

The algebraic expressions $a + b$ and $a - b$ are irreducible over the integers. With algebraic expressions involving exponents, it's harder to tell. For example, $a^2 + b^2$ is irreducible over the integers; but of course, $a^2 - b^2$ can be factored. With experience, it becomes easier to recognize irreducible algebraic expressions.

Example 5. Factor $x^4 - y^4$.

Solution. We treat this expression as a difference of squares:

$$\begin{aligned} x^4 - y^4 &= (x^2)^2 - (y^2)^2 \\ &= (x^2 + y^2)(x^2 - y^2) \\ &= (x^2 + y^2)(x + y)(x - y) \end{aligned}$$

These factors are irreducible over the integers.

Example 6. Factor $x^2 + 2x + 1 - y^2$.

Solution. Here it helps to group the terms.

$$\begin{aligned} x^2 + 2x + 1 - y^2 &= (x^2 + 2x + 1) - y^2 \\ &= (x + 1)^2 - y^2 \\ &= [(x + 1) + y][(x + 1) - y] \\ &= (x + y + 1)(x - y + 1) \end{aligned}$$

Example 7. Factor $32x + 2x^5 + 16x^3$.

Solution.

$$\begin{aligned} 32x + 2x^5 + 16x^3 &= 2x(x^4 + 8x^2 + 16) \\ &= 2x[(x^2)^2 + 8(x^2) + 4^2] \\ &= 2x(x^2 + 4)^2 \end{aligned}$$

These factors are irreducible over the integers.

Exercise Set 2.3 (continued)

In exercises 25–48, factor the given algebraic expression into a product of irreducible factors.

25. $x^3y - xy^3$	26. $2x^3y - 32xy$	27. $x^4y - x^2y^3$
28. $x^3 + 2x^2 + x$	29. $6x^2 - 24x + 24$	30. $x^3y + 2x^2y^2 + xy^3$

31. $x^6 + 2x^3y + y^2$
32. $x^6 + 2x^3y^2 + y^4$
33. $2x^4y^2 + x^7y + xy^3$
34. $x^6 - y^2$
35. $x^6 - y^4$
36. $x^8 - y^2$
37. $x^8 - y^4$
38. $x^8 - 2x^4y^4 + y^8$
39. $x^4 + 2x^2 + 1$
40. $x^4 - 18x^2 + 81$
41. $(x + 3)^2 - 4y^2$
42. $(3x + 2)^2 - 9y^2$
43. $x^2 + 4x + 4 - y^2$
44. $4x^2 + 4x + 1 - 9y^2$
45. $4x^2 - y^2 - 2y - 1$
46. $4x^2 - 4y^2 + 4y - 1$
47. $x^2 + 6x + 9 - y^2 - 2y - 1$
48. $x^2 + 4x - 9y^2 - 6y + 3$

CALCULATOR COMPUTATIONS

Sometimes we must perform a computation that occurs in the form $ab + ac$. It is easier to perform such a computation on an electronic calculator if we factor the monomal factor a and write $(b + c)a$. Then we can enter $\mathbf{b + c \times a =}$, and we do not need to write down intermediate steps. This factoring saves calculation time.

Example 8. CALCULATOR COMPUTATIONS

a. $(3.67)(4.42) + (3.67)(9.12) = (4.42 + 9.12)(3.67) = (13.54)(3.67) = 49.6918$

b. $(6.28)(18.12)^2 + (6.28)(18.12)(26.15) = (18.12 + 26.15)(18.12)(6.28)$
$= (44.27)(18.12)(6.28) = 5037.6426$

Exercise Set 2.3 (continued)

In exercises 49–56, use a calculator to perform the computation.

49. $(1.19)(12.51) + (1.19)(8.62)$
50. $(13.14)(56.21) + (13.14)(43.15)$
51. $(44.63)(17.22) - (29.68)(17.22)$
52. $(53.18)(22.54) - (42.86)(22.54)$
53. $(6.28)(3.9)^2 + (6.28)(3.9)(10.8)$
54. $(6.28)(4.9)^2 + (6.28)(4.9)(8.2)$
55. $(6.28)(8.4)^2 + (6.28)(8.4)(18.8)$
56. $(6.28)(5.2)^2 + (6.28)(5.2)(4.2)$

section 4 • Factoring, II

In this section we discuss a type of factoring that is important in Chapter 5 for solving quadratic equations and inequalities.

FACTORING TRINOMIALS OF THE FORM $x^2 + bx + c$

A trinomial of the form $x^2 + bx + c$ may be irreducible or it may factor. For example, the trinomial $x^2 + x + 1$ is irreducible over the integers, whereas $x^2 + 2x + 1$ factors as $(x + 1)^2$. If a trinomial of this form factors, then it factors as

$$x^2 + bx + c = (x + u)(x + v)$$

for some u and v. When we find this product, we obtain

$$(x + u)(x + v) = x^2 + (u + v)x + uv.$$

So to obtain a factorization, we must be able to find u and v so that

$$b = u + v \qquad \text{and} \qquad c = uv.$$

If it is impossible to find the u and v, the trinomial is irreducible.

Example 1. Factor $x^2 + 6x + 8$.

Solution. We try to find u and v so that

$$x^2 + 6x + 8 = (x + u)(x + v) = x^2 + (u + v)x + uv.$$

We start with the fact that we must have $uv = 8$. Over the integers, there are four possibilities for u and v:

$$8 = 1 \cdot 8, \qquad 8 = (-1)(-8), \qquad 8 = 2 \cdot 4, \qquad 8 = (-2)(-4).$$

Since $2 + 4 = 6$, we try $(x + 2)(x + 4)$ and obtain the factorization

$$x^2 + 6x + 8 = (x + 2)(x + 4).$$

Example 2. Factor $x^2 + x - 12$.

Solution. Again we want to find u and v so that

$$x^2 + x - 12 = (x + u)(x + v) = x^2 + (u + v)x + uv.$$

We must have $uv = -12$; and so the possibilities among the integers are:

$$-12 = 1(-12), \qquad -12 = 2(-6), \qquad -12 = 3(-4),$$
$$-12 = (-1)12, \qquad -12 = (-2)6, \qquad -12 = (-3)4.$$

Since $4 + (-3) = 1 = u + v$, we try $(x + 4)(x - 3)$. We obtain the factorization

$$x^2 + x - 12 = (x + 4)(x - 3).$$

Example 3. Factor $x^2 - 7x + 6$.

Solution. Find u and v so $x^2 - 7x + 6 = (x + u)(x + v)$. Since $uv = 6$, we consider the possibilities:

$$6 = 1 \cdot 6, \qquad 6 = (-1)(-6), \qquad 6 = 2 \cdot 3, \qquad 6 = (-2)(-3).$$

Since $(-1) + (-6) = -7$, we try $(x - 1)(x - 6)$. We obtain, the factorization

$$x^2 - 7x + 6 = (x - 1)(x - 6).$$

Example 4. Factor $x^2 + 2x + 2$.

Solution. Again we try to find u and v so that

$$x^2 + 2x + 2 = (x + u)(x + v) = x^2 + (u + v)x + uv.$$

Since $uv = 2$, we have the possibilities

$$2 = 1 \cdot 2 \quad \text{and} \quad 2 = (-1)(-2).$$

Since $1 + 2 = 3$ and $(-1) + (-2) = -3$, it is impossible to have both $uv = 2$ and $u + v = 2$. Therefore, $x^2 + 2x + 2$ is irreducible over the integers.

Exercise Set 2.4

In exercises 1–20, factor the expression into a product of irreducible factors or specify that it is irreducible over the integers.

1. $x^2 + 3x - 4$
2. $x^2 + 5x + 6$
3. $x^2 - 7x + 12$
4. $x^2 + 11x + 30$
5. $x^2 + 2x - 24$
6. $x^2 - 2x - 24$
7. $x^2 - 2x + 24$
8. $y^2 + 2y - 80$
9. $x^2 - 1$
10. $x^2 - 4$
11. $x^2 + 1$
12. $x^2 + 4$
13. $x^2 - 2$
14. $x^2 - 3$
15. $x^2 + 2$
16. $x^2 + 3$
17. $x^2 + 10x + 21$
18. $x^2 - 13x + 36$
19. $x^2 + x - 1$
20. $x^2 - x + 1$

FACTORING TRINOMIALS OF THE FORM $ax^2 + bx + c$

Previously, we considered trinomials of this form with $a = 1$. If a trinomial of this form factors over the integers, then the factorization has the form

$$ax^2 + bx + c = (sx + u)(tx + v)$$

for some integers s, t, u, v. So now we have four integers to choose. We have

$$(sx + u)(tx + v) = stx^2 + (sv + ut)x + uv,$$

so we must have

$$a = st, \qquad b = sv + ut, \qquad c = uv.$$

Example 5. Factor $2x^2 + 5x - 3$.

Solution. We wish to find s, t, u, v so that

$$2x^2 + 5x - 3 = (sx + u)(tx + v) = stx^2 + (sv + ut)x + uv.$$

Since $2 = st$ and $-3 = uv$, we try the possibilities

$$2 = 1 \cdot 2, \qquad -3 = 1(-3), \qquad -3 = (-1)3.$$

We do not have to try the negative factors of 2, namely $2 = (-1)(-2)$; if there is a factorization, it will follow from the positive factors of 2. Since $2 \cdot 3 + (-1)1 = 5$, we try $(2x - 1)(x + 3)$. When we find the product, we obtain the factorization

$$2x^2 + 5x - 3 = (2x - 1)(x + 3).$$

Example 6. Factor $6x^2 - 11x - 10$.

Solution. We wish to find s, t, u, v so that

$$6x^2 - 11x - 10 = (sx + u)(tx + v) = stx^2 + (sv + ut)x + uv.$$

Since $6 = st$ and $-10 = uv$, we look at the possibilities:

$$\begin{array}{ll} 6 = 1 \cdot 6 & -10 = 1(-10) \\ \boxed{6 = 2 \cdot 3} & -10 = (-1)10 \\ & \boxed{-10 = 2(-5)} \\ & -10 = (-2)5. \end{array}$$

Since $2 \cdot 2 + (-5)3 = -11$, we try $(2x - 5)(3x + 2)$. We obtain the factorization

$$6x^2 - 11x - 10 = (2x - 5)(3x + 2).$$

Example 7. Factor $2x^2 + x - 2$.

Solution. We wish to find s, t, u, v so that

$$2x^2 + x - 2 = (sx + u)(tx + v).$$

So $st = 2$ and $uv = -2$ and the possibilities are

$$2 = 2(1), \qquad -2 = (-2)1, \qquad -2 = 2(-1).$$

Since none of these possibilities gives a factorization, $2x^2 + x - 2$ is irreducible over the integers.

Example 8. Factor $4y^4 + 7y^2 - 2$.

Solution. We treat this as an algebraic expression in y^2, namely $4(y^2)^2 + 7(y^2) - 2$. As an expression in y^2, we find the factorization

$$4(y^2)^2 + 7(y^2) - 2 = (y^2 + 2)(4y^2 - 1).$$

Now $y^2 + 2$ is irreducible over the integers, but

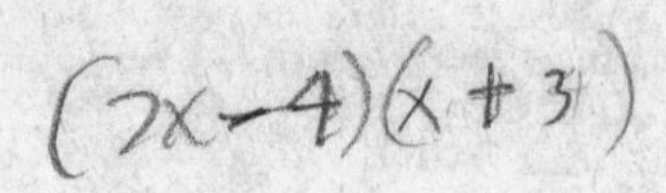

$$4y^2 - 1 = (2y + 1)(2y - 1).$$

Putting these together, we have

$$4y^4 + 7y^2 - 2 = (y^2 + 2)(2y + 1)(2y - 1).$$

Exercise Set 2.4 (continued)

In exercises 21–48, factor the expression into a product of irreducible factors or specify that it is irreducible over the integers.

21. $2x^2 - 5x - 3$
22. $2x^2 + 7x + 5$
23. $3x^2 + 14x + 8$
24. $3x^2 - 13x + 4$
25. $5x^2 - 13x + 6$
26. $7x^2 + 17x - 12$
27. $4x^2 + 8x + 3$
28. $4x^2 + x - 18$
29. $2x^2 - x - 1$
30. $2x^2 + x + 1$
31. $4x^2 - 4x + 1$
32. $4x^2 + 4x + 1$
33. $4x^2 + 21x - 18$
34. $6x^2 - 13x + 6$
35. $4y^2 + 3y - 1$
36. $8y^2 - 17y + 2$
37. $15x^2 + 14x + 3$
38. $12x^2 + 32x + 21$
39. $20x^2 + 9x - 18$
40. $20x^2 - 39x + 18$
41. $10x^2 + 12x - 16$
42. $5x^2 + 5x - 5$
43. $5x^2 + 5x + 4$
44. $7x^2 - 2x + 2$
45. $x^4 + 10x^2 + 21$
46. $x^4 - 3x^2 + 1$
47. $3x^4 - 5x^2 + 2$
48. $x^4 - 13x^2 + 36$

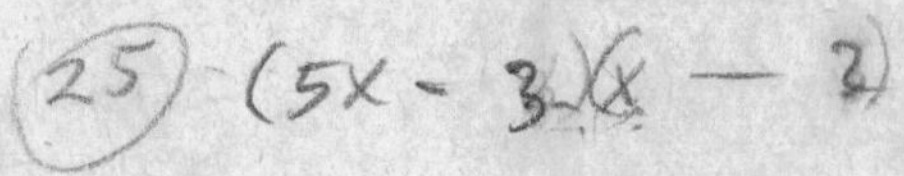

section 5 • Factoring, III

We continue to examine types of factoring.

FACTORING BY GROUPING OF TERMS

Sometimes we can factor an expression by grouping the terms together properly. Suppose we have an algebraic expression of the form $ax + ay + bx + by$. Here we can factor an a from the first two terms and a b from the second two terms to obtain

$$ax + ay + bx + by = a(x + y) + b(x + y).$$

Now we can factor an $x + y$ from each term of the sum to obtain

$$ax + ay + bx + by = (a + b)(x + y).$$

So this type of factoring leads to a product of binomials.

Example 1. Factoring by grouping of terms.

a. $3x + 6y + xz + 2yz = 3(x + 2y) + z(x + 2y) = (3 + z)(x + 2y)$

b. $ab + 2as - bt - 2st = a(b + 2s) - t(b + 2s) = (a - t)(b + 2s)$

c. $x^2y^2 - y^2 + x^2 - 1 = y^2(x^2 - 1) + (x^2 - 1) = (y^2 + 1)(x^2 - 1)$
$= (y^2 + 1)(x + 1)(x - 1)$

Exercise Set 2.5

In exercises 1–8, factor the expression into a product of irreducible factors.

1. $xy + 3x + 2y + 6$
2. $xy + 3x - 2y - 6$
3. $xy - 3x + 2y - 6$
4. $xy - 3x - 2y + 6$
5. $r^2 + rs + rt + st$
6. $2s^2 - rs + 4rs - 2r^2$
7. $abx^2 - bxy + axz - yz$
8. $ax^2 + bxy - axy - by^2$

FACTORING THE CUBE OF A BINOMIAL

Sometimes an algebraic expression of the form $a^3 + 3a^2b + 3ab^2 + b^3$ will occur. We recognize this as a special product, the cube of a binomial. So we have the factorization

$$a^3 + 3a^2b + 3ab^2 + b^3 = (a + b)^3.$$

Example 2. Factoring the cube of a binomial.

a. $x^3 - 3x^2 + 3x - 1 = x^3 + 3x^2(-1) + 3x(-1)^2 + (-1)^3 = (x - 1)^3$

b. $x^6 + 6x^4 + 12x^2 + 8 = (x^2)^3 + 3(x^2)^2(2) + 3(x^2)2^2 + 2^3 = (x^2 + 2)^3$

Exercise Set 2.5 (continued)

In exercises 9–12, factor the expression as the cube of a binomial.

9. $x^3 + 6x^2 + 12x + 8$
10. $x^3 - 9x^2 + 27x - 27$
11. $x^6 + 3x^4 + 3x^2 + 1$
12. $x^6 - 3x^4 + 3x^2 - 1$

FACTORING A DIFFERENCE OF CUBES

If we have a difference of cubes, an expression of the form $a^3 - b^3$, we use a special product to obtain the factorization

$$a^3 - b^3 = (a - b)(a^2 + ab + b^2).$$

Example 3. Factoring a difference of cubes.

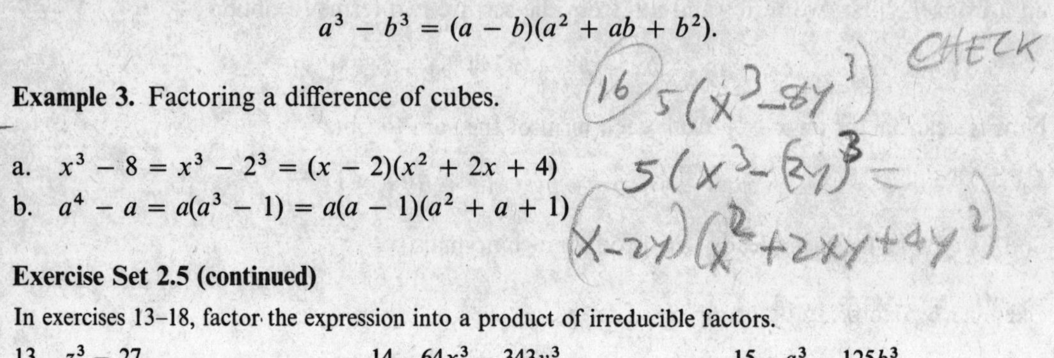

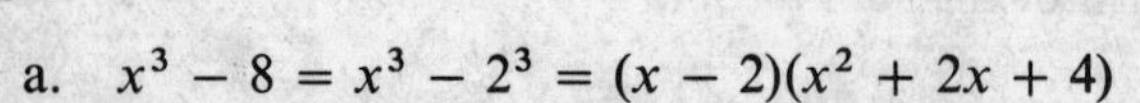

a. $x^3 - 8 = x^3 - 2^3 = (x - 2)(x^2 + 2x + 4)$
b. $a^4 - a = a(a^3 - 1) = a(a - 1)(a^2 + a + 1)$

Exercise Set 2.5 (continued)

In exercises 13–18, factor the expression into a product of irreducible factors.

13. $z^3 - 27$
14. $64x^3 - 343y^3$
15. $a^3 - 125b^3$
16. $5x^3 - 40y^3$
17. $a^4b - ab^4$
18. $a^5 - a^2$

FACTORING A SUM OF CUBES

If we have a sum of cubes, an expression of the form $a^3 + b^3$, we use a special product to obtain the factorization

$$a^3 + b^3 = (a + b)(a^2 - ab + b^2).$$

Example 4. Factoring a sum of cubes.

a. $x^3 + 8 = x^3 + 2^3 = (x + 2)(x^2 - x \cdot 2 + 2^2) = (x + 2)(x^2 - 2x + 4)$
b. $a^5 + a^2 = a^2(a^3 + 1) = a^2(a + 1)(a^2 - a + 1)$

Exercise Set 2.5 (continued)

In exercises 19–24, factor the expression into a product of irreducible factors.

19. $z^3 + 27$
20. $8a^3 + 1$
21. $x + x^4$
22. $x^4y + xy^4$
23. $3a^4b + 24ab$
24. $x^3 + a^6b^6$

FACTORING BY ADDITION AND SUBTRACTION OF A TERM

This factoring technique is tricky, and it takes some practice to recognize when it can be used. It consists of adding and subtracting a term to an algebraic expression to effect a factorization—usually a difference of squares.

Example 5. Factoring by addition and subtraction of a term.

a. $$\begin{aligned} x^4 + x^2y^2 + y^4 &= x^4 + x^2y^2 + y^4 + x^2y^2 - x^2y^2 \\ &= (x^4 + 2x^2y^2 + y^4) - x^2y^2 \\ &= (x^2 + y^2)^2 - x^2y^2 \\ &= [(x^2 + y^2) + xy][(x^2 + y^2) - xy] \\ &= (x^2 + xy + y^2)(x^2 - xy + y^2) \end{aligned}$$

b. $$\begin{aligned} x^4 - 8x^2 + 4 &= x^4 - 8x^2 + 4 + 4x^2 - 4x^2 \\ &= (x^4 - 4x^2 + 4) - 4x^2 = (x^2 - 2)^2 - 4x^2 \\ &= [(x^2 - 2) + 2x][(x^2 - 2) - 2x] \\ &= (x^2 + 2x - 2)(x^2 - 2x - 2) \end{aligned}$$

By the techniques of Section 2.4, both of these factors are irreducible over the integers.

Exercise Set 2.5 (continued)

In exercises 25–32, factor the expression into a product of irreducible factors.

25. $x^4 + x^2 + 1$
26. $x^4 - 3x^2y^2 + y^4$
27. $x^4 - 6x^2y^2 + y^4$
28. $x^4 - 7x^2y^2 + y^4$
29. $x^6 + y^6$
30. $x^2 - y^2 + x - y$
31. $x^3 + x^2 + x + 1$
32. $x^5 + x^3 + x^2 + 1$

REVIEW OF FACTORING

We outline the types of factoring we have considered.

1. Check first for common monomial factors.
2. A binomial expression (two terms)
   - factors if it is a difference of two squares;
   - is irreducible if it is a sum of two squares;
   - factors if it is a sum or difference of two cubes.
3. A trinomial expression (three terms) should be written in decreasing powers of the variable. Check whether it is a binomial square. If it is not, see if it can be factored by trial and error, considering the integer coefficients. In some special cases you may be able to add and subtract a term and then factor.
4. With an expression with four or more terms, try grouping terms.

Exercise Set 2.5 (continued)

In exercises 33–48, factor the expression into a product of irreducible factors.

33. $9x^2 - 49$
34. $x^2 + 10x + 25$
35. $x^2 - 12x + 36$
36. $16x^2 - 81$
37. $x^3 + x^2 - 12x$
38. $x^4 - 10x^3 + 21x^2$
39. $2x^2 - 5x - 12$
40. $3x^2 + 10x - 8$
41. $4x^2 + 8x + 3$
42. $6x^2 - 7x + 2$
43. $-49x + 4x^3$
44. $4x^2 + 9 + 12x$
45. $x^3 + 2x^2 + x + 2$
46. $x^3 + 3x^2 - x - 3$
47. $x^4 + 3x^2 - 4$
48. $x^4 - 3x^2 + 2$

section 6 • Equality of Fractions

In this section we will gain proficiency in dealing with fractions; in particular, we perform the four operations of addition, subtraction, multiplication, and division of fractions. Recall that, if we have a fraction written in the form a/b, then a is the *numerator* and b is the *denominator* of the fraction. Recall also that we must have $b \neq 0$ (Section 1.2); the expression $a/0$ is not defined.

In this section we define equality of two fractions and use this definition to reduce fractions, build fractions, and solve proportions.

DEFINITION OF EQUALITY OF FRACTIONS

A first consideration in dealing with fractions is that two fractions may be equal even though their numerators are not equal and their denominators are not equal. For example, $1/2 = 2/4$, as shown in Figure 2.8. So the first step in understanding fractions is to know when two fractions are equal; that is, we must define equality of fractions.

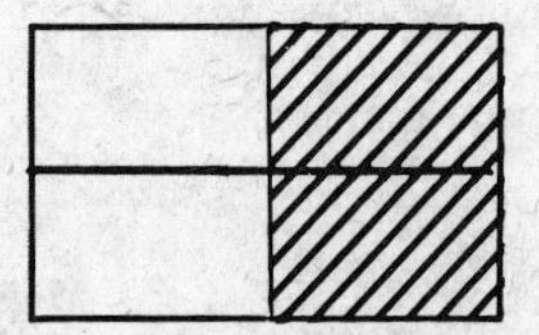

Figure 2.8

DEFINITION OF EQUALITY OF FRACTIONS. If $b \neq 0$ and $d \neq 0$, then

$$a/b = c/d \quad \text{whenever} \quad ad = bc.$$

Example 1. Equality and inequality of fractions.

a. $\dfrac{1}{2} = \dfrac{2}{4}$ since $1 \cdot 4 = 2 \cdot 2$.

b. $\dfrac{1}{2} \neq \dfrac{3}{5}$ since $1 \cdot 5 \neq 2 \cdot 3$.

c. $\dfrac{x}{y} = \dfrac{x^2}{xy}$ since $x \cdot xy = y \cdot x^2$

d. $\dfrac{x+1}{x-2} = \dfrac{5(x+1)}{5(x-2)}$ for $x \neq 2$ since $(x+1) \cdot 5(x-2) = (x-2) \cdot 5(x+1)$.

e. $\dfrac{x+1}{y+1} \neq \dfrac{x}{y}$ for $x \neq y$ since $(x+1)y \neq (y+1)x$.

Exercise Set 2.6

In exercises 1–10, determine whether the two fractions are equal or not equal.

1. $\dfrac{7}{3}, \dfrac{21}{9}$ 2. $\dfrac{7}{6}, \dfrac{10}{12}$ 3. $\dfrac{1}{2}, -\dfrac{1}{2}$ 4. $\dfrac{4}{9}, \dfrac{20}{45}$

5. $\dfrac{8}{10}, \dfrac{12}{15}$ 6. $\dfrac{x^2}{x^3}, \dfrac{1}{x}$ 7. $\dfrac{x^4}{x^3}, \dfrac{1}{x}$ 8. $\dfrac{1}{-x}, \dfrac{-1}{x}$

9. $\dfrac{a}{b}, \dfrac{-a}{-b}$ 10. $\dfrac{-a}{b}, \dfrac{a}{-b}$

REDUCING FRACTIONS

The definition of equality of fractions has an important consequence. Suppose $x \neq 0$; and consider the fraction ax/bx. This fraction has a common factor of x in both numerator and denominator. By the definition of equality of fractions, we see that

$$\frac{ax}{bx} = \frac{a}{b} \qquad \text{since} \qquad ax \cdot b = bx \cdot a.$$

We say that the fraction ax/bx can be *reduced to the* fraction a/b.

REDUCING FRACTIONS. If $b \neq 0$ and $x \neq 0$, then

$$\frac{ax}{bx} = \frac{a}{b}.$$

Frequently we want to reduce a fraction by removing any factors common to both numerator and denominator, because the fraction a/b is usually easier to work with than the more complicated fraction ax/bx. When a fraction has no factors common to the numerator and denominator, we say it is in *lowest terms*.

Example 2. Reducing fractions.

a. $\dfrac{2}{4} = \dfrac{1 \cdot 2}{2 \cdot 2} = \dfrac{1}{2}$

b. $\dfrac{26}{39} = \dfrac{2 \cdot 13}{3 \cdot 13} = \dfrac{2}{3}$

c. $\dfrac{a^2 - 2ab + b^2}{a^2 - b^2} = \dfrac{(a-b)(a-b)}{(a+b)(a-b)} = \dfrac{a-b}{a+b}$ for $a \neq b$ and $a \neq -b$.

d. $\dfrac{x^2yz^3}{xy^2z} = \dfrac{xz^2 \cdot xyz}{y \cdot xyz} = \dfrac{xz^2}{y}$ for $x \neq 0, y \neq 0,$ and $z \neq 0.$

e. $\dfrac{x^2 + 3x + 2}{x^2 - 5x - 6} = \dfrac{(x + 2)(x + 1)}{(x - 6)(x + 1)} = \dfrac{x + 2}{x - 6}$ for $x \neq 6\, x \neq -1.$

It is of the utmost importance to remember that we can only remove common *factors* from the numerator and denominator of a fraction. One of the most frequent mistakes made in algebra is to try to remove from fractions algebraic expressions that occur not as factors of the numerator and denominator but as parts of sums. For example, a common mistake would be to try to reduce the fraction

$$\frac{x + 1}{y + 1} \quad \text{to} \quad \frac{x}{y}.$$

These fractions are not equal, however, if $x \neq y$ (see Example 1e). Only *factors* of both numerator and denominator may be removed from a fraction.

Exercise Set 2.6 (continued)

In exercises 11–26, reduce the fractions to fractions with no factors common to the numerator and denominator.

11. $\dfrac{12}{14}$ 12. $\dfrac{15}{25}$ 13. $\dfrac{66}{77}$ 14. $\dfrac{84}{35}$

15. $\dfrac{axy}{bxy}$ 16. $\dfrac{abx}{aby}$ 17. $\dfrac{x^2y}{xy^2}$ 18. $\dfrac{x^2y^2}{xy}$

19. $\dfrac{x^2y^2}{x^3y - xy^3}$ 20. $\dfrac{x^3 + xy^2}{x^2y + y^3}$ 21. $\dfrac{a^2b + ab^2}{ab}$ 22. $\dfrac{a^3b + ab^3}{a^2b + ab^2}$

23. $\dfrac{x + y}{x^2 - y^2}$ 24. $\dfrac{x^2 + 2xy + y^2}{x^2 - y^2}$ 25. $\dfrac{x^3y - xy^3}{x^2y + xy^2}$ 26. $\dfrac{x^2 - 2xy + y^2}{x^2 - y^2}$

BUILDING FRACTIONS

Sometimes, as when adding or subtracting fractions, we want to put a common factor back into the numerator and denominator of a fraction. That is, we want to start with a/b and obtain an equivalent fraction ax/bx. This procedure is called *building* fractions.

Example 3. Building fractions.

a. Find a fraction equal to 3/2 but with a denominator 12.

Solution. Since $12 = 2 \cdot 6$, we multiply numerator and denominator of the fraction by 6. So

$$\frac{3}{2} = \frac{3 \cdot 6}{2 \cdot 6} = \frac{18}{12}$$

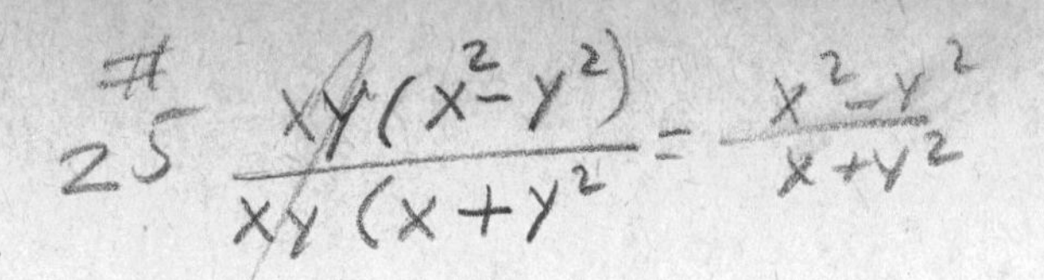

b. Find a fraction equal to x^2/yz but with denominator axy^2z.

Solution. Since $axy^2z = yz \cdot axy$, we multiply numerator and denominator of the fraction by axy. We obtain

$$\frac{x^2}{yz} = \frac{x^2 \cdot axy}{yz \cdot axy} = \frac{ax^3y}{axy^2z}.$$

c. Find a fraction equal to $\dfrac{x+2}{x-6}$ but with denominator $x^2 - 5x - 6$.

Solution. Since $x^2 - 5x - 6 = (x-6)(x+1)$, we multiply numerator and denominator of the fraction by $x + 1$. We obtain

$$\frac{x+2}{x-6} = \frac{(x+2)(x+1)}{(x-6)(x+1)} = \frac{x^2+3x+2}{x^2-5x-6}.$$

Exercise Set 2.6 (continued)

In exercises 27–40, find the fraction that has the new denominator but is still equal to the given fraction.

27. $\dfrac{2}{7}; \dfrac{\quad}{21}$
28. $\dfrac{1}{6}; \dfrac{\quad}{18}$
29. $\dfrac{7}{5}; \dfrac{\quad}{35}$
30. $\dfrac{3}{8}; \dfrac{\quad}{24}$
31. $\dfrac{x^2}{y}; \dfrac{\quad}{xy^2}$
32. $\dfrac{x}{y^2}; \dfrac{\quad}{xy^2}$
33. $\dfrac{xyz}{y}; \dfrac{\quad}{xyz}$
34. $\dfrac{x^2z^2}{y^2}; \dfrac{\quad}{x^2y^2}$
35. $\dfrac{a+b}{a-b}; \dfrac{\quad}{a^2-b^2}$
36. $\dfrac{a+b}{a-b}; \dfrac{\quad}{a^2-2ab+b^2}$
37. $\dfrac{ab}{a-b}; \dfrac{\quad}{a^2-b^2}$
38. $\dfrac{x+1}{x+2}; \dfrac{\quad}{x^2+5x+6}$
39. $\dfrac{x-3}{x+2}; \dfrac{\quad}{x^2+3x+2}$
40. $\dfrac{x-y}{2x-y}; \dfrac{\quad}{4x^2-y^2}$

SOLVING PROPORTIONS

A proportion is an equality of two ratios. Thus, *ratio a/b is proportional to ratio c/d* means

$$a/b = c/d.$$

A common problem involving proportions is to *solve a proportion.* In this situation we are given three of the four numbers a, b, c, and d and we wish to find the fourth number.

We can always solve a proportion by using the definition of equality of fractions. For example, to solve for x in a proportion of the form

$$a/b = x/d,$$

we use the definition of equality of fractions twice:

$$bx = ad$$
$$x = ad/b$$

Example 4. Solve the proportion $x/5 = 14/25$.

Solution. By the definition of equality of fractions,

$$25x = 5 \cdot 14$$
$$x = 5 \cdot 14/25 = 5 \cdot 14/5 \cdot 5$$
$$= 14/5 = 2.8$$

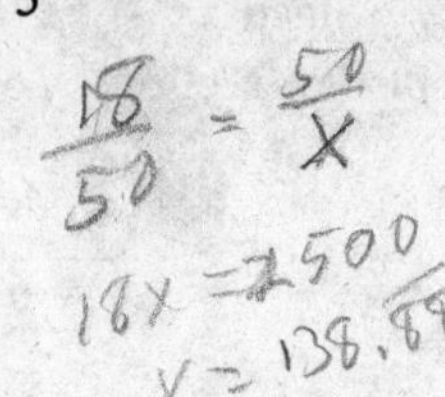

Example 5. Solve the proportion $5.5/2 = 0.75/x$.

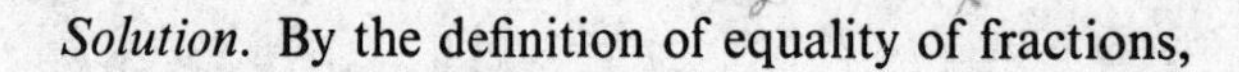

Solution. By the definition of equality of fractions,

$$5.5x = 2(0.75) = 1.5$$
$$x = 1.5/5.5 = 0.273$$

Exercise Set 2.6 (continued)

In exercises 41–60, solve the given proportion.

41. $x/1 = 900/60$	42. $x/600 = 78/150$	43. $x/1 = 30/60$
44. $x/600 = 72/150$	45. $18/50 = 50/x$	46. $96/150 = x/500{,}000$
47. $18/60 = 40/x$	48. $80/150 = x/500{,}000$	49. $200/x = 24/200$
50. $0.2/0.25 = x/1$	51. $200/x = 20/150$	52. $0.1/0.5 = x/2$
53. $x/60 = 16\frac{2}{3}/1$	54. $0.15/0.25 = x/5$	55. $x/60 = 12\frac{1}{2}/1$
56. $12.5/50 = x/1$	57. $108/150 = x/250$	58. $32.5/x = 25.6/13.8$
59. $55/150 = x/250$	60. $14.7/12.8 = 32.5/x$	

section 7 • Addition and Subtraction of Fractions

In this section we consider the operations of addition and subtraction of fractions.

FRACTIONS WITH THE SAME DENOMINATOR

To add or subtract fractions with the same denominator, we keep this denominator and add or subtract the numerators. For two fractions this procedure is expressed by the equations

$$\frac{a}{b} + \frac{c}{b} = \frac{a+c}{b} \quad \text{and} \quad \frac{a}{b} - \frac{c}{b} = \frac{a-c}{b}.$$

Example 1. Adding and subtracting fractions with the same denominator.

a. $\frac{1}{3} + \frac{4}{3} = \frac{1+4}{3} = \frac{5}{3}$ and $\frac{1}{3} - \frac{4}{3} = \frac{1-4}{3} = \frac{-3}{3} = -1$

b. $\frac{x^2}{x^2+y^2} + \frac{y^2}{x^2+y^2} = \frac{x^2+y^2}{x^2+y^2} = 1$ for $x \neq 0$ and $y \neq 0$.

c. $\frac{-x}{x^2-1} + \frac{1}{x^2-1} = \frac{-x+1}{x^2-1} = \frac{-(x-1)}{(x+1)(x-1)} = \frac{-1}{x+1}$

for $x \neq 1$ and $x \neq -1$.

Exercise Set 2.7

In exercises 1–8, perform the operations of addition and subtraction. Reduce the answer to a fraction with no factors common to the numerator and denominator.

1. $\frac{6}{5} + \frac{4}{5}$
2. $\frac{11}{3} - \frac{4}{3}$
3. $\frac{a}{x} + \frac{b}{x}$
4. $\frac{x+1}{3} - \frac{4x+1}{3}$
5. $\frac{x+3}{x} + \frac{x-3}{x}$
6. $\frac{x+1}{2x+1} + \frac{x}{2x+1}$
7. $\frac{a}{a^2-1} - \frac{1}{a^2-1}$
8. $\frac{a^2}{a^2-1} - \frac{a}{a^2-1}$

BUILDING FRACTIONS

Now we wish to add and subtract fractions that don't have the same denominator. We do this by building fractions (Section 2.6). That is, we construct fractions that are equal to the fractions we want to add or subtract, but that have the same denominators.

Example 2. Addition and subtraction of fractions.

a. $\frac{2}{5} + \frac{7}{4} = \frac{2 \cdot 4}{5 \cdot 4} + \frac{7 \cdot 5}{4 \cdot 5} = \frac{8}{20} + \frac{35}{20} = \frac{43}{20}$

b. $\frac{4}{6} - \frac{7}{8} = \frac{4 \cdot 4}{6 \cdot 4} - \frac{7 \cdot 3}{8 \cdot 3} = \frac{16}{24} - \frac{21}{24} = \frac{-5}{24}$

c. $\dfrac{x+1}{x+2} - \dfrac{x+2}{x+3} = \dfrac{(x+1)(x+3)}{(x+2)(x+3)} - \dfrac{(x+2)(x+2)}{(x+3)(x+2)}$

$= \dfrac{(x^2+4x+3)-(x^2+4x+4)}{x^2+5x+6} = \dfrac{-1}{x^2+5x+6}$

for $x \neq -2$ and $x \neq -3$.

d. $\dfrac{x}{x+1} + \dfrac{x}{x-1} - \dfrac{1}{x^2-1} = \dfrac{x(x-1)}{(x+1)(x-1)} + \dfrac{x(x+1)}{(x-1)(x+1)} - \dfrac{1}{(x-1)(x+1)}$

$= \dfrac{(x^2-x)+(x^2+x)-1}{x^2-1} = \dfrac{2x^2-1}{x^2-1}$ for $x \neq 1$ and $x \neq -1$.

Exercise Set 2.7 (continued)

In exercises 9–28, perform the operations of addition and subtraction. Reduce the answer to a fraction with no factors common to the numerator and denominator.

9. $\dfrac{4}{3} + 7$
10. $\dfrac{2}{5} - 6$
11. $\dfrac{2x+3}{x+3} - 1$
12. $\dfrac{3x+1}{x+3} + 1$
13. $x + \dfrac{x^2}{x+1}$
14. $x - \dfrac{x^2}{x+1}$
15. $\dfrac{z}{xy} + xyz$
16. $1 + \dfrac{1}{x^2}$
17. $\dfrac{1}{2} + \dfrac{3}{7}$
18. $\dfrac{1}{4} - \dfrac{5}{3}$
19. $\dfrac{x}{y} + \dfrac{y}{x}$
20. $\dfrac{x}{y^2} + \dfrac{y}{x^2}$
21. $\dfrac{1}{a^2} - \dfrac{1}{b^2}$
22. $\dfrac{a}{b} + \dfrac{a}{c}$
23. $\dfrac{x}{x-1} - \dfrac{x}{x+1}$
24. $\dfrac{5}{x} + \dfrac{12}{x-1}$
25. $\dfrac{2}{x} + \dfrac{4}{x+6}$
26. $\dfrac{2}{x} + \dfrac{x-1}{x-3}$
27. $\dfrac{x+2}{x-1} - \dfrac{x-2}{x+1}$
28. $\dfrac{x-1}{x+2} - \dfrac{x+1}{x-2}$

LEAST COMMON DENOMINATORS

To add or subtract fractions, we must find a *common denominator* for all the fractions involved—that is, we express the fractions so they have the same denominator. Naturally, we want to choose a common denominator that is as simple as possible to work with. Now the product of the denominators of all the fractions involved is always a common denominator. Parts b and d of Example 2 show, however, that the product of the denominators is not always the easiest common denominator to work with. If it's possible to find a smaller common denominator than the product of the denominators, use it; don't make more work than necessary.

How do we find the smallest possible common denominator for fractions, the *least common denominator*? A common denominator must contain *all the factors* contained in the denominators separately. So we factor all the denominators involved and take the product of the necessary factors to construct a common denominator.

Example 3. Common denominators.

a. $\frac{13}{12} + \frac{1}{15}$. We factor the denominators:

$$12 = 2^2 \cdot 3 \qquad \text{and} \qquad 15 = 3 \cdot 5.$$

So a common denominator must have factors of 2^2, 3, and 5; and $60 = 2^2 \cdot 3 \cdot 5$ will do. Then

$$\frac{13}{12} + \frac{1}{15} = \frac{13 \cdot 5}{12 \cdot 5} + \frac{1 \cdot 4}{15 \cdot 4} = \frac{65}{60} + \frac{4}{60} = \frac{69}{60} = \frac{23 \cdot 3}{20 \cdot 3} = \frac{23}{20}.$$

b. $\frac{x}{yz} + \frac{y}{xz} + \frac{z}{xy}$. Here the factors in the denominators are x, y, and z. So a common denominator is xyz. Then

$$\frac{x}{yz} + \frac{y}{xz} + \frac{z}{xy} = \frac{x \cdot x}{yz \cdot x} + \frac{y \cdot y}{xz \cdot y} + \frac{z \cdot z}{xy \cdot z} = \frac{x^2 + y^2 + z^2}{xyz}$$

c. $\frac{x + 2}{x^2 + 2x + 1} + \frac{x}{x^2 - 1}$. We factor the denominators:

$$x^2 + 2x + 1 = (x + 1)^2 \qquad \text{and} \qquad x^2 - 1 = (x + 1)(x - 1).$$

For a common denominator, we need two factors of $x + 1$ and one factor of $x - 1$. So a common denominator is $(x + 1)^2(x - 1)$. Then

$$\frac{x + 2}{(x + 1)^2} + \frac{x}{(x + 1)(x - 1)} = \frac{(x + 2)(x - 1)}{(x + 1)^2(x - 1)} + \frac{x(x + 1)}{(x + 1)^2(x - 1)}$$

$$= \frac{x^2 + x - 2 + x^2 + x}{(x + 1)^2(x - 1)} = \frac{2(x^2 + x - 1)}{(x + 1)^2(x - 1)}$$

d. $2 + \frac{3}{x - 2} + \frac{9x}{(x - 2)^2}$. A common denominator is $(x - 2)^2$. We obtain

$$2 + \frac{3}{x - 2} + \frac{9x}{(x - 2)^2} = \frac{2(x - 2)^2}{(x - 2)^2} + \frac{3(x - 2)}{(x - 2)^2} + \frac{9x}{(x - 2)^2}$$

$$= \frac{2x^2 - 8x + 8 + 3x - 6 + 9x}{(x - 2)^2}$$

$$= \frac{2x^2 + 4x + 2}{(x - 2)^2} = \frac{2(x + 1)^2}{(x - 2)^2}$$

Exercise Set 2.7 (continued)

In exercises 29–48, perform the operations of addition and subtraction. Reduce the answer to a fraction with no factors common to the numerator and denominator.

29. $\frac{3}{4} + \frac{7}{8}$

30. $\frac{13}{12} - \frac{1}{15}$

31. $\frac{a}{b} + \frac{a^2}{b^2}$

32. $\frac{1}{ab} - \frac{a}{b}$

33. $\frac{1}{abc} + \frac{b}{ac}$

34. $\frac{x}{ab} + \frac{y}{bc}$

35. $\frac{y-2}{y+2} + \frac{1}{y+1}$

36. $\frac{y+5}{3y-1} - \frac{6y+5}{2y-1}$

37. $\frac{x+2}{x-1} - \frac{3x-4}{3x+2}$

38. $\frac{x+2}{x-1} - \frac{3x-5}{3x+2}$

39. $\frac{2x+1}{x+1} + \frac{3x-7}{2x+1}$

40. $\frac{2x-1}{(x-2)(x+1)} - \frac{1}{(x-2)(x-1)}$

41. $\frac{7x-2}{(x+1)(x-2)} - 2\frac{3x-4}{(x-1)(x-2)}$

42. $\frac{3x+1}{(x-1)(x-3)} + 2\frac{2x-3}{(x-1)(x-2)}$

43. $\frac{8x+1}{(x-3)(x+2)} - 2\frac{3x+1}{(x-3)(x+1)}$

44. $3\frac{x-1}{(x+1)(x-2)} - \frac{2}{(x+1)(x+2)}$

45. $x - \frac{1}{x+1} + \frac{1}{(x+1)^2}$

46. $x + \frac{1}{x+1} + \frac{1}{x^2-1}$

47. $\frac{1}{x+1} + \frac{1}{x-1} + \frac{2}{x^2-1}$

48. $\frac{x}{x+1} - \frac{x}{x-1} + \frac{x}{x^2-1}$

section 8 • Multiplication and Division of Fractions

In this section we discuss multiplication and division of fractions.

MULTIPLICATION OF FRACTIONS

We start with a familiar rule: to multiply two fractions, we form the fraction whose numerator is the product of the two numerators and whose denominator is the product of the two denominators.

PRODUCT OF FRACTIONS. If $b \neq 0$ and $d \neq 0$, then

$$\frac{a}{b} \cdot \frac{c}{d} = \frac{ac}{bd}.$$

This rule is illustrated in Figure 2.9, where we find

$$\frac{4}{5} \cdot \frac{2}{3} = \frac{4 \cdot 2}{5 \cdot 3} = \frac{8}{15}$$

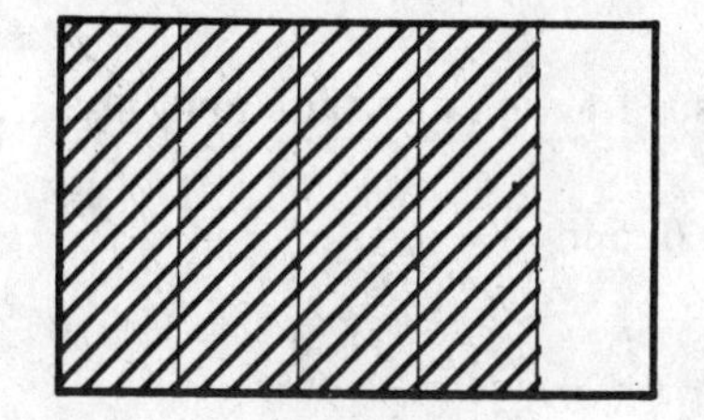

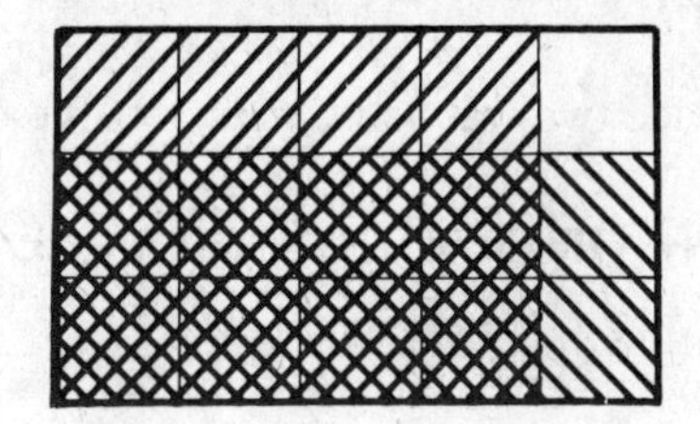

Figure 2.9

Example 1. Product of fractions.

a. $\dfrac{2}{3} \cdot \dfrac{7}{4} = \dfrac{2 \cdot 7}{3 \cdot 4} = \dfrac{7 \cdot 2}{3 \cdot 2 \cdot 2} = \dfrac{7}{6}$

b. $\dfrac{x^2}{yz} \cdot \dfrac{z}{xy} = \dfrac{x^2 \cdot z}{yz \cdot xy} = \dfrac{x \cdot xz}{y^2 \cdot xz} = \dfrac{x}{y^2}$ for $x \neq 0$, $y \neq 0$, and $z \neq 0$.

c. $\dfrac{x+1}{x-2} \cdot \dfrac{x^2+1}{x^2-1} = \dfrac{(x+1)(x^2+1)}{(x-2)(x^2-1)} = \dfrac{(x^2+1)(x+1)}{(x-2)(x-1)(x+1)} = \dfrac{x^2+1}{(x-2)(x-1)}$

for $x \neq 1$, $x \neq -1$, and $x \neq 2$.

d. $\dfrac{5x+25}{2x} \cdot \dfrac{4x}{2x+10} = \dfrac{(5x+25)4x}{2x(2x+10)} = \dfrac{4 \cdot 5 \cdot x(x+5)}{2 \cdot 2 \cdot x(x+5)} = 5$

for $x \neq 0$ and $x \neq -5$.

Exercise Set 2.8

In exercises 1–20, find the product of the fractions and express the answer as a fraction that has no factors common to the numerator and denominator.

1. $\dfrac{1}{3} \cdot \dfrac{2}{5}$
2. $\dfrac{5}{3} \cdot \dfrac{6}{7}$
3. $\dfrac{7}{8} \cdot \dfrac{20}{21}$
4. $\dfrac{22}{15} \cdot \dfrac{35}{55}$
5. $\dfrac{x}{x+1} \cdot \dfrac{x}{x-1}$
6. $\dfrac{x+y}{2x} \cdot \dfrac{x-y}{2y}$
7. $\dfrac{a}{b} \cdot \dfrac{x+y}{x-y}$
8. $\dfrac{a^2}{a+b} \cdot \dfrac{b^2}{a+b}$
9. $\dfrac{x^2+1}{x^2-1} \cdot \dfrac{x+1}{x-1}$

10. $\dfrac{x+1}{x-1} \cdot \dfrac{x^2-1}{(x+1)^2}$

11. $\dfrac{x^2+xy}{xy-y^2} \cdot \dfrac{y^2}{2x^2}$

12. $\dfrac{x^2-xy}{xy-y^2} \cdot \dfrac{y}{x}$

13. $\dfrac{x^2+x}{2x+4} \cdot \dfrac{4x^3}{xy+y}$

14. $\dfrac{ax+ay}{b^2} \cdot \dfrac{bx-by}{x^2-y^2}$

15. $\dfrac{x+1}{x-4} \cdot \dfrac{x^2-16}{x^2-1}$

16. $\dfrac{x^2-9}{x-2} \cdot \dfrac{x^2-4}{x-3}$

17. $\dfrac{x^2-9}{9x^2-1} \cdot \dfrac{3x-1}{x+3}$

18. $\dfrac{x^2+1}{x+1} \cdot \dfrac{x-1}{x^2-1}$

19. $\dfrac{x+y}{2x+y} \cdot \dfrac{4x^2+4xy+y^2}{x^2+2xy+y^2}$

20. $\dfrac{x-y}{x+y} \cdot \dfrac{x^2-y^2}{x^2-2xy+y^2}$

DIVISION OF FRACTIONS

To divide two fractions, we invert the divisor fraction and then multiply.

QUOTIENT OF FRACTIONS. If $b \neq 0$, $c \neq 0$, and $d \neq 0$, then

$$\frac{a}{b} \div \frac{c}{d} = \frac{a}{b} \cdot \frac{d}{c} = \frac{ad}{bc}.$$

Example 2. Division of fractions.

a. $\dfrac{2}{3} \div \dfrac{7}{4} = \dfrac{2 \cdot 4}{3 \cdot 7} = \dfrac{8}{21}$

b. $\dfrac{x^2}{yz} \div \dfrac{z}{xy} = \dfrac{x^2 \cdot xy}{yz \cdot z} = \dfrac{x^3 \cdot y}{z^2 \cdot y} = \dfrac{x^3}{z^2}$ for $x \neq 0$, $y \neq 0$, and $z \neq 0$.

c. $\dfrac{x+1}{x-2} \div \dfrac{x^2+1}{x^2-1} = \dfrac{(x+1)(x^2-1)}{(x-2)(x^2+1)} = \dfrac{(x+1)^2(x-1)}{(x-2)(x^2+1)}$ for $x \neq 1$, $x \neq -1$, and $x \neq 2$.

d. $\dfrac{5x+25}{2x} \div \dfrac{4x}{2x+10} = \dfrac{(5x+25)(2x+10)}{2x \cdot 4x}$

$= \dfrac{5(x+5)2(x+5)}{2 \cdot 4 \cdot x^2} = \dfrac{5(x+5)^2}{4x^2}$ for $x \neq 0$ and $x \neq -5$.

Exercise Set 2.8 (continued)

In exercises 21–48, perform the operations and express the answer as a fraction that has no factors common to the numerator and denominator.

21. $\dfrac{5}{8} \div \dfrac{3}{14}$

22. $\dfrac{9}{10} \div \dfrac{5}{12}$

23. $\dfrac{6}{25} \div \dfrac{18}{35}$

24. $\dfrac{8}{45} \div \dfrac{12}{25}$

25. $\dfrac{xyz}{ab} \div \dfrac{x^2y^2z^4}{a^3b}$

26. $\dfrac{abx}{yz} \div \dfrac{abx^2}{yz}$

27. $\dfrac{x^2}{yz} \div \dfrac{xy^2z}{x^2yz^2}$

28. $\dfrac{abc}{axy} \div \dfrac{ab^2c}{x^2y}$

29. $\dfrac{3x+3}{x^3-x^2} \div \dfrac{2x+2}{x^2-x}$

30. $\dfrac{a^2 - a}{a^2 + a} \div \dfrac{b^2 + b}{b^2 - b}$ 31. $\dfrac{a^3 + b^3}{a - b} \div \dfrac{a + b}{a - b}$ 32. $\dfrac{a^2 - 1}{b^2 - 1} \div \dfrac{a + 1}{b + 1}$

33. $\dfrac{1}{x^2 + x - 2} \div \dfrac{1}{x^2 + 1}$ 34. $\dfrac{1}{x^2 + 3} \div \dfrac{1}{x^2 - 3}$ 35. $\dfrac{y^3}{xy + 3} \div \dfrac{y^2}{1}$

36. $\dfrac{xy}{xy + 1} \div \dfrac{x}{y}$ 37. $\dfrac{x^2 - 3x}{xy - 4y} \div \dfrac{x}{y}$ 38. $\dfrac{x}{x + 1} \div \dfrac{xy}{y + 1}$

39. $\dfrac{x^2 + 2x + 1}{16x^2 - 1} \div \dfrac{x + 1}{4x - 1}$ 40. $\dfrac{9x^2 - 16}{x^2 + 4x + 4} \div \dfrac{3x + 4}{x + 2}$ 41. $\dfrac{x^2 + x - 2}{x^2 + 2x + 1} \div \dfrac{x - 1}{x + 1}$

42. $\dfrac{x^2 + 3x + 2}{x^2 + 4x + 3} \div \dfrac{x + 2}{x + 1}$ 43. $\dfrac{2x + 3}{x + 3} \div \dfrac{2x + 2}{x - 1}$ 44. $\dfrac{x + 1}{x - 1} \div \dfrac{x - 1}{x + 1}$

45. $\dfrac{2}{x} + \dfrac{x - 3}{x} \cdot \dfrac{x + 2}{x + 3}$ 46. $\dfrac{1}{x + 1} + \dfrac{x + 2}{x - 10} \cdot \dfrac{x - 3}{x + 1}$ 47. $\dfrac{3}{x} + \dfrac{x - 1}{x} \div \dfrac{x + 1}{x + 3}$

48. $\dfrac{1}{x + 3} + \dfrac{x + 1}{x + 3} \div \dfrac{x - 3}{x - 6}$

---

section 9 • Simplifying Fractions

Frequently fractions occur whose numerator or denominator (or both) contains fractions. For example,

$$\frac{x + \frac{7}{2}}{x - \frac{3}{4}} \quad \text{and} \quad \frac{\frac{1}{x} + \frac{1}{y}}{\frac{1}{x} - \frac{1}{y}}.$$

Such fractions are called *complex fractions*. A *simple fraction*, then, is a fraction in which neither the numerator nor denominator contains any fractions. Every complex fraction is equal to a simple fraction. This section describes two methods for finding a simple fraction equal to a given complex fraction.

SIMPLIFYING COMPLEX FRACTIONS

The first method of finding a simple fraction equal to a complex fraction is simply to perform the operations denoted in the complex fraction according to the rules governing symbols of grouping. In this method, the main fraction bar corresponds to the division of two fractions.

Example 1. First method of simplifying complex fractions.

a. $$\frac{x + \frac{7}{2}}{x - \frac{3}{4}} = \frac{\frac{2x + 7}{2}}{\frac{4x - 3}{4}} = \frac{2x + 7}{2} \cdot \frac{4}{4x - 3} = \frac{2(2x + 7)}{4x - 3}$$

b. $$\frac{\frac{1}{x} + \frac{1}{y}}{\frac{1}{x} - \frac{1}{y}} = \frac{\frac{y + x}{xy}}{\frac{y - x}{xy}} = \frac{y + x}{xy} \cdot \frac{xy}{y - x} = \frac{y + x}{y - x}$$

The second way to simplify complex fractions is to multiply both numerator and denominator of the complex fraction by an expression that clears the numerator and denominator of any fractions that occur there. This procedure is an application of the process $a/b = ax/bx$, so the value of the complex fraction is not changed (Section 2.6).

Example 2. Second method of simplifying complex fractions.

a. $\dfrac{x + \frac{7}{2}}{x - \frac{3}{4}}$. Multiply numerator and denominator by 4. Then

$$\frac{x + \frac{7}{2}}{x - \frac{3}{4}} = \frac{4\left(x + \frac{7}{2}\right)}{4\left(x - \frac{3}{4}\right)} = \frac{4x + 4\left(\frac{7}{2}\right)}{4x - 4\left(\frac{3}{4}\right)} = \frac{4x + 14}{4x - 3}.$$

b. $\dfrac{\frac{1}{x} + \frac{1}{y}}{\frac{1}{x} - \frac{1}{y}}$. Multiply numerator and denominator by xy. Then

$$\frac{\frac{1}{x} + \frac{1}{y}}{\frac{1}{x} - \frac{1}{y}} = \frac{xy\left(\frac{1}{x} + \frac{1}{y}\right)}{xy\left(\frac{1}{x} - \frac{1}{y}\right)} = \frac{xy \cdot \frac{1}{x} + xy \cdot \frac{1}{y}}{xy \cdot \frac{1}{x} - xy \cdot \frac{1}{y}} = \frac{y + x}{y - x}$$

Exercise Set 2.9

In exercises 1–14, simplify the fractions and reduce the answer to a fraction with no factors common to the numerator and denominator.

1. $\dfrac{x - \frac{1}{3}}{x - \frac{2}{3}}$

2. $\dfrac{x - \frac{5}{6}}{x + \frac{1}{6}}$

3. $\dfrac{x + \frac{5}{6}}{x + \frac{1}{2}}$

4. $\dfrac{x - \frac{1}{4}}{x - \frac{1}{6}}$

5. $\dfrac{\frac{1}{x}+1}{\frac{1}{x}-3}$ 6. $\dfrac{\frac{1}{x^2}-2}{\frac{1}{x^2}+4}$ 7. $\dfrac{\frac{1}{x}+1}{\frac{1}{x^2}-1}$ 8. $\dfrac{\frac{1}{x^2}-1}{\frac{1}{x^3}-1}$

9. $\dfrac{x^2-y^2}{\frac{1}{x}+\frac{1}{y}}$ 10. $\dfrac{\frac{1}{x}-\frac{1}{y}}{x^3-y^3}$ 11. $\dfrac{\frac{a+b}{a-b}+\frac{a-b}{a+b}}{a^2-b^2}$ 12. $\dfrac{\frac{a^2-b^2}{a^2+b^2}}{\frac{1}{a}-\frac{1}{b}}$

13. $\dfrac{\frac{x}{yz}+\frac{y}{xz}+\frac{z}{xy}}{\frac{1}{x}+\frac{1}{y}+\frac{1}{z}}$ 14. $\dfrac{\frac{x}{yz}+\frac{y}{xz}+\frac{z}{xy}}{\frac{yz}{x}+\frac{xz}{y}+\frac{xy}{z}}$

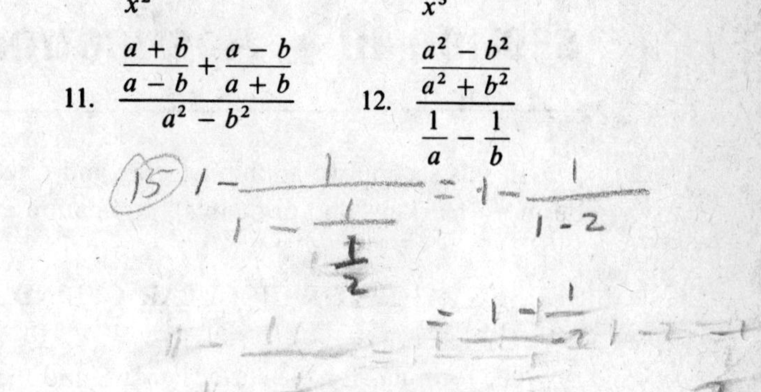

CONTINUED FRACTIONS

Example 3. Simplifying continued fractions.

$$1+\cfrac{1}{1+\cfrac{1}{1+\cfrac{1}{2}}}=1+\cfrac{1}{1+\cfrac{1}{\frac{3}{2}}}$$

$$=1+\cfrac{1}{1+\frac{2}{3}}=1+\cfrac{1}{\frac{5}{3}}=1+\frac{3}{5}=\frac{8}{5}.$$

STUDY

Exercise Set 2.9 (continued)

In exercises 15–20, simplify the continued fractions.

15. $1-\cfrac{1}{1-\cfrac{1}{1-\cfrac{1}{2}}}$ 16. $3+\cfrac{1}{3+\cfrac{1}{3+\cfrac{1}{3}}}$ 17. $5+\cfrac{1}{5-\cfrac{1}{5+\cfrac{1}{5-\cfrac{1}{5}}}}$

18. $1+\cfrac{2}{3+\cfrac{4}{5+\cfrac{6}{7}}}$ 19. $1+\cfrac{1}{1+\cfrac{1}{1+\cfrac{1}{x}}}$ 20. $1-\cfrac{1}{1-\cfrac{1}{1-\cfrac{1}{x}}}$

section 10 • Applications

In this section we apply factoring and fractions to geometry (right circular cylinders), business (markups and discounts), population estimates, and medical dosages.

GEOMETRY (RIGHT CIRCULAR CYLINDERS)

The formulas for the surface area S and the volume V of a right circular cylinder with height h and base radius r (see Figure 2.10) are:

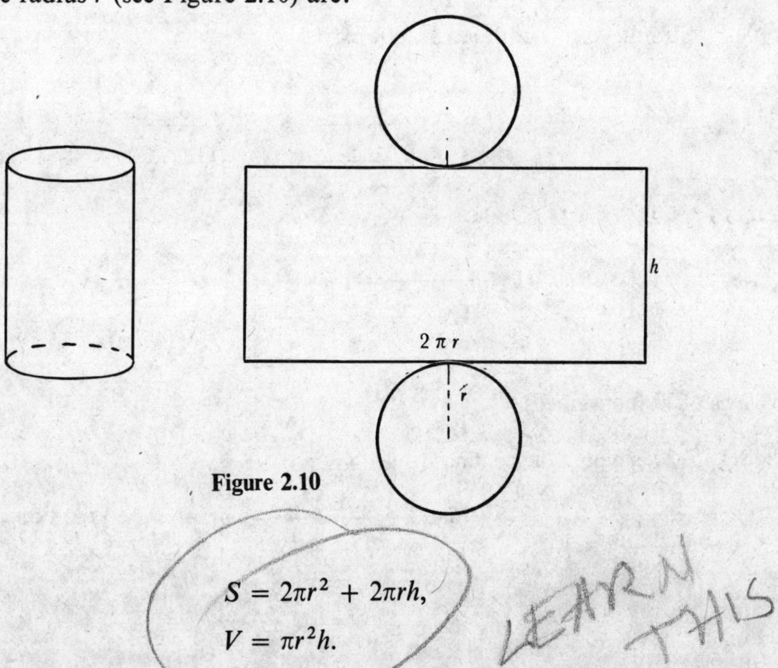

Figure 2.10

$$S = 2\pi r^2 + 2\pi rh,$$
$$V = \pi r^2 h.$$

Notice that, for computational purposes, it is easier to write the formula for surface area (by factoring) as

$$S = (r + h)r(6.28),$$

using 3.14 as an approximation to π.

Example 1. A standard-size soup can has a height of 9.8 centimeters and the radius of its base is 3.3 centimeters. Find the surface area and volume of the can.

Solution. Using 3.14 as an approximation to π, we find

$$S = (3.3 + 9.8)(3.3)(6.28) = 271.5 \text{ square centimeters}$$
$$V = (3.14)(3.3)^2(9.8) = 335.1 \text{ cubic centimeters.}$$

We have rounded off to one decimal place.

Exercise Set 2.10

1. A standard-size can of sweet peas has a height of 10.8 centimeters and the radius of its base is 3.9 centimeters. Find its surface area and volume.
2. A can of vegetable shortening is 8.2 centimeters high, and the radius of its base is 4.9 centimeters. Find its surface area and volume.
3. A can of frozen grape juice is 12 centimeters high, and the radius of its base is 3.2 centimeters. Find its surface area and volume.
4. A cylindrical box of salt is 13.8 centimeters high, and the radius of its base is 4.2 centimeters. Find its surface area and volume.
5. A cylindrical 3.276-liter can of paint is 18.8 centimeters high, and the radius of its base is 8.4 centimeters. Find its surface area and volume.
6. A cylindrical can of car wax is 4.2 centimeters high, and the radius of its base is 5.2 centimeters. Find its surface area and volume.

BUSINESS (MARKUPS AND DISCOUNTS)

A retailer buys a product from a wholesaler for p dollars and determines the selling price by marking up the wholesale price $100r$ percent. Thus the retailer's selling price is

$$p + rp \quad \text{dollars.}$$

For computational purposes, we factor this expression and write

$$(1 + r)p.$$

Similarly, if the selling price of an article is p dollars and a discount of $100r$ percent is offered, the new selling price is

$$p - rp = (1 - r)p.$$

Example 2. A retailer buys transistor radios at \$22.50 and determines the selling price by marking up 18%. What is the retailer's selling price?

Solution. We have $r = 0.18$, so the price is

$$(1 + 0.18)(22.50) = (1.18)(22.50) = 26.55 \text{ dollars.}$$

Exercise Set 2.10 (continued)

7. A retailer buys LP records at \$5.45 and determines the selling price by marking up 20%. What is the retailer's selling price?

8. A retailer buys transistor radios at \$22.50 and determines the selling price by marking up 20%. What is the retailer's selling price?

9. A retailer buys automobiles at \$3280 and determines the selling price by marking up 22%. What is the retailer's selling price?

10. A retailer buys cameras at \$38.50 and determines the selling price by marking up 18%. What is the retailer's selling price?

11. A store advertises a sale of 10% off on every article in the store. What is the sale price of an article costing \$14.95? An article costing \$19.95?

12. A store advertises a sale of 20% off on every article in the store. What is the sale price of an article costing \$14.95? An article costing \$19.95?

13. A mail order house offers a discount of 1% for payments in cash. What is the discount price of an article costing \$58.00? An article costing \$112?

14. A retailer offers a discount of 4% on large volume orders. What is the discount price of an article costing \$19.95? An article costing \$29.95?

POPULATION ESTIMATES

Proportions are used to estimate animal populations in cases in which the whole population is not accessible or is too large to count individually. For example, proportions can be used to count the number of bears in a national park or the number of fish in a stocked pool.

Suppose, for example, that we wish to count the number of bears in a national park. Since it would be difficult to actually find and count all the bears, we tag a sample—a definite number of bears taken at random from the park. We allow time for the tagged bears to mix freely with other bears in the park; then we select another sample of bears at random from the park. In the second sample we have both tagged and untagged bears, but we *assume* that the ratio of tagged bears to total bears in the second sample is proportional to the ratio of tagged bears in the first sample to the entire population.

Example 3. Suppose we tag 100 bears in the first sample. In the second sample we select 100 bears and find that 32 are tagged. How many bears are in the park?

Solution. If we let x denote the number of bears, we have the proportion

$$32/100 = 100/x$$

$$32x = (100)(100) = 10000$$

$$x = 10000/32 = 312.5$$

There are (approximately) 312.5 bears.

Exercise Set 2.10 (continued)

15. We wish to count the number of bears in a park. We tag 50 bears and, in the second sample of 50 bears, find 18 tagged. How many bears are in the park?
16. We wish to count the number of bears in a park. We tag 40 bears and, in the second sample of 60 bears, find 18 tagged. How many bears are in the park?
17. We wish to find the number of fish in a stocked pond. We tag 200 fish and then take a second sample. In the second sample of 200 fish, we find 24 tagged. How many fish are in the pond?
18. We wish to find the number of fish in a stocked pond. We tag 200 fish and then take a second sample. In the second sample of 150 fish, we find 20 tagged. How many fish are in the pond?

MEDICAL DOSAGES

A common nursing task is to administer medication prescribed by a doctor. Frequently this involves changing units of measurement. Both the metric and apothecary systems are used in medicine. The conversion for units of weight is

$$60 \text{ milligrams} = 1 \text{ grain}.$$

Example 4. The doctor prescribes 600 milligrams of aspirin; on hand are 5-grain tablets. How many tablets should be given to the patient?

Solution. Let x denote the number of grains of aspirin prescribed. We have the proportion

$$\frac{600}{60} = \frac{x}{1}$$

$$60x = 600 \cdot 1 = 600$$

$$x = 600/60 = 10$$

Since the prescription calls for 10 grains, we give $10/5 = 2$ tablets.

Exercise Set 2.10 (continued)

19. The doctor prescribes 900 milligrams of aspirin; on hand are 5-grain tablets. How many tablets should be given to the patient?
20. The doctor prescribes 30 milligrams of Valium; on hand are 1/6-grain tablets. How many tablets should be given to the patient?
21. The doctor prescribes $16\frac{2}{3}$ grains of aminophylline; on hand are 250-milligram tablets. How many tablets should be given to the patient?
22. The doctor prescribes $12\frac{1}{2}$ grains of tetracycline; on hand are 250-milligram capsules. How many capsules should be given to the patient?

23. The doctor prescribes 0.15 milligram of digoxin. The label reads "0.25 milligram in 5 cc." How many cc of this liquid should be administered?
24. The doctor prescribes $12\frac{1}{2}$ milligrams of Demerol hydrochloride. The label reads "50 milligrams = 1 cc." How many cc of this liquid should be administered?
25. The doctor prescribes 0.2 milligram of digoxin. The label reads "0.25 milligram = 1 cc." How many cc of this liquid should be administered?
26. The doctor prescribes 0.1 milligram of digoxin. The label reads "0.5 milligram = 2 cc." How many cc of this liquid should be administered?

MEDICAL DOSAGES FOR CHILDREN

Medical dosages for infants and children must be different than those for adults. To find the proper dosage for infants and children, we use proportions based on one adult dosage. We can use either age or weight as the basis of these conversions. The conversion factors are:

$$150 \text{ months} = 1 \text{ adult dosage}$$

$$150 \text{ pounds} = 1 \text{ adult dosage}$$

Example 5. If the adult dosage of penicillin is 500,000 u., find the dosage for a $10\frac{1}{2}$-year-old child.

Solution. We know that $10\frac{1}{2}$ years is $12(10.5) = 126$ months. We solve the proportion

$$\frac{126}{150} = \frac{x}{500{,}000}$$

$$150x = 126(500{,}000)$$

$$x = 126(500{,}000)/150 = 420{,}000$$

The dosage is 420,000 u.

Exercise Set 2.10 (continued)

27. If the adult dosage of penicillin is 500,000 u., find the dosage for an 8-year-old child.
28. If the adult dosage of penicillin is 500,000 u., find the dosage for an 80-pound child.
29. If the adult dosage of tetracycline is 250 milligrams, find the dosage for a 9-year-old child.
30. If the adult dosage of tetracycline is 250 milligrams, find the dosage for a 55-pound child.
31. If the adult dosage of aspirin is 600 milligrams, find the dosage for a $6\frac{1}{2}$-year-old child.
32. If the adult dosage of aspirin is 600 milligrams, find the dosage for a 72-pound child.

Review Exercises

In exercises 1–26, perform the indicated operations on the algebraic expression.

1. $ab(ax + by)$
2. $ab^2(a^2 - b)$
3. $ab(ab + ac + bc)$
4. $abc(a^2 + b^2 + c^2)$
5. $(x + 2)(x + 4)$
6. $(x + 3)(x + 5)$
7. $(x + 2)(x - 4)$
8. $(x - 3)(x + 5)$
9. $(x - 2)(x - 4)$
10. $(x - 3)(x - 5)$
11. $(x + 5)^2 - (x - 5)^2$
12. $(x + a)^2 - (x - a)^2$
13. $(x + 4)(x - 4) + (x + 2)(x - 2)$
14. $(x + a)(x - a) + (x + b)(x - b)$
15. $(x + 3)(x - 3) - (x + 6)(x - 6)$
16. $(x + a)(x - a) - (x + b)(x - b)$
17. $(x - 1)^2 + (x + 2)^2 - (x - 1)(x + 2)$
18. $(x + a)^2 + (x + b)^2 - (x + a)(x + b)$
19. $(x + 3)^2 + (x + 4)^2 - (x - 3)(x - 4)$
20. $(x + a)^2 + (x + b)^2 - (x - a)(x - b)$
21. $(x + 6)^2 + (x - 6)^2 - (x + 6)(x - 6)$
22. $(x + a)^2 + (x - a)^2 - (x + a)(x - a)$
23. $(x + 2)(x + 3) - (x - 2)(x - 3)$
24. $(x + a)(x + b) - (x - a)(x - b)$
25. $(c - b)(x - a) - (c - a)(x - b)$
26. $(ae - bd)(c - f) + (cd - af)(b - e) + (bf - ce)(a - d)$

In exercises 27–80, factor the expression into a product of irreducible factors.

27. $2xy + y^2$
28. $5x + x^2$
29. $x^2y + xy^2 - x^2y^2$
30. $abc - ab - ac$
31. $2\pi r^2 + 2\pi rh$
32. $\pi r^2 + \pi rs$
33. $x^2 + 8x + 16$
34. $x^2 + 18x + 81$
35. $4x^2 - 4x + 1$
36. $9x^2 - 12x + 4$
37. $4x^2 - 9$
38. $9x^2 - 25$
39. $x^4 - 16$
40. $x^4 - 9$
41. $x^4 - 4x^2 + 4$
42. $x^4 - 6x^2 + 9$
43. $x^2 + 6x + 5$
44. $x^2 + 7x + 6$
45. $x^2 - 6x + 8$
46. $x^2 - 6x + 5$
47. $x^2 - 9x + 20$
48. $x^2 - 8x + 15$
49. $x^2 - 3x + 1$
50. $x^2 - 4x + 2$
51. $x^2 + 2x - 15$
52. $x^2 + 3x - 4$

53. $x^2 - 4x - 5$

54. $x^2 - 2x - 8$

55. $x^2 + 3x - 28$

56. $x^2 + 12x - 28$

57. $x^2 - 27x - 28$

58. $x^2 - 12x - 28$

59. $2x^2 + 3x + 1$

60. $2x^2 + 5x + 2$

61. $2x^2 - 7x + 3$

62. $2x^2 - 5x + 3$

63. $2x^2 + 2x - 1$

64. $2x^2 - 2x + 1$

65. $2x^2 - 5x - 3$

66. $2x^2 + x - 1$

67. $2x^2 + x - 3$

68. $2x^2 - x - 6$

69. $x^4 + 8x^2 + 12$

70. $x^4 + 12x^2 + 27$

71. $x^4 + 6x^2 - 7$

72. $x^4 + 4x^2 - 32$

73. $xy + x + y + 1$

74. $xy - x - y + 1$

75. $x^3 + 3x^2 + 3x + 1$

76. $x^3 - 6x^2 + 12x - 8$

77. $x^3 - 1$

78. $x^4 - x$

79. $x^3 + 1$

80. $x^4 + x$

In exercises 81–96, solve the given proportion.

81. $x/5 = 20/30$

82. $x/6 = 50/80$

83. $7/x = 21/32$

84. $9/x = 15/35$

85. $40/82 = x/50$

86. $36/25 = x/45$

87. $30/150 = 100/x$

88. $40/140 = 200/x$

89. $x/14 = 1/4000$

90. $x/18 = 1/5000$

91. $108/150 = x/500{,}000$

92. $100/150 = x/500{,}000$

93. $0.3/6 = 0.15/x$

94. $1.5/4 = 0.85/x$

95. $x/0.3 = 0.12/0.75$

96. $x/0.4 = 0.16/0.6$

In exercises 97–128, perform the operations on the given fractions. Reduce the answer to a fraction with no factors common to the numerator and denominator.

97. $\dfrac{x}{x^2 - 1} + \dfrac{1}{x^2 - 1}$

98. $\dfrac{2x + 1}{x + 2} - \dfrac{x - 1}{x + 2}$

99. $\dfrac{x + 1}{x - 1} + 1$

100. $\dfrac{x - 1}{x + 1} - 1$

101. $\dfrac{1}{x} + \dfrac{1}{y}$

102. $\dfrac{1}{x} - \dfrac{x}{y}$

103. $\dfrac{1}{x + 1} + \dfrac{1}{x - 1}$

104. $\dfrac{1}{x + 1} - \dfrac{1}{x - 1}$

105. $\dfrac{1}{a} - \dfrac{1}{a^2}$

106. $\dfrac{1}{a} - \dfrac{1}{ab}$

107. $\frac{c}{ab} + \frac{a}{bc}$

108. $\frac{d}{abc} + \frac{a}{bcd}$

109. $\frac{x}{x+1} + \frac{x}{x^2-1}$

110. $\frac{1}{x-1} - \frac{x}{x^2-1}$

111. $\frac{a^2b}{c} \cdot \frac{c^2}{ab^2}$

112. $\frac{ab}{cd} \cdot \frac{c^2d}{ab^2}$

113. $\frac{abx}{cy} \div \frac{bx^2}{c^2y}$

114. $\frac{a^2b^3}{xy} \div \frac{a^3b^2}{x^2y^2}$

115. $\frac{2x+1}{2x} \cdot \frac{x}{x+1}$

116. $\frac{x-1}{x+1} \cdot \frac{2x+2}{3x-3}$

117. $\frac{x^2+x}{x^2-x} \div \frac{x+1}{x^2-1}$

118. $\frac{x^2-x}{x+1} \div \frac{x-1}{x^2+x}$

119. $\frac{x^2-9}{x^2-1} \cdot \frac{x-1}{x+3}$

120. $\frac{x^2-1}{x^2-4} \cdot \frac{x-2}{x-1}$

121. $\frac{1}{x+1} + \frac{2}{x-2} + \frac{3}{x^2-x-2}$

122. $\frac{3}{x+3} + \frac{2}{x-2} + \frac{15}{x^2+x-6}$

123. $\frac{2}{x-3} + \frac{1}{x-4} - \frac{1}{x^2-7x+12}$

124. $\frac{3}{x+2} + \frac{2}{x-6} - \frac{16}{(x+2)(x-6)}$

125. $x - 5 + \frac{2x+15}{x+3}$

126. $\frac{x+2}{2x+2} + \frac{x+3}{3x+3} + \frac{x+4}{4x+4}$

127. $\frac{x^2+a^2}{x^2-a^2} + \frac{x+a}{x-a}$

128. $\frac{a+b}{(b-c)(c-a)} + \frac{b+c}{(c-a)(a-b)} + \frac{c+a}{(a-b)(b-c)}$

129. Find the surface area and volume of a cylindrical can which has height 6 centimeters and radius of base 4 centimeters.

130. A retailer buys transistor radios at $28.40 and determines the selling price by marking up 20%. What is the retailer's selling price?

131. A store advertises a sale of 25% off on every article in the store. What is the sale price of an article costing $16.50? An article costing $19.95?

132. We wish to find the number of fish in a stocked pond. We tag 100 fish and then take a second sample. In the second sample of 150 fish, we find 30 tagged. How many fish are in the pond?

133. The doctor prescribes 0.15 milligram of digoxin. The label reads "0.3 milligram in 6 cc." How many cc of this liquid should be administered?

134. If the adult dosage of penicillin is 500,000 u., find the dosage for a 9-year-old child.

chapter 3

Exponents, Roots, and Complex Numbers

1. *Integer Exponents*
2. *Scientific Notation*
3. *Roots*
4. *Properties of Radicals*
5. *Rational Exponents*
6. *Properties of Exponents*
7. *Definition of Complex Numbers*
8. *Operations for Complex Numbers*
9. *Applications*

The concepts discussed in this chapter arise from extending the concept of exponents. A wide variety of information can be expressed in exponential notation and is systematized by using the five properties of exponents.

Basic to this chapter are the definition of exponential notation and the properties of exponents. We define integer exponents in Section 3.1, and we define rational number exponents in Section 3.5. We examine the properties of exponents in Sections 3.1 and 3.6. In Section 3.2 we show how to express real numbers in an exponential notation called scientific notation.

In Sections 3.3 and 3.4 we examine radicals—that is, n^{th} roots of numbers. We define rational exponents in terms of radicals and use radicals in subsequent chapters. To define radicals for negative numbers, we examine complex numbers in Sections 3.7 and 3.8.

In Section 3.9 we apply exponents and roots to geometry (areas and volumes), direct variation, inverse variation, Kepler's Third Law, and compound interest.

section 1 • Integer Exponents

In this section we define integer exponents. That is, we define powers x^n where n is an integer (positive, zero, or negative). We compute powers and show how to write expressions in exponential notation. We also consider the computation of powers on an electronic calculator.

POSITIVE INTEGER EXPONENTS

Recall the definition of positive integer exponents.

DEFINITION OF EXPONENTS. For any positive integer n,

$$x^n = x \cdot x \cdot \cdot \cdot x \qquad (n \text{ factors of } x).$$

This is read "x to the nth" or "the nth power of x." Notice that there are two parts to the notation x^n: n is called the *exponent* and x is called the *base*.

Example 1. Powers of 10.

a. $10^2 = 100$; that is, 100 is the second power of 10.
b. $10^5 = 100{,}000$; that is, 100,000 is the fifth power of 10.

Example 2. Calculate $(1.06)^n$ for $n = 1, 2, 3, 4, 5$.

Solution.

$$\begin{aligned}(1.06)^1 &= 1.06\\ (1.06)^2 &= (1.06)(1.06) = 1.1236\\ (1.06)^3 &= (1.06)^2(1.06) = 1.191016\\ (1.06)^4 &= (1.06)^3(1.06) = 1.262477\\ (1.06)^5 &= (1.06)^4(1.06) = 1.338226\end{aligned}$$

We have rounded off to six decimal places.

Notice that in Example 2 we have used the property that, for all positive integers n,

$$x^n = x^{n-1} \cdot x.$$

Exercise Set 3.1

1. a. Find 10^n for $n = 1, 2, 3, 4, 5, 6$.
   b. State a rule that gives 10^n for a positive integer n.
2. a. Find 0^n for $n = 1, 2, 3, 4, 5, 6$.
   b. State a rule that gives 0^n for a positive integer n.
3. a. Find $(-1)^n$ for $n = 1, 2, 3, 4, 5, 6$.
   b. State a rule that gives $(-1)^n$ for a positive integer n.
4. a. Find 1^n for $n = 1, 2, 3, 4, 5, 6$.
   b. State a rule that gives 1^n for a positive integer n.
5. Find $(1.08)^n$ for $n = 1, 2, 3, 4, 5$. If necessary, round off to six decimal places.
6. Find $(1.05)^n$ for $n = 1, 2, 3, 4, 5$. If necessary, round off to six decimal places.
7. Find $(1.01)^n$ for $n = 1, 2, 3, 4, 5$. If necessary, round off to six decimal places.
8. Find $(1.1)^n$ for $n = 1, 2, 3, 4, 5$. If necessary, round off to six decimal places.

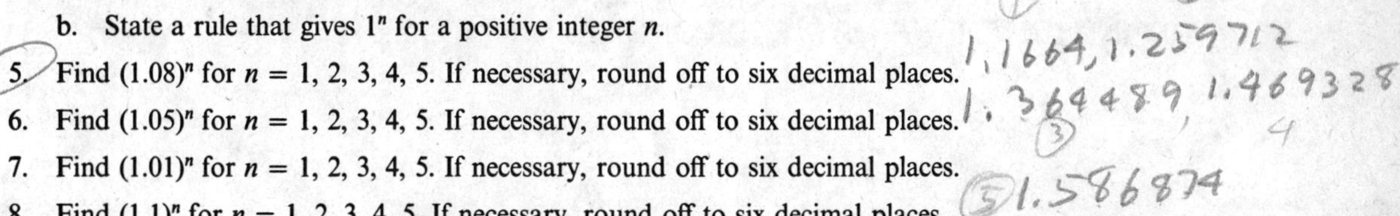

ZERO AND NEGATIVE EXPONENTS

Now we extend the exponent concept by defining zero and negative integer exponents. This extension of exponents is suggested by considering patterns of the following type:

$$\begin{array}{cc}
\vdots & \vdots \\
2^4 = 16 & 3^4 = 81 \\
2^3 = 8 & 3^3 = 27 \\
2^2 = 4 & 3^2 = 9 \\
2^1 = 2 & 3^1 = 3 \\
2^0 = ? & 3^0 = ? \\
2^{-1} = ? & 3^{-1} = ? \\
2^{-2} = ? & 3^{-2} = ? \\
\vdots & \vdots
\end{array}$$

The first column contains powers of 2, and we divide by 2 as we go down the column. The second column contains powers of 3, and we divide by 3 as we go down the column. These patterns suggest that we should define

$$\begin{array}{cc}
2^0 = 1 & 3^0 = 1 \\
2^{-1} = 1/2 & 3^{-1} = 1/3 \\
2^{-2} = 1/4 & 3^{-2} = 1/9
\end{array}$$

In general, we should define

$$x^0 = 1, \qquad x^{-1} = 1/x, \qquad x^{-2} = 1/x^2.$$

We now see how to define zero and negative integer exponents.

DEFINITION OF ZERO EXPONENT. If $x \neq 0$, then

$$x^0 = 1.$$

DEFINITION OF NEGATIVE EXPONENTS. If $x \neq 0$, then

$$x^{-n} = \frac{1}{x^n}.$$

Example 3. Zero and negative integer powers of 5.

a. $5^0 = 1$

b. $5^{-1} = 1/5 = 0.20$

c. $5^{-2} = 1/5^2 = 1/25 = 0.04$

d. $5^{-3} = 1/5^3 = 1/125 = 0.008$

Example 4. Zero and negative integer powers of 10.

a. $10^0 = 1$

b. $10^{-2} = 1/10^2 = 1/100 = 0.01$

c. $10^{-5} = 1/10^5 = 1/100{,}000 = 0.00001$

Example 5. Calculate $(1.06)^{-n}$ for $n = 1, 2, 3$.

Solution. $(1.06)^{-1} = 1/1.06 = 0.943396$
$(1.06)^{-2} = 1/(1.06)^2 = 1/1.1236 = 0.889996$
$(1.06)^{-3} = 1/(1.06)^3 = 1/1.191016 = 0.839619$

We have rounded off to six decimal places.

Exercise Set 3.1 (continued)

9. Find $4^0, 4^{-1}, 4^{-2}, 4^{-3}, 4^{-4}$.
10. Find 6^0 $6^{-1}, 6^{-2}, 6^{-3}, 6^{-4}$.
11. Find $8^0, 8^{-1}, 8^{-2}, 8^{-3}$.
12. Find $9^0, 9^{-1}, 9^{-2}, 9^{-3}$.
13. a. Find 10^n for $n = 0, -1, -2, -3, -4, -5, -6$.
    b. State a rule that gives 10^n for a negative integer n.

14. a. Find $(-1)^n$ for $n = 0, -1, -2, -3, -4, -5, -6$.
    b. State a rule that gives $(-1)^n$ for a negative integer n.
15. Find $(1.08)^n$ for $n = -1, -2, -3, -4$. If necessary, round off to six decimal places.
16. Find $(1.05)^n$ for $n = -1, -2, -3, -4$. If necessary, round off to six decimal places.
17. Find $(1.01)^n$ for $n = -1, -2, -3, -4$. If necessary, round off to six decimal places.
18. Find $(1.1)^n$ for $n = -1, -2, -3, -4$. If necessary, round off to six decimal places.
19. Find $(1/2)^0$, $(1/2)^{-1}$, $(1/2)^{-2}$, $(1/2)^{-3}$.
20. Find $(1/3)^0$, $(1/3)^{-1}$, $(1/3)^{-2}$, $(1/3)^{-3}$.

CALCULATOR COMPUTATIONS

For a positive integer n, the exponential notation x^n has a simple interpretation on the electronic calculator. The base x is the number we enter in the calculator, and the exponent n is the number of times we enter x as a product. Although we can always calculate powers in this way, some calculators have special features for finding powers. For example, on some calculators we can find x^n by entering the number x, pressing $\times$, and then pressing the $=$ key $n - 1$ times. For a calculator with a constant key, you can probably find x^n by entering the number x, pressing $\times$, and then pressing the constant key $n - 1$ times. You should experiment with your calculator to find the easiest way to find x^n.

To find x^{-n} it is frequently easiest first to find x^n and then to find the reciprocal $1/x^n$.

Exercise Set 3.1 (continued)

21. a. Find 2^n for $n = 1, 2, 3$, and so on. What finally happens on your calculator?
    b. Find 2^{-n} for $n = 1, 2, 3$, and so on. What finally happens on your calculator?
22. a. Find 3^n for $n = 1, 2, 3$, and so on. What finally happens on your calculator?
    b. Find 3^{-n} for $n = 1, 2, 3$, and so on. What finally happens on your calculator?

In exercises 23–26, find $(1 + r)^n$ for the given value of r and for $n = 10, 20, -5, -18$, and -20. Round off to six decimal places if necessary.

23. $r = 0.06$ 24. $r = 0.05$ 25. $r = 0.08$ 26. $r = 0.09$
27. Find $4(3.14)r^3/3$ for $r = 2.62$; $r = 7.51$; $r = 8.09$; $r = 21.13$; $r = 101.22$
28. Find $a^3(365.25)^2$ for $a = 0.3871$; $a = 0.7233$; $a = 1.5237$; $a = 5.2027$

PROPERTIES OF EXPONENTS

Exponents are useful because they behave in a simple way with respect to the operations of multiplication and division. This simplicity is due to the following five properties of exponents.

PROPERTIES OF EXPONENTS. In these laws, m and n are integers.

1. $x^m \cdot x^n = x^{m+n}$ 2. $\dfrac{x^m}{x^n} = x^{m-n}$

3. $(x^m)^n = x^{mn}$

4. $(xy)^n = x^n y^n$

5. $\left(\frac{x}{y}\right)^n = \frac{x^n}{y^n}$ if $y \neq 0$

Notice that the first three properties tell what happens if we have the same base and different exponents, and the last two properties tell what happens if we have the same exponent and different bases. We summarize these five properties in the following scheme:

same base

1. product of powers	add exponents
2. quotient of powers	subtract exponents
3. power of powers	multiply exponents

same exponents

4. multiply bases	product of powers
5. divide bases	quotient of powers

Example 6. Properties of exponents.

a. $x^{10}(xy)^7y^4 = x^{10} \cdot x^7 \cdot y^7 \cdot y^4 = x^{10+7}y^{7+4} = x^{17}y^{11}$

b. $(-a^6)^4 = (-1)^4(a^6)^4 = 1 \cdot a^{6 \cdot 4} = a^{24}$

c. $x^{-6} \cdot x^3 = x^{-6+3} = x^{-3} = \frac{1}{x^3}$

d. $\left(\frac{a^{-2}}{a^{-5}}\right)^4 = (a^{-2-(-5)})^4 = (a^{-2+5})^4 = (a^3)^4 = a^{12}$

Exercise Set 3.1 (continued)

In exercises 29–60, use the properties of exponents to simplify each expression; express the result using positive exponents.

29. $x^7 \cdot x^5$
30. $a^6 \cdot a$
31. $b^4 \cdot b^{15} \cdot b^2$
32. $y^5 \cdot y^{13} \cdot y^4$
33. $\frac{x^6}{x^3}$
34. $\frac{x^{12}}{x^{14}}$
35. $\frac{a^2b^4}{a^6b}$
36. $\frac{a^6b^4}{a^6b^{12}}$
37. $a^{-4}a^7$
38. a^5a^{-8}
39. $x^2y^{-1}x^{-2}y^6$
40. $x^{-3}y^{-2}x^{-4}y^3$
41. $a^{-2}ba^{-3}a^4b^{-4}$
42. $a^3a^6a^{-12}b^7a^{-1}$
43. $(a^{-2}b^2)^5$
44. $(a^{-2}b^2)^{-5}$
45. $\frac{a^4b^2}{a^{-3}b^4}$
46. $\frac{a^{-7}b^6}{a^6b^{-3}}$
47. $\frac{(a^2x^{-3})^2}{(a^4b^{-1})^{-3}}$
48. $\frac{(x^6y^3)^{-2}}{x^5y^{-6}}$
49. $\left(\frac{a^3}{b^2}\right)^4$
50. $\left(\frac{x}{y^5}\right)^7$
51. $(4a^2b)^3(3a^3b^2)^2$
52. $(a^2b^2c^2)^3(-ab)^4$

53. $\dfrac{(xy)^2(x^2y)^4}{(xy^2)^3}$

54. $\left(\dfrac{x^2y^3}{z^4}\right)^{11}\left(\dfrac{z^2}{xy^2}\right)^{12}$

55. $x^{4n} \cdot x^{n+3}$

56. $(x^{3n})^{n+3}$

57. $\dfrac{a^{5n}}{a^{6n+2}}$

58. $\dfrac{a^n \cdot a^n \cdot a^n}{a^{5n}}$

59. $\dfrac{(y^{n+1})^n}{y^n}$

60. $\dfrac{(x^{3n-1})^n}{(x^n)^{2n+2}}$

---

section 2 • Scientific Notation

In all sciences, relatively large and relatively small numbers occur; this fact has led to the use of an exponential notation called *scientific notation.* Scientific notation is useful in comparing magnitudes and in performing multiplication and division of large or small numbers. We will find it essential in working with logarithms (Chapter 8).

In this section we see how to write numbers in scientific notation, how to use scientific notation in computations, and how to use scientific notation with an electronic calculator.

EXPRESSING NUMBERS IN SCIENTIFIC NOTATION

We examine multiplication by powers of 10.

Example 1. Multiplication by powers of 10.

a. $2.13 \times 10 = 21.3$ Move the decimal point one place to the right.
b. $2.13 \times 10^2 = 213$ Move the decimal point two places to the right.
c. $2.13 \times 10^3 = 2{,}130$ Move the decimal point three places to the right.

If we multiply a number in decimal notation by 10^n and n is a positive integer, we obtain the product by moving the decimal point n places to the right.

Example 1 (continued). Multiplication by powers of 10.

d. $2.13 \times 10^{-1} = 0.213$ Move the decimal point one place to the left.
e. $2.13 \times 10^{-2} = 0.0213$ Move the decimal point two places to the left.
f. $2.13 \times 10^{-3} = 0.00213$ Move the decimal point three places to the left.

If we multiply a number in decimal notation by 10^{-n} and $-n$ is a negative integer, we obtain the product by moving the decimal point n places to the left. This operation is the same as dividing the number by 10^n.

Scientific notation is a matter of writing a number in a particular form.

DEFINITION OF SCIENTIFIC NOTATION. Let N be a real number. If we write

$$N = n \times 10^k,$$

where $1 \le n < 10$ and k is an integer, then N is said to be expressed in *scientific notation.*

Example 2. Scientific notation.

a. $31{,}250{,}000 = 3.125 \times 10^7$

b. $67{,}413 = 6.7413 \times 10^4$

c. $3020 = 3.02 \times 10^3$

d. $0.0125 = 1.25 \times 10^{-2}$

e. $0.0000362 = 3.62 \times 10^{-5}$

f. $0.0072 = 7.2 \times 10^{-3}$

Exercise Set 3.2

In exercises 1–20, the given number is written in scientific notation; write it in ordinary decimal notation.

1. 1.334×10^5
2. 6.96×10^{10}
3. 5.29×10^{-9}
4. 2.116×10^{-8}
5. 3.1×10^{13}
6. 1.4×10^{11}
7. 1.36×10^{-2}
8. 7.98×10^{-3}
9. 4.37×10^{-5}
10. 6.91×10^{-4}
11. 2.032×10^8; the approximate population of the United States at the time of the 1970 census.
12. 10^{11}; the approximate number of stars in our galaxy.
13. 4.1×10^{13}; the distance from the earth to *a* Centuri, the nearest star, in kilometers.
14. 3.5×10^9; the approximate population of the world in 1970.
15. 2.5×10^8; age of the oceans on the earth in years.
16. 1.99×10^{33}; mass of the sun in grams.
17. 3.53×10^{-2}; the number of ounces in a gram.
18. -1.6×10^{-19}; the charge of an electron in coulombs.
19. 4.0×10^{-7}; the average spacing between molecules of oxygen at STP in centimeters.
20. 1.47×10^7; the area of the moon in square miles.

In exercises 21–40, each given number is written in ordinary decimal notation; write it in scientific notation.

21. 252,000
22. 47,000
23. 2,400,000
24. 367,000,000
25. 0.0212
26. 0.00415
27. 0.000066
28. 0.00000028
29. 0.00000000451
30. 0.0000000333
31. 186,000; the velocity of light in miles per second.
32. 93,000,000; the approximate average distance from the earth to the sun in miles.
33. 5,980,000,000,000,000,000,000,000,000; the mass of the earth in grams.
34. 73,500,000,000,000,000,000,000,000; the mass of the moon in grams.
35. 0.00000000000000000000000166; the mass of a hydrogen atom in grams.
36. 0.00000000000000000000000000091; the mass of an electron in grams.
37. 1.61; the number of kilometers in one mile.

38. 10,000,000,000; the probable age of the universe in years.
39. 637,800,000; the radius of the earth in centimeters.
40. 174,000,000; the radius of the moon in centimeters.

COMPUTATIONS WITH SCIENTIFIC NOTATION

Scientific notation can be useful in doing computations that involve multiplication and division because it helps keep track of the decimal point.

Example 2. Find (3,400,000)(0.06).

Solution. First write each number in scientific notation:

$$3{,}400{,}000 = 3.4 \times 10^{6}$$
$$0.06 = 6.0 \times 10^{-2}$$

Now form the product

$$\begin{aligned}(3.4 \times 10^{6})(6.0 \times 10^{-2}) &= (3.4)(6.0) \times (10^{6})(10^{-2})\\ &= 20.4 \times 10^{4}\\ &= 204{,}000\end{aligned}$$

Notice that, to find the product of two powers of 10, we add the exponents.

Example 3. Find 0.0054/37,000.

Solution. Write each number in scientific notation:

$$0.0054 = 5.4 \times 10^{-3}$$
$$37{,}000 = 3.7 \times 10^{4}$$

Now form the quotient:

$$\begin{aligned}(5.4 \times 10^{-3})/(3.7 \times 10^{4}) &= (5.4/3.7) \times (10^{-3}/10^{4})\\ &= 1.5 \times 10^{-7}\\ &= 0.00000015\end{aligned}$$

Notice that, to find the quotient of two powers of 10, we subtract the exponents.

Exercise Set 3.2 (continued)

In exercises 41–50, perform the computations using scientific notation and express the answer in scientific notation.

41. (56,000)(420,000)
42. (2100)(3,000,000)
43. (0.0006)(4,800,000)
44. (0.00002)(960,000)
45. 480/240,000,000
46. 6,400,000/160,000
47. 0.0006/0.0000012
48. 0.044/0.00000011
49. $(20,000,000,000)^3$
50. $(300,000,000,000)^4$

CALCULATOR COMPUTATIONS

Scientific notation provides a means for extending the capacity of an electronic calculator. That is, we can perform operations on numbers whose total number of digits exceeds the number of digits in the display of the calculator. Scientific notation consists in writing a number as a product of a decimal between 1 and 10 by a power of 10. We enter the decimal part of the number on the calculator; separately, on a sheet of paper, we keep track of the power of 10. To multiply or divide numbers written in scientific notation, we multiply or divide the decimal parts on the calculator. Separately, we use the laws of exponents to multiply or divide the powers of 10.

Example 4. Find $4(3.14)(6.378 \times 10^8)^3/3$

Solution. We write this number as

$$4(3.14)(6.378)^3/3 \times (10^8)^3$$

We perform $4(3.14)(6.378)^3/3$ on the calculator and find 1086.23 (rounded off). So the number is

$$1086.23 \times (10^8)^3 = 1086.23 \times 10^{24}$$
$$= 1.08623 \times 10^{27}$$

Exercise Set 3.2 (continued)

In exercises 51–60, use a calculator and scientific notation to perform each computation; express the answer in scientific notation.

51. $(19.2)^3(365.25)^2$
52. $(30.09)^3(365.25)^2$
53. $4(3.14)(3.4 \times 10^8)^3/3$
54. $4(3.14)(7.14 \times 10^9)^3/3$
55. $(3.14)(5.29 \times 10^{-9})^2$
56. $(3.14)(2.116 \times 10^{-8})^2$
57. $(3.1 \times 10^{13})/0.37$
58. $(3.1 \times 10^{13})/0.751$
59. $(3.75 \times 10^6)^4/(1.21 \times 10^{-5})^3$
60. $(8.18 \times 10^{-5})^6/(7.42 \times 10^5)^7$

In exercises 61–70, perform the computations using scientific notation. The numbers needed are all given in exercises 1–10. Round off answers to three decimal places.

61. How many miles does light travel in an hour?
62. How many miles does light travel in a day?
63. How many miles does light travel in a year?
64. How long does it take light from the sun to reach the earth?
65. Find the distance from the earth to *a* Centuri in miles.
66. How long (in years) does it take light from *a* Centuri to reach the earth?
67. How many times heavier than an electron is a hydrogen atom?
68. How many times heavier than the earth is the sun?
69. If it is assumed that the mass of the sun is entirely hydrogen atoms, how many hydrogen atoms would there be in the sun?
70. How many times longer than the radius of the moon is the radius of the earth?

section 3 • Roots

We now define radicals—that is, n^{th} roots of numbers. We single out square roots for special consideration. We discuss finding approximations to roots.

SQUARE ROOTS

We begin by defining square roots, a special case of nth roots. Finding a square root of a number is the "opposite" of finding the square of a number.

DEFINITION OF SQUARE ROOT. If a is a number, then r is a *square root* of a if $r^2 = a$.

Note: We have seen that, if we have a square whose side has length x, then the area of the square is x^2; that is, we square the length of the side to find the area. If, however, we are given the area a, then a square root of a is the length of the side. For example, if the side of a square is 3, then the area is $3^2 = 9$. Conversely, if the area of a square is 9, then the side is 3. Square roots have many applications besides to squares.

Example 1. Square roots.

a. 3 is a square root of 9 since $3^2 = 9$.
b. -3 is a square root of 9 since $(-3)^2 = 9$.
c. 10 is a square root of 100 since $10^2 = 100$.

d. -10 is a square root of 100 since $(-10)^2 = 100$.

e. 0 is a square root of 0 since $0^2 = 0$.

We see from Example 1 that both 3 and -3 are square roots of 9, and both 10 and -10 are square roots of 100. In fact, if r is a square root of a, then $-r$ is also a square root of a because $(-r)^2 = r^2$. Therefore, square roots occur in positive and negative pairs.

Some real numbers do not have square roots among the real numbers. For example, no real number is the square root of -1. Which real numbers have square roots among the real numbers and which do not? If a is a real number and r is a real number square root of a, then $r^2 = a$. By the rules for multiplying negatives, no matter whether r is positive or negative or 0, we must have $r^2 \geq 0$. So a real number a has a real number square root if $a \geq 0$ and does not have a real number square root if $a < 0$.

If a is a negative real number, there is no real number that is a square root of a. The remedy to this difficulty is to enlarge the system of real numbers to include the system of *complex numbers*. We shall do this in Sections 3.7 and 3.8.

REAL NUMBER SQUARE ROOTS. If a is a positive real number, then a possesses two real number square roots—a positive square root and a negative square root. If a is a negative real number, then a does not possess real number square roots. If $a = 0$, then a possesses one square root, namely 0.

DEFINITION OF POSITIVE AND NEGATIVE SQUARE ROOTS. Suppose a is a positive real number. The positive real number that is a square root of a is called *the positive square root of a* and is written $\sqrt{a}$. The negative real number which is a square root of a is called *the negative square root of a* and is written $-\sqrt{a}$. The positive square root of a is also called the *principal square root of a*. For $a = 0$, we have $\sqrt{0} = -\sqrt{0} = 0$.

Note: Frequently we say "square root of a" when we mean "positive square root of a." The square root of a is understood to mean the positive square root. Also, don't confuse the symbolism $-\sqrt{a}$ with $\sqrt{-a}$. The symbol $-\sqrt{a}$ means the negative square root of a; the symbol $\sqrt{-a}$ means the square root of the negative of a.

Example 2. Notation for square roots.

a. $\sqrt{9} = 3$ and $-\sqrt{9} = -3$.

b. $\sqrt{100} = 10$ and $-\sqrt{100} = -10$.

c. $\sqrt{144} = 12$ and $-\sqrt{144} = -12$.

d. $\sqrt{1.44} = 1.2$ and $-\sqrt{1.44} = -1.2$.

There is a paradox involving positive and negative square roots. We may expect that for all a, $\sqrt{a^2} = a$. If this were the case, then we would have $\sqrt{(-2)^2} = -2$. But when we do the computations,

$$\sqrt{(-2)^2} = \sqrt{4} = 2 \neq -2.$$

So it is *not* true that for all a we have $\sqrt{a^2} = a$. What we can say is

$$\sqrt{a^2} = a \quad \text{if} \quad a \geq 0 \quad \text{and} \quad \sqrt{a^2} = -a \quad \text{if} \quad a < 0.$$

On the other hand, it is true that, if a is a non-negative real number, then

$$(\sqrt{a})^2 = a.$$

Exercise Set 3.3

In exercises 1–20, find the given square root or specify that it is not a real number.

1. $\sqrt{1}$ 2. $\sqrt{4}$ 3. $-\sqrt{1}$ 4. $-\sqrt{4}$ 5. $\sqrt{-1}$

6. $\sqrt{-4}$ 7. $\sqrt{81}$ 8. $\sqrt{25}$ 9. $-\sqrt{81}$ 10. $-\sqrt{25}$

11. $\sqrt{64}$ 12. $\sqrt{144}$ 13. $\sqrt{-64}$ 14. $\sqrt{-144}$ 15. $-\sqrt{64}$

16. $-\sqrt{144}$ 17. $\sqrt{36}$ 18. $\sqrt{49}$ 19. $-\sqrt{36}$ 20. $-\sqrt{49}$

N^{th} ROOTS

Now we define the nth root of a number.

DEFINITION OF nth ROOT. If n is a positive integer and a is a number, then r is an *n^{th} root of a* if $r^n = a$. The case $n = 2$ is, of course, the square root; the case $n = 3$ is called *cube root*; the case $n > 3$ is called n^{th} root.

Example 3. n^{th} roots.

a. 3 is a cube root of 27 since $3^3 = 27$.
b. -3 is a cube root of -27 since $(-3)^3 = -27$.
c. 2 is a fourth root of 16 since $2^4 = 16$.
d. -2 is a fourth root of 16 since $(-2)^4 = 16$.
e. -2 is a fifth root of -32 since $(-2)^5 = -32$.

Note: We have $(-r)^n = (-1)^n r^n$. Also,

$$\text{if } n \text{ is even, } (-1)^n = 1; \text{ if } n \text{ is odd, } (-1)^n = -1.$$

Therefore,

$$\text{if } n \text{ is even, } (-r)^n = r^n; \text{ if } n \text{ is odd, } (-r)^n = -r^n.$$

With square roots we have to distinguish between positive and negative; we have both $\sqrt{a}$ and $-\sqrt{a}$. The basic reason is that $(-r)^2 = r^2$. When we consider cube roots, however, this difficulty does not arise because $(-r)^3 = -r^3 \neq r^3$ (unless $r = 0$). What happens is that, if a is a real number, there is only one real number r that is a cube root of a instead of two as for square roots. But what about fourth roots? Here $(-r)^4 = r^4$. We again have the same situation as for square roots—if r is a fourth root of a, then $-r$ is also a fourth root of a. So we have to distinguish between positive and negative fourth roots.

What can we say in general? If n is even, then $(-r)^n = r^n$; whenever r is an nth root of a, so is $-r$. In this case, we must distinguish between positive and negative roots. On the other hand, if n is odd, then $(-r)^n = -r^n \neq r^n$ (unless $r = 0$). In this case, there is only one real number r which is an nth root of the real number a.

NOTATION FOR Nth ROOTS.

a. If n is an even positive integer and a is a positive real number, then there is a positive real number that is an nth root of a denoted by $\sqrt[n]{a}$ and there is a negative real number that is an nth root of a denoted by $-\sqrt[n]{a}$.

b. If n is an odd positive integer and a is a real number, then there is a single n^{th} root of a denoted by $\sqrt[n]{a}$.

The notation $\sqrt[n]{a}$ is called a *radical*. The radical notation has two parts—a is called the *radicand* and n is called the *index* of the radical.

Example 4. Notation for n^{th} roots.

a. $\sqrt[4]{16} = 2$ and $-\sqrt[4]{16} = -2$.

b. $\sqrt[6]{1} = 1$ and $-\sqrt[6]{1} = -1$.

c. $\sqrt[3]{27} = 3$ and $-\sqrt[3]{27} = -3$.

d. $\sqrt[3]{-27} = -3$ and $-\sqrt[3]{-27} = -(-3) = 3$.

With square roots we ran into the difficulty that the square root of a negative real number is not a real number. This is because $r^2 \geq 0$ for any real number r. However, cube roots of negative numbers present no difficulty because r^3 is positive if r is positive and r^3 is negative if r is negative. However, $r^4 \geq 0$ for all real numbers r, so the fourth root of a negative real number is not a real number.

In general, if n is even, then the nth root of a negative real number is not a real number. If n is odd, then every real number a has an nth root; the nth root is positive if a is positive and negative if a is negative.

Exercise Set 3.3 (continued)

In exercises 21–36, find the given n^{th} root.

21. $\sqrt[3]{8}$
22. $\sqrt[3]{-8}$
23. $-\sqrt[3]{8}$
24. $-\sqrt[3]{-8}$
25. $\sqrt[3]{125}$
26. $-\sqrt[3]{125}$
27. $\sqrt[3]{64}$
28. $\sqrt[3]{-64}$
29. $\sqrt[4]{81}$
30. $-\sqrt[4]{81}$
31. $\sqrt[4]{1}$
32. $-\sqrt[4]{1}$
33. $\sqrt[5]{32}$
34. $\sqrt[5]{-32}$
35. $\sqrt[10]{0}$
36. $\sqrt[13]{-1}$

APPROXIMATING n^{th} ROOTS

If a is a positive real number, then $\sqrt[n]{a}$ is a positive real number. We now consider decimal representations of nth roots. The results stated in Example 5 are proven in more advanced algebra courses.

Example 5. Decimal representations of square roots.

a. Every decimal representation of $\sqrt{2}$ is nonterminating; furthermore, $\sqrt{2}$ is irrational (Section 1.1).

b. Every decimal representation of $\sqrt{3}$ is nonterminating; furthermore, $\sqrt{3}$ is irrational.

c. Since $\sqrt{4} = 2$, there is a terminating decimal representation of $\sqrt{4}$; that is, $\sqrt{4}$ is rational.

d. Every decimal representation for $\sqrt{10}$ is nonterminating; furthermore, $\sqrt{10}$ is irrational.

The point of Example 5 is that $\sqrt[n]{a}$ is frequently an irrational number. Thus, $\sqrt{2}$, $\sqrt{3}$, and $\sqrt{10}$ are irrational. When we perform algebraic computations involving irrational radicals, it is usually easiest to write the radicals in radical notation as $\sqrt{2}$, $\sqrt{3}$, $\sqrt{10}$, and so on. However, in practical applications we are usually interested in a decimal representation of a radical. For practical applications we must find decimal approximations to $\sqrt{2}$, $\sqrt{3}$, $\sqrt{10}$, and so on.

Appendix C is a table of decimal approximations of square roots of integers from 1 to 100. The approximations in the table are accurate to three decimal places.

Using an electronic calculator we can easily find decimal approximations to radicals. Some calculators have a square root key that gives a decimal approximation. Even without a square root key, the trial-and-error method of Example 6 is quick and easy to apply.

Example 6. Find a decimal approximation to $\sqrt{3.14}$ which is accurate to two decimal places.

Calculator Solution. We proceed by trial and error, finding one digit at a time. We know that

$$1^2 = 1 < 3.14 \qquad \text{and} \qquad 2^2 = 4 > 3.14$$

Therefore,

$$1 < \sqrt{3.14} < 2$$

By trial and error we find the next digit (first decimal place). We find

$$1.7^2 = 2.89 < 3.14 \qquad \text{and} \qquad 1.8^2 = 3.24 > 3.14$$

Therefore,

$$1.7 < \sqrt{3.14} < 1.8$$

By trial and error we find the next digit (second decimal place). We find

$$1.77^2 = 3.1329 < 3.14 \qquad \text{and} \qquad 1.78^2 = 3.1684 > 3.14$$

now we can write

$$1.77 < \sqrt{3.14} < 1.78$$

Since $(1.77)^2$ is closer to 3.14 than $(1.78)^2$, the best decimal approximation to $\sqrt{3.14}$ accurate to two places is 1.77.

Example 7. Find a decimal approximation to $\sqrt{1 - x^2}$ which is accurate to three decimal places when $x = 0.637$.

Calculator Solution. First we square x:

$$\begin{aligned} x^2 &= (0.637)^2 = 0.405769 \\ 1 - x^2 &= 1 - 0.405769 = 0.594231 \\ \sqrt{1 - x^2} &= \sqrt{0.594231} \end{aligned}$$

Using either a square root key or trial and error, we obtain 0.771. We also find

$$0.770 < \sqrt{1 - x^2} < 0.771$$

Exercise Set 3.3 (continued)

In exercises 37–52, find a decimal approximation to the given radical which is accurate to two decimal places.

37. $\sqrt{2.5}$	38. $\sqrt{6.5}$	39. $\sqrt{12.74}$	40. $\sqrt{19.75}$	41. $\sqrt{24.18}$
42. $\sqrt{27.62}$	43. $\sqrt{88.15}$	44. $\sqrt{172.08}$	45. $\sqrt{958.60}$	46. $\sqrt{2484.08}$

47. $\sqrt{7738.38}$ 48. $\sqrt{50{,}481.92}$ 49. $\sqrt[3]{3.57}$ 50. $\sqrt[3]{5.55}$ 51. $\sqrt[3]{17.05}$

52. $\sqrt[3]{13.09}$

In exercises 53–60, find a decimal approximation to $\sqrt{1 - x^2}$ for the given value of x, accurate to three decimal places.

53. $x = 0.587$ 54. $x = 0.422$ 55. $x = 0.822$ 56. $x = 0.931$ 57. $x = 0.289$

58. $x = 0.198$ 59. $x = 0.375$ 60. $x = 0.100$

---

section 4 • Properties of Radicals

In this section we discuss two properties of radicals. They tell what happens if the radicand of a radical is a product or a quotient. We apply the properties to simplifying the form of a radical, to rationalizing the denominator, and to scientific notation.

THE TWO PROPERTIES

PROPERTIES OF RADICALS. Suppose n is a positive integer and a and b are positive real numbers. Then

$$\sqrt[n]{ab} = \sqrt[n]{a} \cdot \sqrt[n]{b}$$ (The nth root of a product is the product of the nth roots.)

$$\sqrt[n]{\frac{a}{b}} = \frac{\sqrt[n]{a}}{\sqrt[n]{b}}$$ (The nth root of a quotient is the quotient of the nth roots.)

Notice that these properties are stated only when the radicand is positive, to assure that the radicals exist as positive real numbers. Suppose we look at the first property in more detail to see how it follows from the fourth property of exponents. If we set $x = \sqrt[n]{a}$ and $y = \sqrt[n]{b}$, by the definition of nth root, $x^n = a$ and $y^n = b$. Taking the product, by the fourth property of exponents, we obtain

$$ab = x^n \cdot y^n = (xy)^n.$$

From this equation and the definition of nth root,

$$\sqrt[n]{ab} = xy = \sqrt[n]{a} \cdot \sqrt[n]{b}.$$

In a similar way, the second property follows from the fifth property of exponents.

Example 1. Simplifying radicals.

a. $\sqrt{a^6b^2} = \sqrt{a^6}\sqrt{b^2} = \sqrt{(a^3)^2}\sqrt{b^2} = a^3b$ for $a > 0, b > 0$

b. $\sqrt{x^5y^3} = \sqrt{x^4y^2 \cdot xy} = \sqrt{x^4y^2} \cdot \sqrt{xy} = x^2y\sqrt{xy}$ for $x > 0, y > 0$.

c. $\sqrt{2xy}\sqrt{8x^3y} = \sqrt{2xy \cdot 8x^3y} = \sqrt{16x^4y^2} = 4x^2y$ for $x > 0, y > 0$

It is important to notice that the two properties of radicals pertain to taking the root of a product or quotient. There are no corresponding laws pertaining to the root of a sum or difference. Yet many students of algebra make the *mistake* of thinking that

$$\sqrt[n]{a+b} = \sqrt[n]{a} + \sqrt[n]{b} \quad \text{or} \quad \sqrt[n]{a-b} = \sqrt[n]{a} - \sqrt[n]{b}.$$

Neither of these is correct. Simple examples show that these "laws" do not hold for all a and b.

Example 2. An example in which $\sqrt{a+b} \neq \sqrt{a} + \sqrt{b}$.

Solution. Take $a = 9$ and $b = 16$. Then $a + b = 25$. So we see

$$\sqrt{a+b} = \sqrt{25} = 5 \text{ and } \sqrt{a} + \sqrt{b} = \sqrt{9} + \sqrt{16} = 3 + 4 = 7.$$

Exercise Set 3.4

1. Give an example of positive real numbers a and b so that

$$\sqrt{a-b} \neq \sqrt{a} - \sqrt{b}.$$

2. Show how the second property for radicals follows from the fifth property of exponents.

In exercises 3–16, simplify each radical as in Example 1. Assume that all letters denote positive real numbers.

3. $\sqrt{a^4}$ 4. $\sqrt{16a^4}$ 5. $\sqrt{a^8}$ 6. $\sqrt{81a^8b^8}$ 7. $\sqrt{a^6b^{12}}$

8. $\sqrt{a^{10}b^{18}}$ 9. $\sqrt[3]{x^6}$ 10. $\sqrt[3]{-x^6}$ 11. $\sqrt{a^2b^3}$ 12. $\sqrt{a^5b^7}$

13. $\sqrt{20ab^6}$ 14. $\sqrt{28a^8b^7}$ 15. $\sqrt{3ab}\sqrt{27ab^5}$ 16. $\sqrt{5a^2b}\sqrt{5ab}$

CHANGING THE FORM OF A RADICAL

In Chapter 5 we will use the laws for radicals to change the form of radical expressions.

#10 $\sqrt[3]{-x^6} = -\sqrt[3]{x^6} = -(x^6)^{\frac{1}{3}} = -x^{\frac{6}{3}} = -x^2$

this is OK FOR ODD ROOT NOT FOR POS. ROOT

Example 3. Consider $\sqrt{8}$. The number 8 has a perfect square as a factor: $8 = 4 \cdot 2 = 2^2 \cdot 2$. Using the property for the square root of a product, we have

$$\sqrt{8} = \sqrt{4 \cdot 2} = \sqrt{4} \cdot \sqrt{2} = 2\sqrt{2}.$$

We have written $\sqrt{8}$ as $2\sqrt{2}$; one advantage of this form is that we can simplify a fraction such as

$$\frac{-2 + \sqrt{8}}{2} = \frac{-2 + 2\sqrt{2}}{2} = \frac{2(-1 + \sqrt{2})}{2} = -1 + \sqrt{2}.$$

Example 4. Consider $\sqrt{48}$. The number 48 has a perfect square as a factor: $48 = 16 \cdot 3 = 4^2 \cdot 3$. Using the property for the square root of a product, we have

$$\sqrt{48} = \sqrt{16 \cdot 3} = \sqrt{16} \cdot \sqrt{3} = 4\sqrt{3}.$$

One advantage of writing $\sqrt{48}$ as $4\sqrt{3}$ is that we can simplify a fraction such as

$$\frac{4 - \sqrt{48}}{2} = \frac{4 - 4\sqrt{3}}{2} = \frac{2(2 - 2\sqrt{3})}{2} = 2 - 2\sqrt{3}.$$

Example 5. Consider $\sqrt{200}$. The number 200 has a perfect square as a factor: $200 = 100 \cdot 2 = 10^2 \cdot 2$. Using the property for the square root of a product, we have

$$\sqrt{200} = \sqrt{100 \cdot 2} = \sqrt{100}\sqrt{2} = 10\sqrt{2}$$

An advantage of the form $10\sqrt{2}$ is that we can use the square root table of Appendix C to find $\sqrt{200}$ (which does not occur in the table). From the table we find $\sqrt{2} = 1.414$ Therefore,

$$\sqrt{200} = 10\sqrt{2} = 10(1.414) = 14.14$$

Exercise Set 3.4 (continued)

In exercises 17–32, remove all possible perfect-square factors from under the square root; then simplify the given fraction.

17. $\sqrt{20}$; $(4 + \sqrt{20})/2$

18. $\sqrt{40}$; $(2 + \sqrt{40})/2$

19. $\sqrt{24}$; $(-2 - \sqrt{24})/2$

20. $\sqrt{12}$; $(-2 - \sqrt{12})/2$

21. $\sqrt{32}$; $(-4 + \sqrt{32})/2$

22. $\sqrt{80}$; $(-4 + \sqrt{80})/2$

23. $\sqrt{18}$; $(3 + \sqrt{18})/6$
24. $\sqrt{27}$; $(3 + \sqrt{27})/6$
25. $\sqrt{45}$; $(3 - \sqrt{45})/6$
26. $\sqrt{54}$; $(3 - \sqrt{54})/6$
27. $\sqrt{1444}$; $(2 + \sqrt{1444})/10$
28. $\sqrt{1296}$; $(4 + \sqrt{1296})/10$
29. $\sqrt{136}$; $(-2 - \sqrt{136})/4$
30. $\sqrt{184}$; $(-2 - \sqrt{184})/4$
31. $\sqrt{176}$; $(-4 + \sqrt{176})/4$
32. $\sqrt{1984}$; $(-4 + \sqrt{1984})/4$

In exercises 33–40, remove all possible perfect-square factors from under the square root; then use the square root table of Appendix C to find a decimal approximation to the square root.

33. $\sqrt{800}$
34. $\sqrt{600}$
35. $\sqrt{2400}$
36. $\sqrt{720}$
37. $\sqrt{512}$
38. $\sqrt{120{,}000}$
39. $\sqrt{1225}$
40. $\sqrt{3969}$

RATIONALIZING THE DENOMINATOR

Sometimes when we have fractions involving radicals, we wish to remove the radicals from the denominator. This procedure is called *rationalizing the denominator*. It consists of multiplying the numerator and denominator of the fraction by the same expression; this transformation of the form $a/b = ax/bx$ (Section 2.6) doesn't change the value of the fraction. If we choose the expression x appropriately, the radical is removed from the denominator.

Example 6. Rationalizing the denominator.

a. $$\frac{1}{\sqrt{2}} = \frac{1 \cdot \sqrt{2}}{\sqrt{2} \cdot \sqrt{2}} = \frac{\sqrt{2}}{\sqrt{2 \cdot 2}} = \frac{\sqrt{2}}{2}$$

b. $$\frac{\sqrt{2}}{\sqrt{3}} = \frac{\sqrt{2} \cdot \sqrt{3}}{\sqrt{3} \cdot \sqrt{3}} = \frac{\sqrt{2 \cdot 3}}{\sqrt{3 \cdot 3}} = \frac{\sqrt{6}}{3}$$

c. $$\frac{1}{\sqrt[3]{6}} = \frac{1 \cdot \sqrt[3]{6} \cdot \sqrt[3]{6}}{\sqrt[3]{6} \cdot \sqrt[3]{6} \cdot \sqrt[3]{6}} = \frac{\sqrt[3]{6 \cdot 6}}{\sqrt[3]{6 \cdot 6 \cdot 6}} = \frac{\sqrt[3]{36}}{6}$$

If the denominator of a fraction is a binomial expression, then rationalizing the denominator is a little trickier. If the denominator is of the form $\sqrt{a} - \sqrt{b}$, then we multiply both numerator and denominator by $\sqrt{a} + \sqrt{b}$, which rationalizes the denominator because

$$(\sqrt{a} - \sqrt{b})(\sqrt{a} + \sqrt{b}) = (\sqrt{a})^2 - (\sqrt{b})^2 = a - b.$$

So the trick is that the product of a difference and a sum is a difference of squares. What should we do if the denominator is of the form $\sqrt{a} + \sqrt{b}$?

Example 7. Rationalizing the denominator.

a. $$\frac{1}{\sqrt{2}-\sqrt{3}} = \frac{1 \cdot (\sqrt{2}+\sqrt{3})}{(\sqrt{2}-\sqrt{3})(\sqrt{2}+\sqrt{3})} = \frac{\sqrt{2}+\sqrt{3}}{(\sqrt{2})^2-(\sqrt{3})^2} = \frac{\sqrt{2}+\sqrt{3}}{2-3}$$
$$= \frac{\sqrt{2}+\sqrt{3}}{-1} = -\sqrt{2}-\sqrt{3}$$

b. $$\frac{3}{4+\sqrt{5}} = \frac{3(4-\sqrt{5})}{(4+\sqrt{5})(4-\sqrt{5})} = \frac{12-3\sqrt{5}}{(4)^2-(\sqrt{5})^2} = \frac{12-3\sqrt{5}}{16-5} = \frac{12-3\sqrt{5}}{11}$$

Exercise Set 3.4 (continued)

In exercises 41–48, rationalize the denominator.

41. $\frac{1}{\sqrt{3}}$ 42. $\frac{2}{\sqrt{6}}$ 43. $\frac{4}{\sqrt[3]{3}}$ 44. $\frac{6}{\sqrt[3]{9}}$

45. $\frac{1}{\sqrt{5}-\sqrt{6}}$ 46. $\frac{\sqrt{2}-\sqrt{3}}{\sqrt{2}+\sqrt{3}}$ 47. $\frac{1-\sqrt{3}}{1+\sqrt{3}}$ 48. $\frac{4}{2-\sqrt{7}}$

RADICALS AND SCIENTIFIC NOTATION (CALCULATOR)

In many applications we must find the nth root of a number written in scientific notation.

Example 8. Find $\sqrt{4.7193 \times 10^5}$

Solution. By the property for square roots,

$$\sqrt{4.7193 \times 10^5} = \sqrt{4.7193} \times \sqrt{10^5}$$

Note that $\sqrt{10^5} = \sqrt{10^4}\sqrt{10} = 10^2\sqrt{10}$. Now we must find *two* square roots, $\sqrt{4.7193}$ and $\sqrt{10}$. We realize that we can save work if we rewrite the original number:

$$\sqrt{4.7193 \times 10^5} = \sqrt{47.193 \times 10^4} = \sqrt{47.193} \times \sqrt{10^4}$$
$$= \sqrt{47.193} \times 10^2 = 6.870 \times 10^2$$

Therefore, by rewriting the power of 10 so that it is a perfect square, we only have to find one square root.

In general, to find the nth root of a number written in scientific notation, you can save time by rewriting the power of 10 to be a perfect n^{th} power.

Exercise Set 3.4 (contd.)

In exercises 49–60, find the indicated n^{th} root.

49. $\sqrt{9.4424 \times 10^8}$
50. $\sqrt{8.2219 \times 10^9}$
51. $\sqrt{4.6828 \times 10^{12}/1.3621 \times 10^5}$
52. $\sqrt{9.5763 \times 10^{19}/4.3626}$
53. $\sqrt{5.6783 \times 10^{-11}}$
54. $\sqrt{8.2754 \times 10^{-15}}$
55. $\sqrt{(5.2027)^3(365.25)^2}$
56. $\sqrt{(9.546)^3(365.25)^2}$
57. $\sqrt[3]{1.7042 \times 10^{-23}}$
58. $\sqrt[3]{1.3079 \times 10^{-23}}$
59. $\sqrt[3]{27/4(3.14159)(6.0235 \times 10^{23})}$
60. $\sqrt[3]{42/4(3.14159)(6.0235 \times 10^{23})}$

section 5 • Rational Exponents

We return now to the extension of the definition of exponents. So far we have defined exponents x^n for every integer n—positive, negative, and 0. Now we wish to extend the definition of exponents to cover all rational numbers. That is, if r is any rational number, we wish to attach a meaning to x^r. Recall (Section 1.1) that if r is a rational number, then r is a quotient of two integers, say $r = p/q$ for integers p and q. How shall we assign a meaning to $x^{p/q}$? It should come as no surprise that the definition is in terms of roots.

EXPONENTS 1/q

We shall take as a starting point the third property of exponents. To simplify the work, we first suppose we want to define $x^{1/q}$ and then later look for a definition of the general case $x^{p/q}$. So for the time being, we look for a definition of $x^{1/2}, x^{1/3}, x^{1/4}, \ldots$.

Suppose we take $x^{1/q}$ and raise it to the qth power. The third property of exponents tells what we want to have happen when we raise to powers—we multiply the exponents. So

$$(x^{1/q})^q = x^{(1/q)q} = x^1 = x.$$

This says that $x^{1/q}$ raised to the qth power is x. Since $x^{1/q}$ is thus a qth root of x, this is how we define $x^{1/q}$.

DEFINITION OF $x^{1/q}$. If q is a positive integer, then

$$x^{1/q} = \sqrt[q]{x}.$$

In this definition we assume that x is a positive real number if q is even and that the qth root denotes the positive q^{th} root (Section 3.3).

Example 1. Rational exponents.

a. $4^{1/2} = \sqrt{4} = 2$

b. $8^{1/2} = \sqrt{8} = 2\sqrt{2}$

c. $8^{1/3} = \sqrt[3]{8} = 2$

d. $(81)^{1/4} = \sqrt[4]{81} = 3$

Exercise Set 3.5

In exercises 1–16, find the given number.

1. $9^{1/2}$
2. $16^{1/2}$
3. $81^{1/2}$
4. $100^{1/2}$
5. $12^{1/2}$
6. $27^{1/2}$
7. $27^{1/3}$
8. $64^{1/3}$
9. $(-27)^{1/3}$
10. $(-64)^{1/3}$
11. $16^{1/4}$
12. $(100)^{1/4}$
13. $625^{1/4}$
14. $625^{1/2}$
15. $32^{1/5}$
16. ~~$245^{1/5}$~~ $243^{1/5}$

DEFINITION OF RATIONAL EXPONENTS

With this definition of $x^{1/q}$, we now proceed to a definition of $x^{p/q}$. We look at the third law of exponents again and obtain

$$x^{p/q} = x^{(1/q)p} = (x^{1/q})^p = (\sqrt[q]{x})^p.$$

DEFINITION OF RATIONAL EXPONENTS. If r is a rational number, write $r = p/q$ where p is an integer and q is a positive integer. Then

$$x^{p/q} = (\sqrt[q]{x})^p.$$

We note two things about the definition of rational exponents. First, the rational exponent p/q can be either positive or negative. If it's positive, we can assume that both p and q are positive. If it's negative, we can assume that the negative sign is in the numerator, so q is positive. We assume q is positive when we apply the definition of rational exponents; negative exponents are taken care of in the numerator p. Second, rational exponents are not defined for all bases. The definition depends on the existence of the radicals involved. If x is a positive real number, then $x^{p/q}$ is always defined.

Example 2. Rational exponents.

a. $27^{2/3} = (\sqrt[3]{27})^2 = 3^2 = 9$

b. $9^{3/2} = (\sqrt{9})^3 = 3^3 = 27$

c. $32^{4/5} = (\sqrt[5]{32})^4 = 2^4 = 16$

d. $27^{-2/3} = (\sqrt[3]{27})^{-2} = \dfrac{1}{3^2} = \dfrac{1}{9}$

e. $32^{-2/5} = (\sqrt[5]{32})^{-2} = \dfrac{1}{2^2} = \dfrac{1}{4}$

f. $8^{-1/2} = (\sqrt{8})^{-1} = \dfrac{1}{\sqrt{8}} = \dfrac{\sqrt{2}}{4}$

Exercise Set 3.5 (continued)

In exercises 17–32, find the given number.

17. $16^{3/2}$
18. $4^{3/2}$
19. $9^{5/2}$
20. $4^{5/2}$
21. $8^{2/3}$
22. $125^{2/3}$
23. $8^{4/3}$
24. $125^{4/3}$
25. $16^{-3/2}$
26. $4^{-3/2}$
27. $9^{-3/2}$
28. $4^{-5/2}$
29. $16^{-1/4}$
30. $32^{-3/5}$
31. $(-1)^{3/5}$
32. $(-1)^{-3/7}$

WRITING EXPRESSIONS AS RADICALS

If an expression is written in exponential notation, we can use the definition of rational exponents to rewrite it in radical notation.

Example 3. Exponential notation.

a. $a^{5/3} = (\sqrt[3]{a})^5 = (\sqrt[3]{a})^3(\sqrt[3]{a})^2 = a(\sqrt[3]{a})^2$

b. $x^{1/2}y^{-1/2} = \dfrac{x^{1/2}}{y^{1/2}} = \dfrac{\sqrt{x}}{\sqrt{y}} = \sqrt{\dfrac{x}{y}}$

c. $x^{3/2}(x+1)^{-1/2} = \dfrac{x^{3/2}}{(x+1)^{1/2}} = \dfrac{(\sqrt{x})^3}{\sqrt{x+1}} = \dfrac{x\sqrt{x}}{\sqrt{x+1}}$

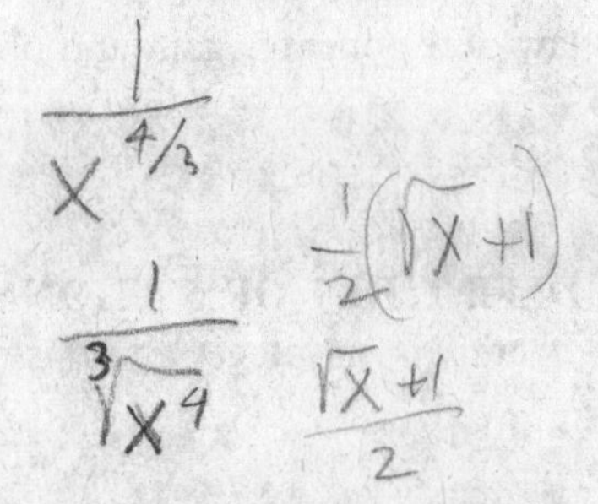

Exercise Set 3.5 (continued)

In exercises 33–44, write the algebraic expressions in terms of radicals and positive integer exponents. Assume that all letters represent positive real numbers.

33. $a^{4/5}$
34. $b^{4/3}$
35. $x^{-4/3}$
36. $y^{-5/2}$
37. $x^{2/3}y^{-2/3}$
38. $a^{-3/2}b^{5/2}$
39. $\dfrac{1}{2}(x+1)^{-1/2}$
40. $\dfrac{1}{2}(x^{1/2}+1)$
41. $\dfrac{1}{2}(x^{-1/2}+1)$
42. $x(x^2+1)^{-3/2}$
43. $x^{-3/2}(x^2+1)^{1/2}$
44. $x^{-3/2}(x^2+1)^{-1/2}$

WRITING RADICALS IN EXPONENTIAL NOTATION

Any expression written in radical notation can be written in exponential notation using fractional exponents.

Example 4. Exponential notation.

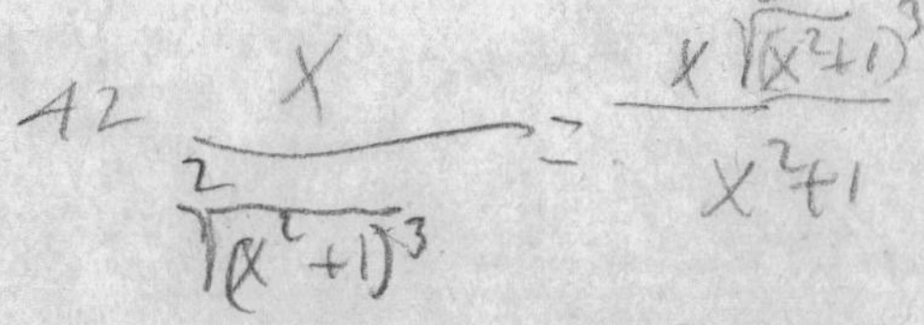

a. $(\sqrt[n]{a})^2 = a^{2/n}$.

b. $\dfrac{\sqrt{a}}{\sqrt[4]{b}} = \dfrac{a^{1/2}}{b^{1/4}} = a^{1/2}b^{-1/4}$

c. $\sqrt{x^2+1} \cdot \sqrt[3]{x^2-1} = (x^2+1)^{1/2}(x^2-1)^{1/3}$

Exercise Set 3.5 (continued)

In exercises 45–56, write the algebraic expressions in terms of rational exponents. Assume that all letters represent positive real numbers.

45. $(\sqrt{a})^3$
46. $\sqrt[n]{a}$
47. $(\sqrt[3]{a})^2$
48. $(\sqrt{a})^n$
49. $\dfrac{\sqrt[m]{a}}{\sqrt[n]{b}}$
50. $\dfrac{1}{\sqrt[m]{ab}}$
51. $x^2\sqrt{x^2+1}$
52. $\dfrac{x^2}{\sqrt{x^2+1}}$
53. $\dfrac{x}{(\sqrt{x}+1)^3}$
54. $\sqrt{x}(\sqrt[3]{x^3}-1)^2$
55. $\dfrac{\sqrt{x}}{(\sqrt{x}-1)^n}$
56. $\sqrt{\sqrt{x}+1}$

---

section 6 • Properties of Exponents

In this chapter we have extended the definition of exponents to cover rational number exponents. We now re-examine the properties of exponents. The properties of exponents are important because we can express a wide variety of computations in exponential notation. We apply the properties for exponents to radicals and practice simplifying algebraic expressions.

EXTENSION OF PROPERTIES OF EXPONENTS

It can be shown that all five properties of exponents extend to rational exponents when the base is restricted to positive real numbers.

PROPERTIES OF EXPONENTS. In these properties, r and s denote rational numbers and x and y denote positive real numbers.

1. $x^r \cdot x^s = x^{r+s}$
2. $\dfrac{x^r}{x^s} = x^{r-s}$
3. $(x^r)^s = x^{rs}$
4. $(xy)^r = x^r y^r$
5. $\left(\dfrac{x}{y}\right)^r = \dfrac{x^r}{y^r}$

In these properties of rational exponents, we restrict the *base* to being a positive real number; there is no such restriction on the base for integer exponents. The properties for rational exponents do not necessarily hold when the base is negative. Example 1, shows a case in which the third property of exponents does not hold for a negative base.

Example 1. The third property of exponents does not necessarily hold for a negative base. We take the base $x = -2$ and compute $[(-2)^2]^{1/2}$ in two ways. First

$$[(-2)^2]^{1/2} = [4]^{1/2} = \sqrt{4} = 2.$$

Second, if we use the third property of exponents,

$$[(-2)^2]^{1/2} = (-2)^{2(1/2)} = (-2)^1 = -2.$$

Exercise Set 3.6

1. Compute $(2^2)^{1/2}$ and $2^{2(1/2)}$ to see if they are equal.
2. a. Compute $(3^2)^{1/2}$ and $3^{2(1/2)}$ to see if they are equal.
   b. Compute $[(-3)^2]^{1/2}$ and $(-3)^{2(1/2)}$ to see if they are equal.
3. Compute $(64^{1/2})(64)^{1/3}$ and $64^{(1/2)+(1/3)}$ to see if they are equal.
4. Compute $64^{1/2}/64^{1/3}$ and $64^{(1/2)-(1/3)}$ to see if they are equal.

COMPUTATIONS WITH RADICALS

We can use the properties of exponents to simplify expressions containing radicals.

Example 2. Changing the index of a radical.

a. $\sqrt[5]{\sqrt[3]{x}} = \sqrt[5]{x^{1/3}} = (x^{1/3})^{1/5} = x^{1/3(1/5)} = x^{1/15} = \sqrt[15]{x}.$

b. $\sqrt[4]{x^2} = (x^2)^{1/4} = x^{2(1/4)} = x^{2/4} = x^{1/2} = \sqrt{x}$

c. $\sqrt{x}\sqrt[3]{x} = x^{1/2}x^{1/3} = x^{(1/2)+(1/3)} = x^{5/6} = \sqrt[6]{x^5}$

Exercise Set 3.6 (continued)

In exercises 5–20, change the index of the radical by using the properties of exponents (as in Example 2); write the result in radical notation. Assume that the letters denote positive real numbers.

5. $\sqrt{\sqrt{x}}$
6. $\sqrt[3]{\sqrt[5]{x}}$
7. $\sqrt[4]{\sqrt{x}}$
8. $\sqrt[3]{\sqrt[3]{x}}$
9. $\sqrt[6]{x^3}$
10. $\sqrt[6]{x^2}$
11. $\sqrt[10]{x^2}$
12. $\sqrt[10]{x^4}$
13. $\sqrt{a} \cdot \sqrt[3]{a}$
14. $\sqrt[4]{a} \cdot \sqrt{a}$
15. $\sqrt[3]{a}\sqrt[4]{a}$
16. $\sqrt{a}\sqrt[5]{a}$

17. $\sqrt[3]{a}/\sqrt[4]{a}$ 18. $\sqrt{a}/\sqrt[5]{a}$ 19. $\sqrt[5]{ab} \cdot \sqrt[3]{ab}$ 20. $\sqrt[3]{ab} \cdot \sqrt[4]{ab}$

SIMPLIFYING EXPRESSIONS

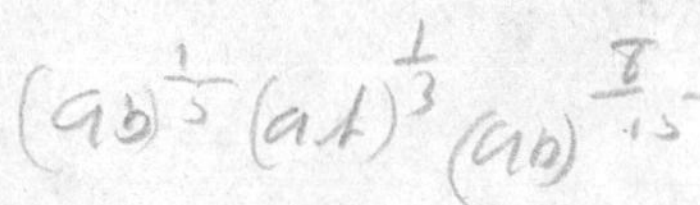

We can use the five properties of exponents to simplify algebraic expressions that contain rational exponents.

Example 3. Simplifying algebraic expressions. We assume that the radicands involved are positive, so the radicals are defined.

a. $\sqrt{a}(\sqrt[8]{a})^3 = a^{1/2} \cdot a^{3/8} = a^{1/2+3/8} = a^{7/8}$

b. $\left(\dfrac{\sqrt{a}}{\sqrt[3]{a}}\right)^4 = \left(\dfrac{a^{1/2}}{a^{1/3}}\right)^4 = (a^{1/2-1/3})^4 = (a^{1/6})^4 = a^{1/6(4)} = a^{2/3}$

c. $\dfrac{a\sqrt[3]{a}}{b^2\sqrt[3]{b}} = \dfrac{a \cdot a^{1/3}}{b^2 \cdot b^{1/3}} = \dfrac{a^{1+1/3}}{b^{2+1/3}} = \dfrac{a^{4/3}}{b^{7/3}} = a^{4/3}b^{-7/3}$

d. $x^{1/3}(x^{2/3} + x^{5/3}) = x^{1/3} \cdot x^{2/3} + x^{1/3} \cdot x^{5/3} = x^{1/3+2/3} + x^{1/3+5/3} = x + x^2$

e. $(x^{-1/2} + y^{-1/2})(x^{-1/2} - y^{-1/2}) = (x^{-1/2})^2 - (y^{-1/2})^2 = x^{-1/2(2)} - y^{-1/2(2)}$

$$= x^{-1} - y^{-1} = \frac{1}{x} - \frac{1}{y} = \frac{y - x}{xy}$$

f. $\dfrac{x^{1/3}}{y^{2/3}} + \dfrac{y^{1/3}}{x^{2/3}} = \dfrac{x^{1/3} \cdot x^{2/3} + y^{1/3} \cdot y^{2/3}}{x^{2/3}y^{2/3}} = \dfrac{x^{1/3+2/3} + y^{1/3+2/3}}{(xy)^{2/3}} = \dfrac{x + y}{(xy)^{2/3}}$

Exercise Set 3.6 (continued)

In exercises 21–48, simplify each given algebraic expression by using the properties of exponents; write the result using exponential notation. Assume that the radicands are positive.

21. $x^{1/2}(x^{2/3})^6$

22. $x^{-3/4}(x^{-1/2})^{-6}$

23. $\dfrac{(x^{2/5})^4}{(x^3)^{1/5}}$

24. $\dfrac{(x^{1/3})^{1/4}}{(x^{3/4})^{1/3}}$

25. $(a^{5/2})^{-3}(a^2b)^{1/4}$

26. $(a^{7/2}b^{-5/2})^4(a^{-3/2}b^4)^{-1}$

27. $\dfrac{(a^{1/3}b^{1/6})^{-2}}{(a^3b^{-2})^{1/6}}$

28. $\dfrac{(a^{12}b^6)^{-1/3}}{(a^{-2/3}b^{4/9})^{3/2}}$

29. $\dfrac{x^{1/3}}{y^{4/5}} \cdot \dfrac{x^{5/3}}{y^{-1/5}}$

30. $\dfrac{a^4}{b^{3/2}} \cdot \dfrac{a^{1/2}}{b^{7/2}}$

31. $\dfrac{x^{1/3}y^2}{z} \cdot \dfrac{x^{10/3}z}{y^{4/3}}$

32. $\dfrac{y^{3/4}z^{-1/2}}{x^{1/4}} \cdot \dfrac{x^{3/4}z^{1/2}}{y^{5/4}}$

33. $\dfrac{a^7}{b^6} \div \dfrac{a^{2/3}}{b^{4/3}}$

34. $\dfrac{a^2x^{1/5}}{b^{3/5}} \div \dfrac{a^{3/5}y^{1/5}}{b^{8/5}}$

35. $\dfrac{a^{1/2}x^{1/2}}{b} \div \left(\dfrac{ax}{b}\right)^{1/2}$

36. $\left(\dfrac{x^2y^4}{z}\right)^{1/2} \div \left(\dfrac{xy}{z^2}\right)^{1/4}$

37. $x^{1/5}(x^{4/5} + x^{-1/5})$

38. $x^{-3/4}(x^{7/4} - x^{3/4})$

39. $x^{1/2}y^{-1/2}\left(x^{1/2} + \dfrac{y^{1/2}}{x^{1/2}}\right)$

40. $\dfrac{x^{1/3}}{y^{1/3}}\left(\dfrac{x^{2/3}}{y^{5/3}} + \dfrac{y^{1/3}}{x^{4/3}}\right)$

41. $\left(\dfrac{x}{y}\right)^{1/2} + \dfrac{x}{y}$

42. $\left(\dfrac{x}{y}\right)^{1/2} + \left(\dfrac{x}{y}\right)^{-1/2}$

43. $\dfrac{x^{1/2}}{y^{1/2}} + \dfrac{x^{3/2}}{y^{3/2}}$

44. $\dfrac{x}{y^{1/3}} + \dfrac{y}{x^{1/3}}$

45. $(x^{1/2} + y^{1/2})(x^{1/2} - y^{1/2})$

46. $(a^{1/3}b^{1/3} + 1)(a^{1/3}b^{1/3} - 1)$

47. $(x^{1/2} + y^{1/2})^2$

48. $(x^{1/4} + 1)^2$

section 7 • Definition of Complex Numbers

In this section we define complex numbers and other concepts relating to the system of complex numbers. In Chapter 5 we will apply the system of complex numbers to the solution of quadratic equations.

THE IMAGINARY UNIT

It is important to review the fact that the square root of a negative real number is not a real number.

SQUARE ROOT OF A NEGATIVE REAL NUMBER. If a is a negative real number, then $\sqrt{a}$ is not a real number.

To see that this is true, consider the following statements:

$$\text{If } \sqrt{a} = b, \text{ then } a = b^2.$$
$$\text{If } b \text{ is a real number, then } b^2 \geq 0.$$
$$\text{If } a < 0, \text{ then } a \neq b^2.$$

These statements show that $\sqrt{a}$ is not a real number for $a < 0$.

In particular, $\sqrt{-1}$ is not a real number. We now define a number (which is not a real number) to be the square root of -1.

DEFINITION OF THE IMAGINARY UNIT. Let $i = \sqrt{-1}$. Then i is a number called the *imaginary unit*; it has the basic property that $i^2 = -1$.

Note: The fact that "real" and "imaginary" are used to name numbers shows the distrust mathematicians had of imaginary numbers when they were invented in the sixteenth century.

Now suppose x is any positive real number, so $-x$ is negative. We claim that $\sqrt{x}i$ is a square root of $-x$; to see this we square the number $\sqrt{x}i$ to obtain

$$(\sqrt{x}i)^2 = (\sqrt{x})^2 i^2 = x(-1) = -x.$$

Therefore we now have square roots of negative real numbers. We express the fact that $\sqrt{x}i$ is a square root of $-x$ by writing

$$\sqrt{-x} = \sqrt{x}i.$$

Example 1. Square roots of negative real numbers.

a. $\sqrt{-4} = \sqrt{4}i = 2i$ b. $\sqrt{-25} = \sqrt{25}i = 5i$

c. $\sqrt{-3} = \sqrt{3}i$ d. $\sqrt{-6} = \sqrt{6}i$

e. $\sqrt{-8} = \sqrt{8}i = 2\sqrt{2}i$ f. $\sqrt{-20} = \sqrt{20}i = 2\sqrt{5}i$

Exercise Set 3.7

In exercises 1–16, write the given square root in the form of a real number times i as in Example 1.

1. $\sqrt{-9}$ 2. $\sqrt{-81}$ 3. $\sqrt{-16}$ 4. $\sqrt{-49}$ 5. $\sqrt{-7}$

6. $\sqrt{-5}$ 7. $\sqrt{-10}$ 8. $\sqrt{-14}$ 9. $\sqrt{-12}$ 10. $\sqrt{-50}$

11. $\sqrt{-18}$ 12. $\sqrt{-27}$ 13. $\sqrt{-40}$ 14. $\sqrt{-24}$ 15. $\sqrt{-800}$

16. $\sqrt{-40{,}000}$

THE COMPLEX NUMBER SYSTEM

Now we define the system of complex numbers. There are two conditions we want the system of complex numbers to satisfy—it must contain all the real numbers, and it must contain the square roots of negative real numbers. We have just seen that the square root of a negative real number can be written in the *form* of a real number times i, so we want to include all numbers of the form bi for any real number b. The numbers of the form bi do not include any real numbers. To include real numbers, say a, we have to include all numbers of the form $a + bi$. We now define a complex number.

DEFINITION OF A COMPLEX NUMBER. A *complex number* is any number of the form $a + bi$, where a and b are real numbers and i satisfies the basic property $i^2 = -1$.

We emphasize that every real number is a complex number—the real number a is the complex number of the form $a + 0i$. So the set of real numbers is a subset of the set of complex numbers.

Usually we write a complex number in the form $a + bi$, which is called the *standard form.* That is, we write a complex number as a real number plus a real number times i. Written in the form $a + bi$, a complex number has two parts; and these parts have names.

DEFINITION OF REAL AND IMAGINARY PARTS. If $a + bi$ is a complex number, then a is called the *real part* and b is called the *imaginary part.* The complex number is a real number if $b = 0$. The complex number is called *imaginary* or *purely imaginary* if $a = 0$.

Example 2. Real and imaginary parts of complex numbers.

a. $2 + 3i$ has real part 2 and imaginary part 3
b. $4 - 5i$ has real part 4 and imaginary part -5
c. $-2 - i$ has real part -2 and imaginary part -1
d. $7i$ is purely imaginary.

DEFINITION OF EQUALITY OF COMPLEX NUMBERS. If $a + bi$ and $c + di$ are complex numbers, then $a + bi = c + di$ only if $a = c$ and $bi = di$. That is, two complex numbers are equal only if their real parts are equal and their imaginary parts are equal.

In the next section and in Chapter 5 we will use the definition of the conjugate of a complex number.

DEFINITION OF THE CONJUGATE OF A COMPLEX NUMBER. If $a + bi$ is a complex number, the *conjugate* of $a + bi$ is $a - bi$.

Example 2 (continued). Conjugates of complex numbers.

a. The conjugate of $2 + 3i$ is $2 - 3i$.
b. The conjugate of $4 - 5i$ is $4 + 5i$.
c. The conjugate of $-2 - i$ is $-2 + i$.
d. The conjugate of $7i$ is $-7i$.

Exercise Set 3.7 (continued)

In exercises 17–32, specify the real part and the imaginary part of the given complex number and write its conjugate.

17. $1 + 2i$ 18. $4 + i$ 19. $1 - 2i$ 20. $4 - i$ 21. $2 + i$

22. $2 + 3i$ 23. $4 - 3i$ 24. $3 - 4i$ 25. $3 + i$ 26. $2 + 2i$

27. $1 - i$ 28. $4 - 2i$ 29. $1 - \sqrt{2}i$ 30. $2 - \sqrt{3}i$ 31. $3 + \sqrt{3}i$
32. $1 + \sqrt{5}i$

SOME COMPUTATIONS

We will use complex numbers in Chapter 5 in solving quadratic equations. Now we do some computations with complex numbers to prepare for the computations we will encounter in Chapter 5.

Example 3. If $a = 1$, $b = 4$, and $c = 5$, calculate $D = b^2 - 4ac$ and $x = (-b + \sqrt{D})/2a$.

Solution. When we substitute we find

$$D = b^2 - 4ac = 4^2 - 4(1)(5) = -4$$

$$\sqrt{D} = \sqrt{-4} = \sqrt{4}i = 2i$$

$$\begin{aligned} x &= (-b + \sqrt{D})/2a = (-4 + 2i)/2(1) \\ &= -2 + i \end{aligned}$$

Example 4. If $a = 3$, $b = -2$, and $c = 3$, calculate $D = b^2 - 4ac$ and $x = (-b + \sqrt{D})/2a$.

Solution. When we substitute we find

$$D = b^2 - 4ac = (-2)^2 - 4(3)(3) = -32$$

$$\sqrt{D} = \sqrt{-32} = \sqrt{32}i = 4\sqrt{2}i$$

$$\begin{aligned} x &= (-b + \sqrt{D})/2a = (-(-2) + 4\sqrt{2}i)/2(3) \\ &= (2 + 4\sqrt{2}i)/2(3) = (1 + 2\sqrt{2}i)/3 \end{aligned}$$

Exercise Set 3.7 (continued)

In exercises 33–44, calculate $D = b^2 - 4ac$ and $x = (-b + \sqrt{D})/2a$ for the given values of a, b, and c.

33. $a = 1, b = -2, c = 2$
34. $a = 1, b = 2, c = 2$
35. $a = 1, b = 4, c = 5$
36. $a = 1, b = -6, c = 13$
37. $a = 1, b = 4, c = 4$
38. $a = 2, b = -11, c = 12$
39. $a = 2, b = 3, c = -2$
40. $a = 3, b = -1, c = -2$
41. $a = 1, b = 2, c = -4$
42. $a = 1, b = 2, c = 4$
43. $a = 2, b = 2, c = 2$
44. $a = 2, b = -2, c = -2$

section 8 • Operations for Complex Numbers

Now we consider the four operations of addition, subtraction, multiplication, and division for complex numbers. These operations are extensions of the same operations for real numbers, with the basic property that $i^2 = -1$.

ADDITION AND SUBTRACTION

Addition and subtraction of complex numbers follows the ordinary rules for addition and subtraction.

$$(a + bi) + (c + di) = (a + c) + (bi + di) = (a + c) + (b + d)i$$

$$(a + bi) - (c + di) = (a - c) + (bi - di) = (a - c) + (b - d)i$$

To add two complex numbers, we add the real parts and add the imaginary parts. To subtract two complex numbers, we subtract the real parts and subtract the imaginary parts.

Example 1. Addition and subtraction of complex numbers.

a. $(2 + 3i) + (4 - 4i) = (2 + 4) + (3i - 4i) = 6 - i$

b. $(2 + 3i) - (4 - 4i) = (2 - 4) + (3i + 4i) = -2 + 7i$

c. $(4 - i) + (3 + i) = (4 + 3) + (-i + i) = 7$

d. $(4 - i) - (3 + i) = (4 - 3) + (-i - i) = 1 - 2i$

Exercise Set 3.8

In exercises 1–16, find the sum of the two given complex numbers, and then subtract the second complex number from the first.

1. $1 + i; 2 + 3i$
2. $2 + 4i; 3 + i$
3. $2 - 2i; 3 + i$
4. $5 - 3i; 2 + 4i$
5. $6 - 3i; 3 - 2i$
6. $4 - 2i; 5 - i$
7. $4 + 3i; 4 - 3i$
8. $2 + i; 2 - i$
9. $1 - 2i; 1 + 2i$
10. $3 - i; 3 + i$
11. $3 + 3i; 1 - 3i$
12. $2 - 4i; 1 + 4i$
13. $5 - 4i; 2 + 3i$
14. $6 - i; 2 + 3i$
15. $5 - i; 5 + 3i$
16. $4 - 2i; 4 + 4i$

MULTIPLICATION

Multiplication of complex numbers proceeds just as for ordinary algebraic expressions, with the exception that we always replace i^2 by -1.

$$(a + bi)(c + di) = ac + adi + bci + bdi^2 = (ac - bd) + (ad + bc)i.$$

Example 2. Multiplication of complex numbers.

a. $$\begin{aligned}(2 + 3i)(1 + 4i) &= 2(1) + 2(4i) + 3i(1) + (3i)(4i)\\ &= 2 + 8i + 3i + 12(-1)\\ &= -10 + 11i\end{aligned}$$

b. $$\begin{aligned}(2 - i)(1 + 3i) &= 2(1) + 2(3i) - i(1) - i(3i)\\ &= 2 + 6i - i - 3(-1)\\ &= 5 + 5i\end{aligned}$$

c. $$\begin{aligned}(1 + 2i)(1 - 2i) &= 1(1) + 1(-2i) + 2i(1) + 2i(-2i)\\ &= 1 - 2i + 2i - 4(-1)\\ &= 5\end{aligned}$$

Exercise Set 3.8 (continued)

In exercises 17–26, find the product of the two complex numbers.

17. $(5 + 5i)(1 + 2i)$
18. $(2 + 3i)(4 + i)$
19. $(1 + i)(3 + i)$
20. $(2 - i)(2 + 2i)$
21. $(2 - 3i)(1 + 2i)$
22. $(3 - 2i)(2 + i)$
23. $(1 - i)(2 - i)$
24. $(3 - 2i)(1 - i)$
25. $(1 - 4i)(4 + i)$
26. $(2 + i)(1 - 2i)$

In exercises 27–34, find the product of the given complex number with its conjugate (Example 2c).

27. $1 + 2i$
28. $4 + i$
29. $2 - 3i$
30. $3 - i$
31. $2 + i$
32. $2 + 3i$
33. $1 - i$
34. $4 - 3i$

DIVISION

Division of complex numbers is most easily performed by using a trick. Suppose we wish to divide $a + bi$ by $c + di$. The idea is to get the i out of the denominator; in other words, we want to have a real number in the denominator without changing the value of the fraction. To see how to do this, consider what happens when we multiply a complex number by its conjugate. We obtain

$$(a + bi)(a - bi) = a^2 - abi + abi - b^2i^2 = a^2 + b^2.$$

So the product of a complex number and its conjugate is always a real number.

Now here's the trick for performing the division. We multiply both numerator and denominator by the conjugate of the denominator. This does not change the value of the fraction (Section 2.6), and we obtain

$$\frac{a+bi}{c+di} = \frac{(a+bi)(c-di)}{(c+di)(c-di)} = \frac{ac - adi + bci - bdi^2}{c^2 - cdi + cdi - d^2i^2}$$

$$= \frac{(ac+bd)+(bc-ad)i}{c^2+d^2} = \frac{ac+bd}{c^2+d^2} + \frac{bc-ad}{c^2+d^2}i.$$

The trick removes the i from the denominator and permits us to write the quotient in the form of a complex number.

Example 3. Division of complex numbers.

a. $$\frac{5+5i}{1+2i} = \frac{(5+5i)(1-2i)}{(1+2i)(1-2i)} = \frac{5+5i-10i-10i^2}{1+2i-2i-4i^2}$$

$$= \frac{15-5i}{5} = 3 - i$$

b. $$\frac{2+3i}{4+i} = \frac{(2+3i)(4-i)}{(4+i)(4-i)} = \frac{2(4)-2i+3(4i)-3i^2}{4(4)-4i+4i-i^2}$$

$$= \frac{11+10i}{17} = \frac{11}{17} + \frac{10}{17}i$$

c. $$\frac{1+i}{2+3i} = \frac{(1+i)(2-3i)}{(2+3i)(2-3i)} = \frac{2-3i+2i-3i^2}{2(2)-2(3i)+2(3i)-3(3i^2)}$$

$$= \frac{5-i}{13} = \frac{5}{13} - \frac{1}{13}i$$

Exercise Set 3.8 (continued)

In exercises 35–42, perform the indicated divisions of complex numbers.

35. $\frac{5+5i}{2+i}$

36. $\frac{13+13i}{2+3i}$

37. $\frac{10-5i}{4-3i}$

38. $\frac{5-5i}{3-4i}$

39. $\frac{1+i}{3+i}$

40. $\frac{2-i}{2+2i}$

41. $\frac{1+i}{1-i}$

42. $\frac{4+3i}{4-3i}$

section 9 • Applications

In this section we apply exponents and roots to geometry (areas and volumes), direct variation, inverse variation, Kepler's Third Law, and compound interest.

GEOMETRY PROBLEMS: AREA

We will use the following formulas for the area of a square and of a circle:

AREA OF A SQUARE. If a square has side of length x and area A, then:

Given x, to find A use $A = x^2$

Given A to find x use $x = \sqrt{A}$

AREA OF A CIRCLE. If a circle has radius of length r and area A, then:

Given r, to find A use $A = \pi r^2$

Given A, to find r use $r = \sqrt{A/\pi}$

Example 1. When designing a building, an architect determines that a circular pipe must carry a certain volume of water. The architect finds that, to carry this volume, the pipe must have a cross-sectional area of 4360 square centimeters. What must be the diameter of the pipe?

Solution. First we find the radius by finding

$$r = \sqrt{A/\pi} = \sqrt{4360/3.14} = \sqrt{1389} = 37$$

We have rounded off to the nearest centimeter. The diameter should be 2(37) = 74 centimeters.

Exercise Set 3.9

1. An architect finds that, to carry a certain volume of water, a circular pipe must have a cross-sectional area of 3010 square centimeters. What must be the diameter of the pipe?

2. An architect finds that, to carry a certain volume of water, a circular pipe must have a cross-sectional area of 7800 square centimeters. What must be the diameter of the pipe?

3. An engineer finds that, to carry a certain number of wires, a circular pipe must have a cross-sectional area of 40 square centimeters. What must be the diameter of the pipe?
4. An engineer finds that, to carry a certain number of wires, a circular pipe must have a cross-sectional area of 62 square centimeters. What must be the diameter of the pipe?
5. What must be the length of the side of a square office room if its area is to be 6200 square meters?
6. What must be the length of the side of a square storage room if its area is to be 10,400 square meters?
7. a. The radius of a standard LP record is 14.9 centimeters. Find its area.
   b. The length of the side of a standard LP record jacket is 31 centimeters. Find its area.
8. a. The diameter of a U.S. dime is 18 millimeters. Find the area of the face.
   b. The diameter of a U.S. quarter is 24 millimeters. Find the area of the face.
9. In Bohr's theory of the hydrogen atom, the electron is assumed to move in a circular path about the nucleus. However, only certain orbits are permissible. The radius of the first permissible orbit is 5.29×10^{-9} centimeters, that of the second permissible orbit is 2.116×10^{-8} centimeters, and that of the third permissible orbit is 4.761×10^{-8} centimeters. Find the area of each permissible circular orbit.
10. L-shell electrons of the magnesium atom have a circular orbit of radius 2.7×10^{-9} centimeters. Find the area of the orbit.

GEOMETRY PROBLEMS: VOLUME

We will use the following formulas for the volumes of a cube and a sphere.

VOLUME OF A CUBE. If each side of a cube has length x and the cube has volume V, then:

$$\text{Given } x\text{, to find } V \text{ use } V = x^3$$

$$\text{Given } V\text{, to find } x \text{ use } x = \sqrt[3]{V}.$$

VOLUME OF A SPHERE. If a sphere has radius r and volume V, then:

$$\text{Given } r\text{, to find } V \text{ use } V = 4\pi r^3/3$$

$$\text{Given } V\text{, to find } r \text{ use } r = \sqrt[3]{3V/4\pi}$$

Example 2. Assuming that the earth is a sphere, find the volume of the earth, given that its radius is 6.378×10^8 centimeters.

Solution. Using the formula for the volume of a sphere,

$$\begin{aligned} V &= 4\pi r^3/3 = 4(3.14)(6.378 \times 10^8)^3/3 \\ &= 4(3.14)(6.378)^3/3 \times (10^8)^3 = 1086.2 \times 10^{24} \\ &= 1.0862 \times 10^{27} \text{ cubic centimeters} \end{aligned}$$

Exercise Set 3.9 (continued)

11. Find the volume of the sun, given that its radius is measured to be 6.96×10^{10} centimeters.
12. Find the volume of the moon, given that its radius is measured to be 1.74×10^{8} centimeters.
13. Find the volume of Mars, given that its radius is measured to be 3.400×10^{8} centimeters.
14. Find the volume of Jupiter, given that its radius is measured to be 7.140×10^{9} centimeters.
15. It is possible to find the radius of any star whose absolute magnitude and temperature are known. Find the volume of Sirius, given that its radius is determined to be 1.4×10^{11} centimeters.
16. Find the volume of the star σ Draconis, given that its radius is determined to be 4.9×10^{10} centimeters.

Many atoms are assumed to be spherical in shape. To find the radius of an atom of a given element, it is easiest to find its volume first. This volume is a/N, where a is called the gram-atomic volume of the element and $N = 6.0235 \times 10^{23}$ is the Avogadro number.

17. Find the radius of a zinc atom, given that its gram-atomic volume is measured to be 9.
18. Find the radius of a magnesium atom, given that its gram-atomic volume is measured to be 14.
19. Find the radius of a xenon atom, given that its gram-atomic volume is measured to be 43.
20. Find the radius of a krypton atom, given that its gram-atomic volume is measured to be 33.

DIRECT VARIATION

Variation is a vocabulary for describing relationships between measurable quantities represented by variables.

DEFINITION OF DIRECT VARIATION. If x and y are variables, then y is said to *vary directly* as x if there is a non-zero constant k so that

$$y = kx.$$

The constant k is called the *constant of variation.*

DEFINITION OF DIRECT VARIATION WITH RESPECT TO A POWER. If x and y are variables, then y *varies directly as the* n^{th} *power of* x if

$$y = kx^n$$

for some non-zero constant k called the *constant of variation.*

Example 3. The distance required to stop a car varies directly as the square of the speed of the car; it takes 187.5 feet to stop a car going 50 mph.

a. Find the constant of variation.

b. What distance is required to stop a car going 70 mph?

Solution. Let d be the distance required to stop a car going at speed s. Then $d = ks^2$ where k is the constant of variation.

a. We wish to find k; we are given $d = 187.5$ when $s = 50$. So

$$k = d/s^2 = 187.5/50^2 = 0.075$$

b. From part a we have $d = 0.075s^2$. Therefore, when $s = 70$,

$$d = 0.075s^2 = 0.075(70)^2 = 367.5 \text{ feet.}$$

Exercise Set 3.9 (continued)

21. a. What distance is required to stop a car going 100 mph?
    b. If the speed of a car is doubled, what effect does this have on the distance required to stop the car?
22. a. What distance is required to stop a car going 25 mph?
    b. If the speed of a car is halved, what effect does this have on the distance required to stop the car?
23. The area of a circle varies directly as the square of its radius.
    a. What is the constant of variation?
    b. If the radius of a circle is doubled, what effect does this have on the area?
24. The volume of a sphere varies directly as the cube of its radius.
    a. What is the constant of variation?
    b. If the radius of a sphere is doubled, what effect does this have on the volume?
25. The distance an object falls from rest varies directly as the square of the time it falls. The constant of variation, frequently denoted by $\frac{1}{2}g$, is an important physical constant called the acceleration due to gravity. What would be your determination of $\frac{1}{2}g$ if you dropped a stone from the Leaning Tower of Pisa, which is 184.5 feet high, and found that it took 3.4 seconds to hit the ground?
26. The caloric intake necessary for mammals varies directly as the cube of the mammal's height. When Gulliver visited the Island of Lilliput, he found that the Lilliputians were only 6 inches tall. If Gulliver was 6 feet tall, how many Lilliputian meals did he require at each eating?
27. A store sells transistor radios priced at \$24.95 apiece. Does the total income I obtained from selling the radios vary directly as the number n of radios sold? If so, specify the constant of variation.
28. A car travels at a constant rate of 55 mph. Does the distance d that it covers vary directly as the time t that it has traveled? If so, specify the constant of variation.
29. Does temperature measured in degrees Fahrenheit F vary directly as the measure in degrees Centigrade C? If so, specify the constant of variation. Recall that $F = 1.8C + 32$.
30. One side of a rectangle is 3 meters longer than the adjacent side. Does the perimeter P vary directly as the length of the shorter side x? If so, specify the constant of variation. In this case $P = 2x + 2(x + 3)$.

INVERSE VARIATION

Variation is a vocabulary for describing relationships between measurable quantities represented by variables. Inverse variation describes the situation in which what happens to one variable affects the other in the "opposite" way.

DEFINITION OF INVERSE VARIATION. If x and y are variables, then y *varies inversely* with respect to x if

$$y = \frac{k}{x},$$

where k is a non-zero constant called the *constant of variation*. Also, y is said to vary inversely as the nth power of x if $y = k/x^n$ for some non-zero constant k.

Example 4. The distance d to a star varies inversely as the parallax p of the star. (The parallax of a star is an angular measurement to the star based on the revolution of the earth about the sun.) If d is measured in kilometers and p is measured in seconds of arc, then

$$d = \frac{3.1 \times 10^{13}}{p};$$

so the constant of variation is 3.1×10^{13}.

Find the distance to α Centauri (the nearest star), given that its parallax has been measured to be 0.751 seconds of arc.

Solution. Since $p = 0.751$, we have

$$d = (3.1 \times 10^{13})/0.751 = 4.1 \times 10^{13} \text{ kilometers.}$$

Exercise Set 3.9 (continued)

31. Find the distance to the star Spica, given that its parallax has been measured to be 0.02 seconds of arc.
32. Find the distance to Sirius (the brightest star), given that its parallax has been measured to be 0.37 seconds of arc.
33. One *parsec* is defined as the distance of a star whose parallax is one second of arc.
    a. How many kilometers is one parsec?
    b. What is the distance of α Centauri in parsecs?
34. Find the parallax of Vega, given that its distance is 8.12 parsecs.

The intensity of solar radiation for a planet varies inversely as the square of the planet's distance from the sun.

35. The mean distance of Mercury from the sun is 0.3781 of the mean distance of Earth from the sun. How many times greater is the intensity of solar radiation on Mercury than on Earth?
36. The mean distance of Venus from the sun is 0.7233 of the mean distance of Earth from the sun. How many times greater is the intensity of solar radiation on Venus than on Earth?

37. The mean distance of Mars from the sun is 1.5237 times the mean distance of Earth from the sun. How many times less is the intensity of solar radiation on Mars than on Earth?

38. The mean distance of Neptune from the sun is 30.09 times the mean distance of Earth from the sun. How many times less is the intensity of solar radiation on Neptune than on Earth?

KEPLER'S THIRD LAW

The sixteenth-century astronomer and mathematician Johann Kepler discovered three laws governing the motion of the planets. Kepler's third law says that the square of the length of a planet's year varies directly as the cube of the semi-major axis of the planet's orbit. If t is the length of the planet's year and d is the semi-major axis of the orbit (basically, the planet's distance from the sun), then

$$t^2 = kd^3$$

for a constant of proportionality k.

Example 5. Find the length of Mars's year, given that the semi-major axis of Mars's orbit is 1.5237 times the semi-major axis of Earth's orbit.

Solution. If d_m is the semi-major axis of Mars's orbit and d_e is the semi-major axis of Earth's orbit, we are given that $d_m = 1.5237\, d_e$. Let t_m be the length of Mars's year and t_e be the length of Earth's year. We consider ratios:

$$\frac{t_m^2}{t_e^2} = \frac{kd_m^3}{kd_e^3} = \frac{d_m^3}{d_e^3} = \frac{(1.5237\, d_e)^3}{d_e^3} = \frac{(1.5237)^3 d_e^3}{d_e^3} = (1.5237)^3$$

$$t_m^2 = (1.5237)^3 t_e^2 = (1.5237)^3(365.25)^2$$

$$t_m = \sqrt{(1.5237)^3(365.25)^2} = (1.5237)(365.25)\sqrt{1.5237}$$

$$= (1.5237)(365.25)(1.2344) = 686.97$$

The length of Mars's year is 687 days.

We see from Example 6 that we do not need to know the semi-major axis of a planet's orbit if we know its *ratio* to the semi-major axis of Earth's orbit. This ratio is measured in *astronomical units*. The following list gives the semi-major axis in astronomical units for each planet:

Mercury	~~0.3871~~ .3781	Saturn	9.546
Venus	0.7233	Uranus	19.20
Earth	1.0000	Neptune	30.09
Mars	1.5237	Pluto	39.5
Jupiter	5.2027		

Exercise Set 3.9 (continued)

39. Find the length of the year of each planet, given that the length of Earth's year is 365.25 days.
40. In Kepler's Third Law, use rational exponents to:
    a. Express t in terms of d.
    b. Express d in terms of t.

COMPOUND INTEREST

Suppose we make an investment that yields an annual rate of interest. For example, we may deposit money in a savings account at an annual interest rate. We wish to know how much money we have after n years.

The original amount invested is called the *principal* and is denoted by P. If the interest is $100r\%$, after one year Pr dollars is added to the account. So the amount after the first year is

$$P + Pr = P(1 + r).$$

After the second year $P(1 + r)r$ dollars is added to the account and the amount is

$$P(1 + r) + P(1 + r)r = P(1 + r)^2.$$

After the third year $P(1 + r)^2r$ dollars is added to the account, and the total amount is

$$P(1 + r)^2 + P(1 + r)^2r = P(1 + r)^3.$$

Continuing in this way, we find that if the amount after n years is denoted by A, then

$$A = P(1 + r)^n.$$

We use this equation to compute the amount A, given the principal P. On the other hand, if we are given the amount A and wish to find the principal P, we use the equation

$$P = A/(1 + r)^n = A(1 + r)^{-n}.$$

The term $(1 + r)^{-n}$ is called the *discount factor*.

Note: Frequently, interest is compounded quarterly rather than annually. We consider this type of compound interest in Section 8.6.

Example 6. Suppose a bank offers 6% interest compounded annually on savings accounts.

a. If we deposit \$1000, how much do we have after 5 years?
b. How much must we deposit now to have \$2000 after 5 years?

Solution.

a. We have $r = 0.06$, $P = 1000$, and $n = 5$. Thus,

$$A = 1000(1 + 0.06)^5 = 1000(1.06)^5$$
$$= 1000(1.33822) = 1338.22 \text{ dollars.}$$

After 5 years we have \$1338.22.

b. We have $r = 0.06$, $n = 5$, and $A = 2000$. Thus,

$$P = 2000(1 + 0.06)^{-5} = 2000(1.06)^{-5}$$
$$= 2000(0.74726) = 1494.52 \text{ dollars}$$

We must deposit \$1494.52.

Exercise Set 3.9 (continued)

41. Suppose we deposit \$3000 in an account that yields 6% interest compounded annually. How much do we have after 1 year? 2 years? 3 years? 4 years? 5 years? If you have a calculator, find the amount after 10 years; after 20 years.

42. Suppose we deposit \$1000 in an account that yields 5% interest compounded annually. How much do we have after 1 year? 2 years? 3 years? 4 years? 5 years? If you have a calculator, find the amount after 10 years; after 20 years.

43. How much must we invest now in an account that yields 6% interest compounded annually to have \$3000 at the end of 3 years?

CALCULATOR EXERCISE: How much must we invest now in an account that yields 6% interest compounded annually to have \$20,000 at the end of 18 years?

44. How much must we invest now in an account that yields 5% interest compounded annually to have \$3000 at the end of 4 years?

CALCULATOR EXERCISE: How much must we invest now in an account that yields 5% interest compounded annually to have \$20,000 at the end of 20 years?

45. Suppose we deposit \$10,000 in an account that yields 8% interest compounded anually. How much do we have after 1 year? 2 years? 3 years? 4 years? 5 years? If you have a calculator, find the amount after 10 years; after 20 years.

46. Suppose we deposit \$20,000 in an account that yields 9% interest compounded annually. How much do we have after 1 year? 2 years? 3 years? 4 years? 5 years? If you have a calculator, find the amount after 10 years; after 20 years.

47. How much must we invest now in an account that yields 8% interest compounded annually to have \$10,000 at the end of 5 years?

CALCULATOR EXERCISE: How much must we invest now in an account that yields 8% interest compounded annually to have $20,000 at the end of 20 years?

48. How much must we invest now in an account that yields 9% interest compounded annually to have $10,000 at the end of 5 years?

CALCULATOR EXERCISE: How much must we invest now in an account that yields 9% interest compounded annually to have $20,000 at the end of 20 years?

Review Exercises

---

1. Find 4^n for $n = 0, 1/2, 1, 3/2, 2, 5/2, 3, -1/2, -1, -3/2, -2$.
2. Find 9^n for $n = 0, 1/2, 1, 3/2, 2, 5/2, 3, -1/2, -1, -3/2, -2$.
3. Find 16^n for $n = 0, 1, 2, -1, -2, 1/4, 1/2, 3/4, -1/4, -1/2, -3/4$.
4. Find 81^n for $n = 0, 1, 2, -1, -2, 1/4, 1/2, 3/4, -1/4, -1/2, -3/4$.
5. Find 8^n for $n = 0, 1, 2, -1, -2, 1/3, -1/3, 2/3, -2/3, 4/3$.
6. Find 27^n for $n = 0, 1, 2, -1, -2, 1/3, -1/3, 2/3, -2/3, 4/3$.

In exercises 7–12, the given number is written in scientific notation; write it in ordinary decimal notation.

7. 3.66×10^{-2}
8. 4.55×10^{-4}
9. 8.14×10^{4}
10. 9.09×10^{3}
11. 5.56×10^{-3}
12. 8.24×10^{-5}

In exercises 13–18, the given number is written in ordinary decimal notation; write it in scientific notation.

13. 6480
14. 213.2
15. 0.0043
16. 0.000061
17. 51.6
18. 54,300

In exercises 19–28, perform each computation and write the answer in scientific notation.

19. $(3 \times 10^{4})(2 \times 10^{5})(6 \times 10^{10})$
20. $(9 \times 10^{12})(2 \times 10^{8})(5 \times 10^{12})$
21. $(6 \times 10^{6})^2/(3 \times 10^{14})$
22. $(8 \times 10^{15})/(4 \times 10^{7})^2$
23. $\sqrt{6.4 \times 10^{13}}$
24. $\sqrt{3.6 \times 10^{-9}}$
25. $\sqrt[3]{1.25 \times 10^{11}}$
26. $\sqrt[3]{2.7 \times 10^{-8}}$
27. $(1.5 \times 10^{4})^2/(5 \times 10^{5})^3$
28. $(2.5 \times 10^{-3})^2/(5 \times 10^{-8})^3$

In exercises 29–36, find the indicated n^{th} root.

29. $\sqrt{400}$
30. $\sqrt{900}$

31. $\sqrt{625}$

32. $\sqrt{256}$

33. $\sqrt[3]{1000}$

34. $\sqrt[3]{8000}$

35. $\sqrt[4]{625}$

36. $\sqrt[4]{256}$

In exercises 37–44, remove all possible perfect-square factors from under the square root; then simplify the given fraction.

37. $\sqrt{8}$; $(4 + \sqrt{8})/2$

38. $\sqrt{20}$; $(4 + \sqrt{20})/2$

39. $\sqrt{12}$; $(2 - \sqrt{12})/2$

40. $\sqrt{24}$; $(2 - \sqrt{24})/2$

41. $\sqrt{32}$; $(4 - \sqrt{32})/4$

42. $\sqrt{48}$; $(4 + \sqrt{48})/4$

43. $\sqrt{18}$; $(9 + \sqrt{18})/6$

44. $\sqrt{27}$; $(9 - \sqrt{27})/6$

In exercises 45–50, rationalize the denominator of the given fraction.

45. $4/\sqrt{2}$

46. $6/\sqrt{3}$

47. $1/\sqrt{6}$

48. $3/\sqrt{6}$

49. $1/(\sqrt{3} - \sqrt{2})$

50. $1/(\sqrt{2} - 1)$

In exercises 51–62, find the given number.

51. $100^{1/2}$

52. $49^{1/2}$

53. $8^{1/3}$

54. $125^{1/3}$

55. $100^{-1/2}$

56. $49^{-1/2}$

57. $8^{-1/3}$

58. $125^{-1/3}$

59. $100^{3/2}$

60. $49^{3/2}$

61. $8^{2/3}$

62. $125^{2/3}$

In exercises 63–98, use the laws for exponents to simplify each given expression; write the result in exponential notation, using negative and rational exponents if necessary.

63. a^2a^8

64. $a^{10}a^5$

65. a^7a^{-3}

66. a^8a^{-5}

67. $(a^4)^3$

68. $(b^6)^5$

69. $(a^5)^{-3}$

70. $(a^{-4})^6$

71. $(xy^{-3}z^2)^3$

72. $(x^{-2}y^2z^4)^{-4}$

73. $x^{-3}(xy^{-2})^2y^{-4}$

74. $x^4y^2(x^{-2}y^3)^{-4}$

75. $\dfrac{a^7 \cdot a^4}{a^{12}}$

76. $\dfrac{b^{13} \cdot b^2}{b^4 \cdot b^5}$

77. $\dfrac{c^7 \cdot c \cdot c^2}{c^5 \cdot c^2 \cdot c^3}$

78. $\dfrac{x^2 \cdot x^7 \cdot x^6}{x^{10}}$

79. $\sqrt{a^2b^6}$

80. $\sqrt{a^8b^{12}}$

81. $\sqrt[3]{a^9b^{12}}$

82. $\sqrt[3]{a^{15}b^9}$

83. $\sqrt[3]{\sqrt{x}}$

84. $\sqrt[5]{\sqrt{x}}$

85. $\sqrt{a}/\sqrt[3]{a}$

86. $\sqrt[3]{a}/\sqrt{a}$

87. $(a^2)^{-3/2}$

88. $(a^6)^{-1/3}$

89. $(a^{2/3})^6$

90. $(a^{5/2})^4$

91. $a^{5/2}a^{-1/2}$

92. $a^{1/3}a^{5/3}$

93. $(x^{2n}y^m)^{-2}$

94. $(x^{-n}y^{-n})^{-1}$

95. $(x^2)^n \cdot x^{2n} \cdot x^4$

96. $(x^n)^3 \cdot (x^3)^n$

97. $\left(\frac{a^{n+1}}{a^{n-1}}\right)^n$

98. $\left(\frac{a^{1+n}}{a^{1-n}}\right)^2$

In exercises 99–104, write the given square root in the form of a real number times i.

99. $\sqrt{-4}$

100. $\sqrt{-25}$

101. $\sqrt{-6}$

102. $\sqrt{-15}$

103. $\sqrt{-12}$

104. $\sqrt{-18}$

In exercises 105–110, calculate $D = b^2 - 4ac$ and $x = (-b + \sqrt{D})/2a$ for the given values of a, b, and c.

105. $a = 1, b = 2, c = 2$

106. $a = 1, b = -2, c = 5$

107. $a = 1, b = -6, c = 10$

108. $a = 1, b = 4, c = 13$

109. $a = 2, b = 2, c = 5$

110. $a = 2, b = 4, c = 5$

In exercises 111–114, find: a. the sum of the complex numbers; b. the first complex number minus the second complex number; c. the product of the complex numbers; d. the first complex number divided by the second complex number.

111. $3 + 2i; 1 + i$

112. $4 - 2i; 1 - i$

113. $2 + i; 2 - i$

114. $-2 + i; 2 + i$

115. The diameter of a U.S. nickel is 20 millimeters. Find the area of the face.

116. An architect finds that, to carry a certain volume of water, a circular pipe must have a cross-sectional area of 3500 square centimeters. What must be the diameter of the pipe?

117. Find the volume of Saturn, given that its radius is measured to be 6.04×10^9 centimeters.

118. The perimeter of a square varies directly as the length of its side. Specify the constant of variation.

119. Suppose we deposit \$1000 in an account that yields 5% interest compounded annually. How much do we have after 1 year? 2 years?

120. How much must we invest now in an account that yields 5% interest compounded annually to have \$1000 at the end of 2 years?

chapter 4

Linear Equations and Inequalities

In Chapters 4, 5, and 6 we solve equations and inequalities in one variable. In this chapter we consider linear (or first-degree) equations and inequalities; in Chapter 5 we consider quadratic (or second-degree) equations and inequalities; and in Chapter 6 we consider polynomial (or n^{th} degree) equations. In each chapter we must develop additional techniques to solve the equations and inequalities.

In Section 4.1 we apply the operations of addition, subtraction, multiplication, and division to solving equations and inequalities. We solve various types of linear equations and inequalities in Sections 4.2 and 4.3. Section 4.4 shows how to solve an equation in more than one variable for a specified variable.

In Section 4.5 we define the concept of absolute value and solve inequalities expressed in absolute value. In Section 4.6 we apply linear equations and inequalities to geometry problems, rate problems, mixing solutions, diet problems, business and management considerations, engineering tolerances, and historical exercises.

section 1 • Solutions

In this section we define the solution of an equation or inequality. We also define equivalent equations and equivalent inequalities. Then we relate the operations of addition, subtraction, multiplication, and division to equations and inequalities.

DEFINITION OF SOLUTIONS

If we have an equation or inequality in one variable, we can substitute various numbers for the variable. A given number may or may not satisfy the equation or inequality when it is substituted for the variable.

Example 1. Consider the equation $(x - 1)(x + 2) = 10$.

a. The number 3 satisfies this equation, since $(3 - 1)(3 + 2) = 10$.
b. The number 1 does not satisfy this equation, since $(1 - 1)(1 + 2) \neq 10$.
c. The number 0 does not satisfy this equation, since $(0 - 1)(0 + 2) \neq 10$.
d. The number -4 satisfies this equation, since $(-4 - 1)(-4 + 2) = 10$.

Example 2. Consider the inequality $3x + 8 \geq 20$.

a. The number 5 satisfies this inequality, since $3(5) + 8 \geq 20$.
b. The number 2 does not satisfy this inequality, since $3(2) + 8 < 20$.
c. The number -1 does not satisfy this inequality, since $3(-1) + 8 < 20$.
d. The number 4 satisfies this inequality, since $3(4) + 8 \geq 20$.

Naturally we are interested in those numbers that satisfy a given equation or inequality; this set of numbers is called the solution.

DEFINITION OF THE SOLUTION OF AN EQUATION OR INEQUALITY. The *solution* of an equation or inequality in one variable is the set of numbers which satisfy the equation or inequality when substituted for the variable.

To *solve* an equation or inequality means to find its solution. We will also use the definition of equivalent equations and inequalities.

DEFINITION OF EQUIVALENT EQUATIONS OR INEQUALITIES. Two equations or inequalities in one variable are *equivalent* if they have the same solution.

Example 3. The equations $x - 6 = 2$ and $3x = 24$ are equivalent.

Suppose we look at the solution of $x - 6 = 2$. First, the number 8 satisfies the equation because $8 - 6 = 2$. Now 8 is the only number from which we can subtract 6 to obtain 2. So 8 is the only number that satisfies the equation $x - 6 = 2$, and the solution is 8.

For the equation $3x = 24$, again 8 satisfies the equation because $3 \cdot 8 = 24$. Also, 8 is the only number which when multiplied by 3 gives 24. So the solution of $3x = 24$ is 8.

These two equations have the same solution, so they are equivalent. For equations to be equivalent, they don't have to *look* the same.

Example 4. The inequalities $4x \geq 6$ and $2x \geq 5$ are *not* equivalent. The number 2 satisfies the inequality $4x \geq 6$ but does not satisfy the inequality $2x \geq 5$. Since they do not have the same solution, the inequalities are not equivalent.

Exercise Set 4.1

In exercises 1–10, substitute the numbers for the variable in each equation or inequality and tell which numbers satisfy the equation or inequality.

1. $(x - 4)(2x - 4) = 48$; 1, 2, 3, 4, 5, 6, 7, 8, 9, −1, −2, −3, −4, −5, −6.
2. $3x + 8 \geq 20$; 0, 1, 2, 3, 4, 5, 6, 7, −1, −2, −3, $\frac{1}{2}, \frac{3}{2}, \frac{5}{2}$.
3. $4y + 3 \geq 5$; 0, 1, 2, 3, 4, 5, 6, 7, −1, −2, −3, $\frac{1}{2}, \frac{3}{2}, -\frac{1}{2}$.
4. $3x + 5 = 2x + 2$; 0, 1, 2, 3, 4, 5, 6, 7, −1, −2, −3, $\frac{1}{2}, \frac{3}{2}, -\frac{1}{2}$.
5. $5x + 3 < 2x - 7$; 0, 1, 2, 3, 4, 5, −1, −2, −3, −4, $\frac{1}{2}, -\frac{1}{2}$.
6. $\frac{-2x + 7}{3} \geq 5$; 0, 1, 2, 3, 4, 5, −1, −2, −3, −4, −5.
7. $\frac{x + 3}{x - 2} = \frac{x + 1}{x - 1}$; 0, 1, 2, 3, 4, 5, −1, −2, −3, −4, −5, $\frac{1}{2}, -\frac{1}{2}$.
8. $\frac{2}{x - 4} + \frac{2x}{x^2 - 16} = \frac{1}{x + 4}$; 0, 1, 2, 3, 4, 5, −1, −2, −3, −4, −5.
9. $\frac{x}{x + 1} \geq 3$; 0, 1, 2, 3, 4, −1, −2, −3, −4, $-\frac{1}{2}, -\frac{3}{2}, -\frac{5}{2}$.
10. $x^2 - x - 6 = 0$; 0, 1, 2, 3, 4, 5, −1, −2, −3, −4, −5.

In exercises 11–16, determine the solution of each equation or inequality and tell whether the pairs of equations or inequalities are equivalent.

11. $x - 6 = 0$ and $x - 2 = 4$
12. $3x = 9$ and $7x = 21$
13. $x + 2 \geq 0$ and $x + 2 \leq 0$
14. $x + 1 \geq 0$ and $2x \geq -2$
15. $x + 1 = 0$ and $x^2 = 1$
16. $x^2 = 4$ and $2x = 4$

ADDITION AND SUBTRACTION

Now we show how to use the operations of addition and subtraction to solve equations and inequalities.

ADDITION OR SUBTRACTION OF AN ALGEBRAIC EXPRESSION TO AN EQUATION OR INEQUALITY.

If $a = b$, then $a + c = b + c$ and $a - c = b - c$.

If $a > b$, then $a + c > b + c$ and $a - c > b - c$.

If $a < b$, then $a + c < b + c$ and $a - c < b - c$.

Remember that whenever we perform an operation on one side of an equation or inequality, we must perform the same operation on the other side.

Example 5. Solve $x - 6 = 2$.

Solution. We want x to appear alone on one side of the equation. If we add 6 to both sides of the equation, we obtain

$$x - 6 + 6 = 2 + 6$$
$$x = 8$$

We have thus found that x must equal 8.

Now we want to find the solution set of the equation $x - 6 = 2$. We have performed an operation on the equation to obtain another equation. But how do we know that the equation $x - 6 = 2$ is equivalent to $x = 8$? What we have shown is that, *if* $x - 6 = 2$, *then* $x = 8$. So the only number that can possibly be a solution of $x - 6 = 2$ is 8. What we want to show now is that 8 actually is a solution. But this is easy; we check 8 in the equation. Substituting 8 for x, $8 - 6 = 2$; so 8 satisfies the equation. We have shown that the solution is 8.

Checking the answer as we did in Example 5 may seem to be unnecessary here. But when we get to more complicated examples, we will see that checking *is* necessary. Performing an operation on an equation does not always lead to an equivalent equation.

Example 6. Solve $x + 5 > 4$.

Solution. We subtract 5 from both sides of the inequality to obtain

$$x + 5 - 5 > 4 - 5$$
$$x > -1$$

We have found x, but again the point comes up—have we actually found the solution? We have shown that *if* $x + 5 > 4$, then $x > -1$. But let us now reverse the operation. Start with $x > -1$ and add 5. We obtain

$$x + 5 > -1 + 5$$
$$x + 5 > 4$$

So if $x > -1$, then $x + 5 > 4$. We have now shown that the two inequalities are equivalent; thus, the solution is the set of numbers x so that $x > -1$.

Exercise Set 4.1 (continued)

In exercises 17–28, solve the given equation or inequality.

17. $x - 5 = 0$ 18. $x + 12 = 0$ 19. $y + 7 = 11$ 20. $y - 12 = -2$
21. $x - 7 > 3$ 22. $x - 3 \leq 4$ 23. $x + 2 \leq 5$ 24. $x + 11 > 14$
25. $2x - 3 = x + 1$ 26. $3x - 1 = 2x - 2$ 27. $2x + 3 > x - 2$
28. $3x + 4 > 2x + 5$

MULTIPLICATION AND DIVISION—EQUATIONS

Now we turn to the operations of multiplication and division. We treat equations and inequalities separately.

MULTIPLYING OR DIVIDING AN EQUATION BY AN ALGEBRAIC EXPRESSION.

If $a = b$, then $ac = bc$ and $a/c = b/c$ for $c \neq 0$.

Example 7. Solve $5x = -15$.

Solution. Dividing both sides of the equation by 5, we obtain

$$5x/5 = -15/5$$
$$x = -3$$

We check to see if -3 satisfies the equation. Since $5(-3) = -15$, the solution is -3.

Example 8. Solve $\dfrac{x}{x-2} = \dfrac{2}{x-2}$.

Solution. Multiplying both sides of the equation by $x - 2$, we obtain

$$(x-2)\frac{x}{x-2} = (x-2)\frac{2}{x-2}$$
$$x = 2$$

However, the number 2 does not satisfy the original equation. This is because, when we substitute 2 for x, we have 0 in the denominator, and we cannot divide by 0. Therefore, there are no values of x that satisfy this equation.

Example 8 shows that an equation may have no solution. It also shows that performing an operation on an equation does not always lead to an equivalent equation. There are two operations that always produce an equivalent equation or inequality.

OPERATIONS ON EQUATIONS AND INEQUALITIES THAT ALWAYS PRODUCE EQUIVALENT EQUATIONS OR INEQUALITIES.

1. Addition of an algebraic expression,
2. Multiplication by a non-zero *constant*.

Exercise Set 4.1 (continued)

In exercises 29–40, solve the given equation.

29. $2x = 12$
30. $4x = 20$
31. $3x = -12$
32. $2x = -6$
33. $-4x = 16$
34. $-5x = 15$
35. $-5x = -20$
36. $-3x = -18$
37. $\dfrac{x}{x-3} = \dfrac{3}{x-3}$
38. $\dfrac{x}{x-4} = \dfrac{4}{x-4}$
39. $\dfrac{x}{x-2} = \dfrac{5}{x-2}$
40. $\dfrac{x}{x+1} = \dfrac{1}{x+1}$

MULTIPLICATION AND DIVISION—INEQUALITIES

When we multiply or divide both sides of an inequality by a number, we must distinguish between the case in which the number is positive and the case in which the number is negative.

Example 9. Consider the inequality $2 < 3$.

a. If we multiply both sides of this inequality by 2 (which is positive), the correct inequality is $2(2) < 2(3)$ or $4 < 6$. We say that we *preserve* the inequality.

b. If we multiply both sides of this inequality by -2 (which is negative), the correct inequality is $-2(2) > -2(3)$ or $-4 > -6$. We say that we *reverse* the inequality.

These basic properties are:

a. If we multiply or divide both sides of an inequality by a positive number, we preserve the inequality.

b. If we multiply or divide both sides of an inequality by a negative number, we reverse the inequality.

MULTIPLYING OR DIVIDING AN INEQUALITY BY A NUMBER.

If $a > b$ and c is positive, then $ac > bc$ and $a/c > b/c$ (preserve)
If $a < b$ and c is positive, then $ac < bc$ and $a/c < b/c$ (preserve)

If $a > b$ and c is negative, then $ac < bc$ and $a/c < b/c$ (reverse)
If $a < b$ and c is negative, then $ac > bc$ and $a/c > b/c$ (reverse)

Example 10. Solve $7x < 21$.

Solution. Dividing both sides of the inequality by 7, we preserve the inequality and obtain

$$x < 3.$$

The solution is the set of numbers x such that $x < 3$.

Example 11. Solve $-4x < 20$.

Solution. Dividing both sides of the inequality by -4, we reverse the inequality and obtain

$$x > -5.$$

The solution is the set of numbers x such that $x > -5$.

Exercise Set 4.1 (continued)

In exercises 41–52, solve the given inequality.

41. $5x < 25$	42. $3x \leq 24$	43. $3x \geq 21$	44. $2x > 8$
45. $-2x > 20$	46. $-5x \geq 20$	47. $-6x \leq -24$	48. $-4x < -16$
49. $15x \leq 5$	50. $12x \geq 6$	51. $-6x \geq 3$	52. $-8x \leq -4$

section 2 • Linear Equations and Inequalities

Now we solve linear equations and inequalities in one variable.

DEFINITION OF A LINEAR EQUATION OR INEQUALITY IN ONE VARIABLE. A *linear equation* or *first-degree equation* in one variable x is any equation equivalent to an equation of the form

$$ax + b = 0,$$

where a and b are constants and $a \neq 0$.

A *linear inequality* or *first-degree inequality* in one variable x is an inequality equivalent to one of the form

$$ax + b > 0 \quad \text{or} \quad ax + b \geq 0$$
$$\text{or } ax + b < 0 \quad \text{or} \quad ax + b \leq 0,$$

where a and b are real number constants and $a \neq 0$.

Both the names *linear* and *first-degree* are used in connection with these equations and inequalities. The name *linear* refers to the fact that these equations and inequalities are related to the graph of a straight line in the Cartesian coordinate system. The name *first-degree* refers to the fact that there is a variable (which we've called x) in the equation or inequality, but it occurs only to the first power—there is no x^2, x^3, etc. This is the essence of the definition: we have an equation or inequality that involves the first power of x but no higher powers. Of course, the variable need not be called x; it may just as well be y or z or t or some other letter.

An equation or inequality is linear if it is *equivalent* to an equation or inequality of the appropriate form. At first sight, the equation or inequality may *look* quite different from the appropriate form; and it may require several operations to put it in this form.

LINEAR EQUATIONS

The technique we use to solve a linear equation is to perform operations on both sides of the equation to cause the variable to occur by itself on one side of the equation.

Example 1. Solve $7x + 5 = -2$.

Solution. Subtracting 5 from both sides of the equation, we have

$$7x = -7.$$

Now we divide both sides of the equation by 7 to obtain

$$x = -1.$$

We check the number -1 in the original equation: $7(-1) + 5 = -2$. Therefore, the solution is -1.

Example 2. Solve $3x + 5 = 2x + 2$.

Solution. We subtract 5 from both sides of the equation to obtain

$$3x = 2x - 3.$$

Now we subtract $2x$ from both sides of the equation:

$$x = -3.$$

We check the number -3 in the original equation: $3(-3) + 5 = 2(-3) + 2$. Therefore, the solution is -3.

Example 3. Solve $2x + 2(x + 12) = 74$.

Solution. First we simplify the left-hand side of the equation.

$$2x + 2x + 24 = 74$$

$$4x + 24 = 74$$

$$4x = 50$$

$$x = 50/4 = 25/2$$

We check $25/2$ in the original equation:

$$2(25/2) + 2((25/2) + 12) = 25 + 2(49/2) = 74.$$

Therefore, the solution is $25/2$.

Exercise Set 4.2

In exercises 1–18, solve the given equation.

1. $3x + 7 = 10$
2. $6y - 4 = -16$
3. $-4x + 3 = -1$
4. $-6y - 3 = 0$
5. $4 + 2z = 18$
6. $-11 - 5x = 4$
7. $2 + 4y = 4$
8. $6x + 3 = 0$
9. $5 - 3t = 6$
10. $2 - 10z = -1$
11. $2y + 5 = 7y + 3$
12. $6x + 11 + x = 5x - 4$
13. $4x + 2 = 4(x + 2)$
14. $3(x + 2) = x + 4 + 2(x + 1)$
15. $2x + 2(x + 8) = 74$
16. $2x + 2(x + 6) = 18$
17. $2x + 2(5x) = 45$
18. $2x + 2(3x) = 48$

LINEAR INEQUALITIES

The technique for solving a linear inequality is the same as that for solving a linear equation—we perform operations on both sides of the inequality to cause the variable to occur by itself on one side of the inequality.

Example 4. Solve $3x + 8 \geq 20$.

Solution. Subtracting 8 from both sides of the inequality, we obtain

$$3x \geq 12.$$

Now we divide both sides of the inequality by 3 to preserve the inequality and obtain

$$x \geq 4.$$

Since we can reverse these steps, the solution of this inequality is the set of numbers x so that $x \geq 4$. The graph of the solution is shown in Figure 4.1.

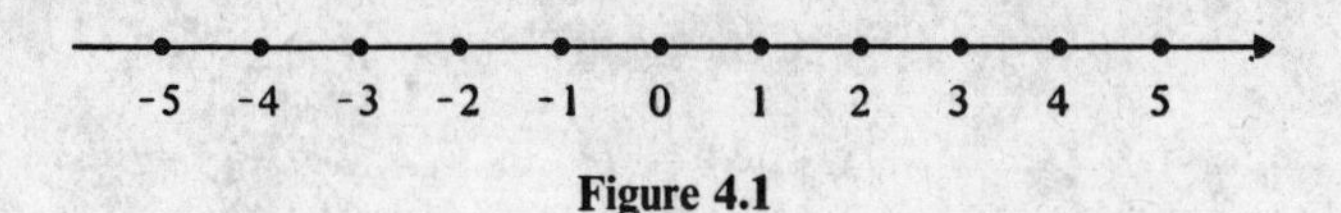

Figure 4.1

Example 5. Solve $5x + 3 < 2x - 7$.

Solution. We subtract 3 from both sides of the inequality:

$$5x < 2x - 10.$$

Now we subtract $2x$ from both sides of the inequality:

$$3x < -10.$$

Dividing both sides of the inequality by 3, we preserve the inequality and obtain

$$x < -10/3.$$

Since we can reverse these steps, the solution is the set of numbers x such that $x < -10/3$. Figure 4.2 shows the graph of the solution.

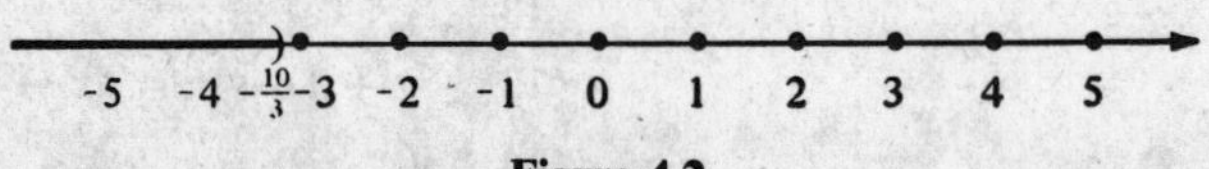

Figure 4.2

Example 6. Solve $\dfrac{-2x + 7}{3} \geq 5$.

Solution. When we multiply both sides of this inequality by 3, we preserve the inequality and find

$$-2x + 7 \geq 15.$$

Subtracting 7 from both sides of the inequality, we obtain

$$-2x \geq 8$$

Dividing both sides of the inequality by -2, we reverse the inequality and obtain

$$x \leq -4.$$

We can reverse these steps, so the solution is the set of numbers x such that $x \leq -4$. The graph of the solution is shown in Figure 4.3.

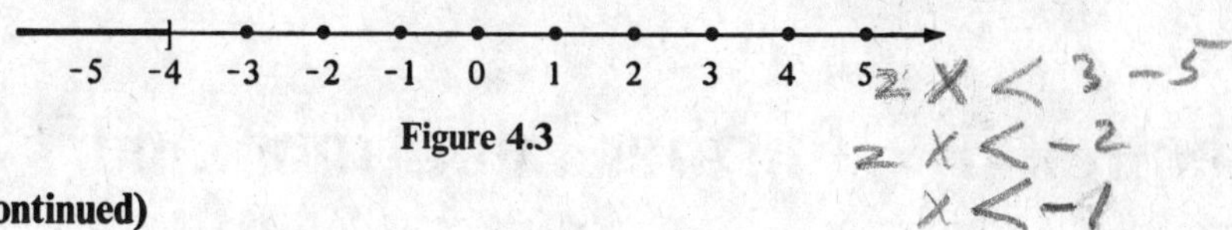

Figure 4.3

Exercise Set 4.2 (continued)

In exercises 19–36, solve the given inequality.

19. $3x - 7 > 2$
20. $2x + 5 < 3$
21. $4y + 3 \geq 5$
22. $-3z - 5 < 1$
23. $1 - 6x > 4$
24. $1 + 6x < -4$
25. $-6x + 1 \geq 13$
26. $4 - 5x \leq -6$
27. $-2 - 4x > 10$
28. $3 + 3x \leq 15$
29. $6x + 1 \geq 2x - 3$
30. $-2x - 5 < 11x + 1$
31. $3t - 2 < -3t + 1$
32. $3x + 5 > 4x - 7$
33. $\dfrac{5x + 2}{3} > 9$
34. $\dfrac{-14y + 4}{3} < 12y$
35. $\dfrac{-2x + 1}{-2} < 3x$
36. $\dfrac{6x - 3}{-4} \geq -6$

SIMPLIFYING EQUATIONS

We give further examples of linear equations in which we must simplify the equations to solve them.

Example 7. Solve $(x + 5)^2 = (x - 2)^2$.

Solution. First we rewrite the equation by squaring out the binomials on each side:

$$\begin{aligned} x^2 + 10x + 25 &= x^2 - 4x + 4 \\ 10x + 25 &= -4x + 4 \\ 10x &= -4x - 21 \\ 14x &= -21 \\ x &= -21/14 = -3/2. \end{aligned}$$

Checking, we have $((-3/2) + 5)^2 = ((-3/2) - 2)^2$. So the solution is $-3/2$.

Exercise Set 4.2 (continued)

In exercises 37–48, solve the given equation.

37. $100x = 5(x + 285)$
38. $60x = 80(x - 2)$
39. $(x + 2)^2 = (x - 1)^2$
40. $(x + 4)^2 = (x - 4)^2$

41. $(2x + 1)^2 = (2x - 1)^2$
42. $(2x + 1)^2 = 4x^2 + 1$
43. $(x + 1)^2 = x^2 + 13$
44. $(x + 6)^2 = x^2 + 76$
45. $(x + 5)(2x - 10) = (x - 5)(2x + 1)$
46. $(x - 1)(2x + 1) = x(2x - 1)$
47. $x(x + 3) + 12 = (x + 1)(x + 4)$
48. $(x - 2)(x + 1) + 38 = x(x + 3)$

section 3 • Further Equations and Inequalities

In this section we consider further topics in solving equations and inequalities: fractional equations, double inequalities, and equations and inequalities with decimal coefficients.

FRACTIONAL EQUATIONS

If we have an equation that contains fractions, it is usually easiest first to multiply both sides of the equation by an expression that will clear the equation of the fractions.

Example 1. Solve $\dfrac{x + 3}{x - 2} = \dfrac{x + 1}{x - 1}$.

Solution. To clear the equation of fractions, we multiply both sides of the equation by $(x - 2)(x - 1)$.

$$(x + 3)(x - 1) = (x + 1)(x - 2)$$
$$x^2 + 2x - 3 = x^2 - x - 2$$
$$2x - 3 = -x - 2$$
$$3x = 1$$
$$x = 1/3$$

We check 1/3 in the original equation:

$$\frac{(1/3) + 3}{(1/3) - 2} = -2 = \frac{(1/3) + 1}{(1/3) - 1}.$$

Therefore, the solution is 1/3.

Example 2. Solve $\dfrac{2}{x-4} + \dfrac{2x}{x^2-16} = \dfrac{1}{x+4}$.

Solution. We notice that $x^2 - 16 = (x-4)(x+4)$, and we can clear of fractions if we multiply by this product.

$$2(x+4) + 2x = 1(x-4)$$
$$2x + 8 + 2x = x - 4$$
$$4x + 8 = x - 4$$
$$3x = -12$$
$$x = -4$$

But -4 does not check in the original equation because we cannot divide by 0. The operations performed on the equations show that *if* there is a solution, it must be -4. But -4 is not a solution, so the equation has no solution.

Exercise Set 4.3

In exercises 1–24, solve the given equation.

1. $\dfrac{5}{x} = \dfrac{15}{2}$

2. $\dfrac{-4}{x} = \dfrac{1}{3}$

3. $\dfrac{1}{x} = \dfrac{1}{2} + \dfrac{1}{3}$

4. $\dfrac{1}{x} = \dfrac{1}{2} - \dfrac{1}{3}$

5. $\dfrac{3}{4}t - 2 = \dfrac{7}{8}t + \dfrac{1}{2}$

6. $\dfrac{1}{5}x + \dfrac{1}{3} = \dfrac{1}{3}x + \dfrac{1}{5}$

7. $\dfrac{x}{x+2} = \dfrac{2x}{2x+2}$

8. $\dfrac{2x}{x+4} = \dfrac{2x-5}{x+4}$

9. $\dfrac{x-1}{2x+1} = \dfrac{x}{2x+1}$

10. $\dfrac{x+5}{x-5} = \dfrac{2x+1}{2x-10}$

11. $\dfrac{1236}{x+400} = \dfrac{36}{x}$

12. $\dfrac{1620}{x+480} = \dfrac{180}{x}$

13. $\dfrac{y}{y+2} = \dfrac{2y}{y+2} - 2$

14. $\dfrac{x}{x+2} = \dfrac{2x}{x+2} - 1$

15. $\dfrac{x}{x-2} = \dfrac{3x}{x-2} + 1$

16. $\dfrac{2x-2}{x-2} = \dfrac{x}{x-2} + 3$

17. $\dfrac{x}{x-1} = \dfrac{x}{x+1}$

18. $\dfrac{t+3}{t-1} = \dfrac{t-1}{t+2}$

19. $\dfrac{x+2}{x+1} = \dfrac{x-1}{x-2}$

20. $\dfrac{2x}{2x+3} = \dfrac{x-1}{x-3}$

21. $\dfrac{1}{x+1} + \dfrac{1}{x-1} = \dfrac{x}{x^2-1}$

22. $\dfrac{2}{x+1} - \dfrac{4}{x-1} = \dfrac{x}{x^2-1}$

23. $\dfrac{1}{x-2} + \dfrac{3}{x+1} = \dfrac{3}{x^2-x-2}$

24. $\dfrac{4}{x+5} - \dfrac{2}{x-3} = \dfrac{-16}{x^2+2x-15}$

DOUBLE INEQUALITIES

Sometimes we encounter inequalities of the form $a \le b \le c$. We call these *double inequalities* because we really have two inequalities: $a \le b$ and $b \le c$.

Example 3. Solve $-1 \le \dfrac{2x+1}{-3} \le 1$.

Solution. We have two inequalities here, but we can deal with them simultaneously. We multiply each side of these inequalities by -3:

$$3 \ge 2x + 1 \ge -3$$

$$2 \ge 2x \ge -4$$

$$1 \ge x \ge -2.$$

Since we can reverse these steps, the solution is the set of numbers x such that

$$-2 \le x \le 1.$$

The graph of the solution is shown in Figure 4.4.

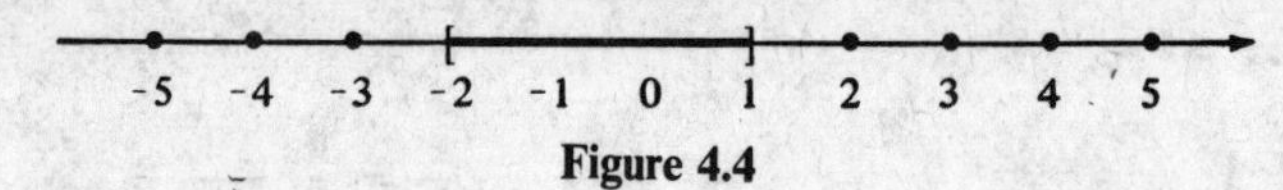

Figure 4.4

Exercise Set 4.3 (continued)

In exercises 25–34, solve the double inequality.

25. $-3 \le 4x - 6 < 7$
26. $-5 \ge 3x + 7 \ge 4$
27. $-4 \le x - 5 \le 4$
28. $-1 < x - 3 < 1$
29. $-6 \le 2x + 6 \le 6$
30. $-7 < 3x + 1 < 7$
31. $1 \le \dfrac{x+6}{-1} \le 2$
32. $-5 \le \dfrac{2z+6}{-4} \le 10$
33. $-2 < \dfrac{4x+1}{2} < 2$
34. $-\dfrac{5}{6} \le \dfrac{x-1}{3} \le \dfrac{5}{6}$

DECIMAL COEFFICIENTS

Frequently we must solve equations and inequalities in which the numbers are decimals rather than integers.

Example 4. Solve $0.04x + 0.05(4) \le 0.72$.

Solution. We have

$$0.04x + 0.2 \le 0.72$$

$$0.04x \le 0.52$$

$$x \leq 0.52/0.04 = 13.$$

The solution is the set of numbers x such that $x \leq 13$.

Example 5. Solve $0.6(10) + 0.2x = 0.35(10 + x)$.

Solution. We have

$$6 + 0.2x = 3.5 + 0.35x$$
$$2.5 = 0.15x$$
$$x = 2.5/0.15 = 16.67.$$

The solution is 16.67.

Exercise Set 4.3 (continued)

In exercises 35–44, solve the given equation or inequality.

35. $0.9 - 0.9x = 0.85$
36. $0.15(6) + x = 0.1(6 + x)$
37. $0.4(15) + x = 0.5(15 + x)$
38. $0.04x + 0.05(6000 - x) = 275$
39. $0.04x + 0.05(4) \leq 0.6$
40. $0.03(4) + 0.03x \leq 0.54$
41. $0.03x + 0.06(5) \leq 0.58$
42. $0.02(5) + 0.04x \leq 0.62$
43. $0.4(30) + 0.1x = 0.3(30 + x)$
44. $0.45(40) + 0.15x = 0.3(40 + x)$

CALCULATOR EXERCISES: In exercises 45–52, use a calculator to solve the given equation. Express the solution as a decimal rounded off to two decimal places.

45. $3.36x + 2.45 = 1.13x - 6.22$
46. $1.15x - 13.11 = 2.45x + 3.67$
47. $3.14x - 3.33 = 1.14(x + 2.5)$
48. $6.28x - 4.45 = 3.65(x + 4.7)$
49. $2.16x + 6.61 = 3.33(x - 5.8)$
50. $5.61x - 13.44 = 5.44(x - 6.6)$
51. $14.45 = 3.14(2.31)^2 + 3.14(2.31)x$
52. $28.28 = 3.14(5.55)^2 + 3.14(5.55)x$

section 4 • Solving for a Variable

In this section we consider equations in more than one variable. If an equation has more than one variable, to obtain information from the equation, it is frequently useful to solve for one variable in terms of the others. In other words, we want to isolate one of the variables on one side of the equation. The way to do this is to perform operations on the equation. The techniques are analogous to those for solving linear equations.

EQUATIONS FROM SCIENCE

We give some examples of equations that occur in various sciences.

Example 1. The equation for converting temperature from degrees Centigrade to degrees Fahrenheit is $F = \frac{9}{5}C + 32$. Find the equation for converting from degrees Fahrenheit to degrees Centigrade.

Solution. We wish to solve the equation for the variable C.

$$F = \frac{9}{5}C + 32$$

$$F - 32 = \frac{9}{5}C$$

$$\frac{5}{9}(F - 32) = C$$

The equation is $C = \frac{5}{9}(F - 32)$.

Example 2. Solve $\frac{1}{s} = \frac{1}{t} - \frac{1}{T}$ for t. (This equation occurs in astronomy; it relates two different sidereal periods with a synodic period.)

Solution. To clear of fractions, we multiply both sides of the equation by stT:

$$tT = sT - st$$

$$ts + tT = sT$$

$$t(s + T) = sT$$

$$t = \frac{sT}{s + T}. \qquad \text{for } s \neq -T$$

Example 3. Solve $S = 2\pi r^2 + 2\pi rh$ for h. (This is the formula for the surface area S of a right circular cylinder in terms of the radius r and the height h.)

Solution.

$$2\pi rh = S - 2\pi r^2$$

Dividing both sides of the equation by $2\pi r$, we obtain

$$h = \frac{S - 2\pi r^2}{2\pi r}.$$

We can also write

$$h = \frac{S}{2\pi r} - r.$$

We now outline a procedure which can be used to solve an equation for a specified variable. You may wish to review Examples 1, 2, and 3 in view of this procedure.

PROCEDURE FOR SOLVING AN EQUATION FOR A SPECIFIED VARIABLE.

1. If necessary, clear the equation of fractions by multiplying both sides of the equation by a common denominator.
2. Take all terms involving the specified variable to one side of the equation and all terms not involving the specified variable to the other side of the equation.
3. If necessary, factor the specified variable as a monomial factor on its side of the equation; note the remaining factor.
4. Divide both sides of the equation by this remaining factor. Simplify the result if necessary.

Exercise Set 4.4

In exercises 1–20, solve for the specified variable.

1. $c = 2\pi r$ for r
2. $F = ma$ for a
3. $F = G\dfrac{mM}{r^2}$ for M
4. $\dfrac{P_1V_1}{T_1} = \dfrac{P_2V_2}{T_2}$ for P_1
5. $l = a + (n - 1)d$ for n
6. $pa = (1 - p)b$ for p
7. $P = (p - c)n - F$ for n
8. $P = A(1 - dt)$ for d
9. $S = \dfrac{a}{1 - r}$ for r
10. $I = \dfrac{E}{R + ar}$ for r
11. $\dfrac{1}{r} = \dfrac{1}{a} + \dfrac{1}{b}$ for r
12. $\dfrac{1}{r} = \dfrac{1}{a} - \dfrac{1}{b}$ for r
13. $d = \dfrac{r}{1 + rt}$ for t
14. $d = \dfrac{r}{1 + rt}$ for r
15. $S = \dfrac{rt - a}{r - 1}$ for r
16. $\dfrac{1}{H - a} + \dfrac{1}{H - b} = \dfrac{2}{H}$ for H
17. $V = \pi r^2 h$ for h
18. $E = mc^2$ for m
19. $\dfrac{1}{1 - x} = 1 - \dfrac{v}{1 - b}$ for x
20. $\dfrac{1}{1 - x} = 1 - \dfrac{v}{1 - x}$ for x

EQUATIONS FROM MATHEMATICS

We give some examples of equations that arise in mathematics. Some of these equations occur later in this book; some occur in calculus.

Example 4. Solve $2x + 3y - 12 = 0$ for y.

Solution. We have

$$3y = -2x + 12$$

$$y = \frac{-2x + 12}{3}$$

$$= -\frac{2}{3}x + 4.$$

Example 5. Solve $2b^2x + 2a^2yy' = 0$ for y'.

Solution. Note that y and y' are two different symbols here.

$$2a^2yy' = -2b^2x$$

$$y' = -2b^2x/2a^2y = -b^2x/a^2y$$

Exercise Set 4.4 (continued)

In exercises 21–40, solve each equation for the specified variable.

21. $3x + 2y + 12 = 0$ for y
22. $3x + 2y + 12 = 0$ for x
23. $6x - 2y + 5 = 0$ for y
24. $6x - 2y + 5 = 0$ for x
25. $4x + 3y - 15 = 0$ for y
26. $3x + 5y - 15 = 0$ for y
27. $2x - 4y = 6$ for y
28. $3x - 6y = 15$ for y
29. $5x - 4y = 20$ for y
30. $Ax + By + C = 0$ for y
31. $mx + b = m'x + b'$ for x
32. $ax + c = a^2x + c^2$
33. $\frac{x}{2} + \frac{y}{3} = 1$ for y
34. $\frac{x}{a} + \frac{y}{b} = 1$ for y
35. $xy' + y = 0$ for y'
36. $1 + y' + xy' + y = 0$ for y'
37. $2x + 2y\,y' = 0$ for y'
38. $\frac{2x}{a^2} - \frac{2y}{b^2}y' = 0$ for y'
39. $3x^2 + 3y^2y' - 6xy' - 6y = 0$ for y'
40. $4y^3y' + 2yx^2y' + 2xy^2 = 0$ for y'

section 5 • Absolute Value

In this section we examine the concept of distance on the real number line and, in particular, the relation between distance and *absolute value*. We define absolute value and solve inequalities involving absolute values.

DEFINITION OF ABSOLUTE VALUE

Recall that when we graph real numbers on the real number line, we graph positive numbers to the right of the origin and negative numbers to the left of the origin. If we let the variable x denote a real number, then we can ask, What is the distance between x and the origin? We consider three cases here: the number x may be positive, negative, or 0.

First, suppose x denotes a positive real number. Then $-x$ is negative and the points x and $-x$ are symmetric about the origin (Figure 4.5). Furthermore, the distance between

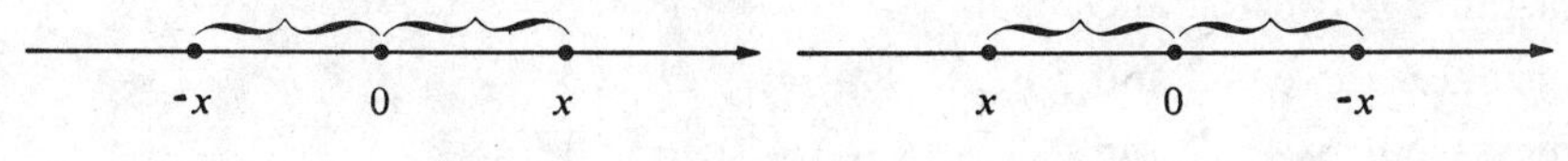

Figure 4.5

x and the origin is just x. Second, suppose x denotes a negative real number. Then $-x$ is positive, and again the points x and $-x$ are symmetric about the origin (Figure 4.5). But now the distance between x and the origin is $-x$. Third, if x denotes the real number 0, then the distance between 0 and the origin is 0.

The distance between x and the origin is what we call the *absolute value* of x.

DEFINITION OF ABSOLUTE VALUE. If x is a real number, then

$$|x| = \begin{cases} x & \text{if } x \geq 0 \\ -x & \text{if } x < 0. \end{cases}$$

We read $|x|$ as the *absolute value* of x.

So $|x|$ measures the distance between the real number x and the origin on the real number line.

Example 1. Absolute value.

a. $|0| = 0$
b. $|5| = 5$ since 5 is positive.
c. $|-5| = -(-5) = 5$ since -5 is negative.
d. $|1.5| = 1.5$ since 1.5 is positive.
e. $|-1.5| = -(-1.5) = 1.5$ since -1.5 is negative.

BASIC PROPERTY OF ABSOLUTE VALUE. If x is a real number, then $|x| \geq 0$; and if $x \neq 0$, then $|x| > 0$.

We have seen the absolute value in Section 3.3 in another connection. In discussing square roots we saw that, if x denotes a real number, then

$$|x| = \sqrt{x^2}.$$

The absolute value $|x|$ measures the distance between x and the origin. What about $|x - a|$? This expression tells us to subtract $x - a$ and then take the absolute value. So $|x - a|$ measures the distance between x and a (Figure 4.6).

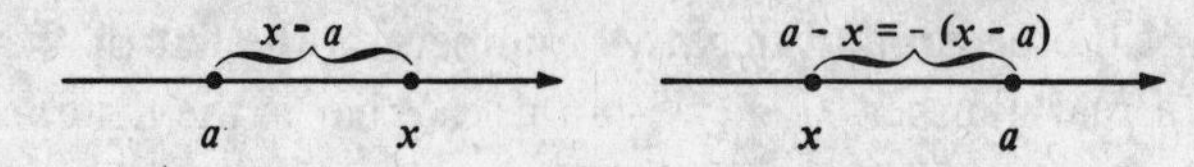

Figure 4.6

Example 2. Distance between two points.

a. The distance between 5 and 2 is $|5 - 2| = |3| = 3$.

b. The distance between 2 and 5 is $|2 - 5| = |-3| = 3$.

c. The distance between 3 and -7 is $|3 - (-7)| = |10| = 10$.

d. The distance between -7 and 3 is $|-7 - 3| = |-10| = 10$.

Exercise Set 4.5

In exercises 1–16, find the given absolute value.

1. $\|6\|$	2. $\|3.7\|$	3. $\|-6\|$	4. $\|-3.7\|$
5. $\|-12\|$	6. $\|-15\|$	7. $\|\pi\|$	8. $\|-\pi\|$
9. $\|8 - 3\|$	10. $\|6 - 2\|$	11. $\|3 - 8\|$	12. $\|2 - 6\|$
13. $\|-1 - 4\|$	14. $\|-3 - 5\|$	15. $\|-4 - 1\|$	16. $\|-5 - 3\|$

EXPRESSING INTERVALS AS ABSOLUTE VALUES

In applications (such as calculus) we use absolute values to express intervals on the real number line. The inequality

$$|x - a| < r$$

is satisfied by the set of points x whose distance from a is less than r. These points are inside the interval with center at the point a and radius r; that is, the interval from $a - r$ to $a + r$ (see Figure 4.7). This interval is the set of numbers x satisfying the double inequality $a - r < x < a + r$.

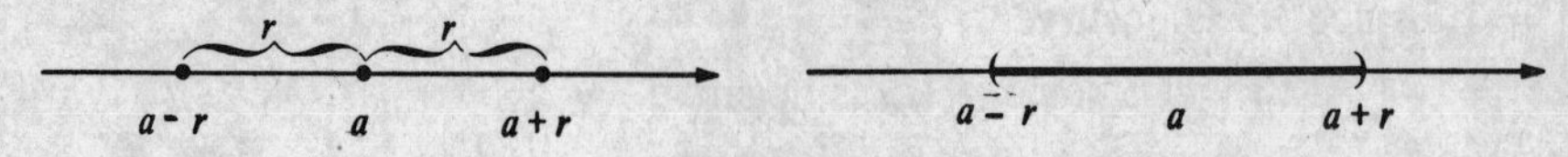

Figure 4.7

Example 3. Express the interval from 2 to 6 (not including the endpoints) as an absolute value inequality.

Solution. The center of this interval is the point $\frac{1}{2}(2 + 6) = 4$ and the radius is $6 - 4 = 2$. Therefore, the interval consists of the set of numbers x satisfying the inequality

$$|x - 4| < 2.$$

We can also express this interval using the double inequality

$$2 < x < 6.$$

The inequality

$$|x - a| > r$$

is satisfied by the set of points x outside the interval from $a - r$ to $a + r$. We can express these points as the set of numbers x satisfying

$$x < a - r \quad \text{or} \quad x > a + r$$

(see Figure 4.8). We cannot combine these two inequalities as a double inequality.

Figure 4.8

Example 4. Express the points outside the interval from -4.3 to 2.1 (including the endpoints) as an absolute value inequality.

Solution. The center of the interval is the point $\frac{1}{2}(-4.3 + 2.1) = -1.1$ and the radius is $2.1 - (-1.1) = 3.2$. Therefore, the points x outside the interval satisfy the inequality

$$|x + 1.1| \geq 3.2$$

We can also express these points as the set of numbers x satisfying

$$x \leq -4.3 \quad \text{or} \quad x \geq 2.1$$

Exercise Set 4.5 (continued)

In exercises 17–22, express the given interval (not including the endpoints) as an absolute value inequality.

17. From 3 to 7
18. From 2 to 4
19. From 1.3 to 3.7
20. From 2.2 to 3.2
21. From -3 to 5
22. From -2 to 2

In exercises 23–28, express the points outside the given interval (including the endpoints) as an absolute value inequality.

23. From -4.6 to 1.5
24. From -3.7 to 4.2

25. From -6 to -1
26. From -5 to -2
27. From 0.6 to 3.4
28. From 0.2 to 1.7

SOLVING INEQUALITIES

We can use the following method to solve absolute value inequalities.

METHOD FOR SOLVING ABSOLUTE VALUE INEQUALITIES. If r is a positive number,

replace $|y| < r$ by the double inequality $-r < y < r$.

replace $|y| > r$ by the two inequalities $y < -r$ or $y > r$.

Example 5. Solve $\left|\dfrac{3x + 2}{5}\right| \leq 1$.

Solution. Replace the absolute value inequality by the double inequality

$$-1 \leq \frac{3x + 2}{5} \leq 1$$

$$-5 \leq 3x + 2 \leq 5$$

$$-7 \leq 3x \leq 3$$

$$-\frac{7}{3} \leq x \leq 1.$$

The solution is the set of numbers x satisfying $-7/3 \leq x \leq 1$; its graph is Figure 4.9.

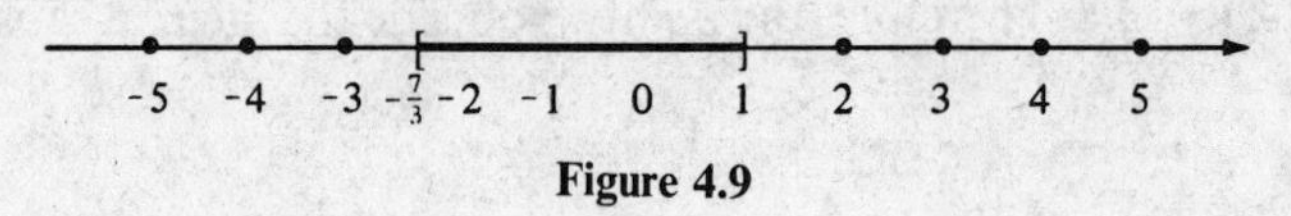

Figure 4.9

Example 6. Solve $|x + 3| > 2$.

Solution. Replace the absolute value inequality by the two inequalities

$$x + 3 < -2 \quad \text{or} \quad x + 3 > 2$$

$$x < -5 \quad \text{or} \quad x > -1$$

The solution is the set of numbers x satisfying $x < -5$ or $x > -1$; its graph is Figure 4.10.

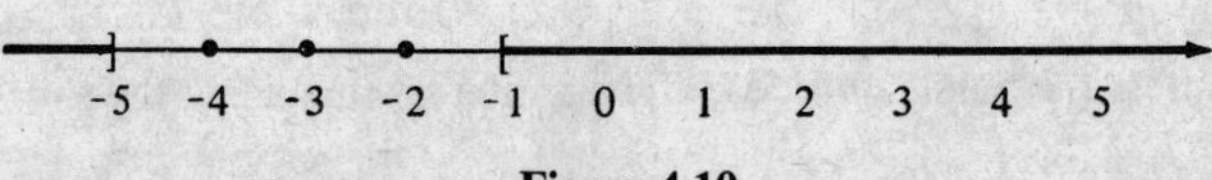

Figure 4.10

Exercise Set 4.5 (continued)

In exercises 29–44, solve each given inequality.

29. $|x - 5| \leq 4$
30. $|x - 3| < 1$
31. $|x - 2| < 3$
32. $|x + 1| \leq \frac{1}{2}$
33. $|2x + 6| \leq 6$
34. $|3x + 1| < 7$
35. $\left|\frac{4x + 1}{2}\right| < 2$
36. $\left|\frac{x - 1}{3}\right| \leq \frac{5}{6}$
37. $|x + 5| > 12$
38. $|2x + 2| \geq 3$
39. $\left|\frac{x + 6}{3}\right| \geq 1$
40. $\left|\frac{x - 4}{5}\right| \geq \frac{4}{5}$
41. $|4x - 14.28| < 0.06$
42. $|4x - 9.32| < 0.03$
43. $|4x - 8.24| < 0.04$
44. $|4x - 5.48| < 0.01$

section 6 • Applications

This section applies linear equations and inequalities to geometry problems, rate problems, mixing solutions, diet problems, business and management considerations, engineering tolerances, and historical exercises.

GEOMETRY PROBLEMS

Example 1. One side of a rectangle is 12 meters longer than the other side. If the perimeter is 74 meters, what are the lengths of the sides?

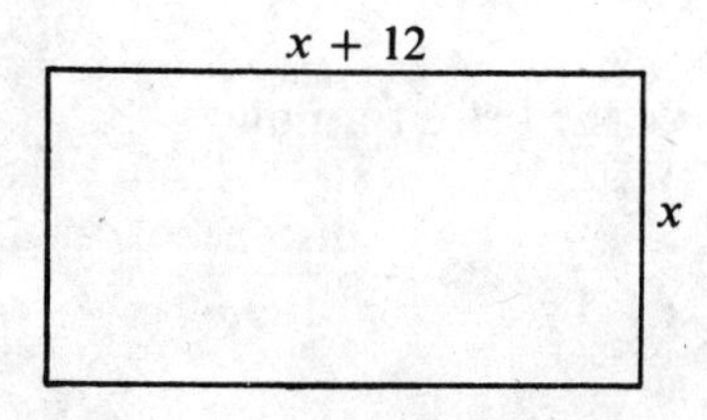

Figure 4.11

Solution. It may help you to draw a picture of a rectangle like that shown in Figure 4.11. We want to know the length of each side. If we let x be the length of the shorter side, then $x + 12$ is the length of the longer side. Now we look for an equation involving the variable x. On the one hand we know that the perimeter is $2x + 2(x + 12)$, and on the other hand we know that the perimeter is 74. So we have the equation

$$2x + 2(x + 12) = 74$$

$$2x + 2x + 24 = 74$$

$$4x = 50$$

$$x = 50/4 = 25/2$$

The shorter side is $12\frac{1}{2}$ meters and the longer side is $12\frac{1}{2} + 12 = 24\frac{1}{2}$ meters. (Check for yourself that this solution results in a perimeter of 74 meters.)

Exercise Set 4.6

1. One side of a rectangle is 8 meters longer than the other side. If the perimeter is 74 meters, what is the length of each side?
2. One side of a rectangle is 6 meters longer than the other side. If the perimeter is 18 meters, what is the length of each side?
3. One side of a rectangle is 5 times as long as the other. If the perimeter is 45 meters, how long is each side?
4. One side of a rectangle is 3 times as long as the other side. If the perimeter is 48 meters, how long is each side?
5. A farmer wishes to enclose a rectangular area with a fence on three sides and the side of his barn as the fourth side. He wants the rectangle to be 40 feet long, and he has 90 feet of fencing. How wide should the rectangle be?
6. When the sides of a square are increased by 6 meters, the area is increased by 76 square meters. What is the side of the square?
7. One side of a rectangle is 3 inches longer than the other. If each side is increased by 1 inch, the area is increased by 12 square inches. What is the length of each side?
8. A rectangular picture has a 2-inch frame around it. The length of the frame is 3 inches greater than the width. If the area of the frame is 38 square inches, what are the outside length and width of the frame?

RATE PROBLEMS

Problems involving rate use the basic formula

$$\text{rate} = \frac{\text{distance}}{\text{time}}.$$

Example 2. An airplane goes 1275 miles in the same time a boat goes 25 miles. If the airplane goes 500 mph faster than the boat, find the rate of each.

Solution. Let r denote the rate of the boat, so $r + 500$ denotes the rate of the airplane. We're given that both *times* are equal, so we use the basic formula in the form $\text{time} = \dfrac{\text{distance}}{\text{rate}}$. The time for the airplane is then $\dfrac{1275}{r + 500}$, and the time for the boat is $\frac{25}{r}$. Since these are equal, the equation is

$$\frac{1275}{r + 500} = \frac{25}{r}$$

$$1275r = 25(r + 500)$$

$$1275r = 25r + 12500$$

$$1250r = 12500$$

$$r = 10$$

The rate of the boat is 10 mph and the rate of the airplane is 10 + 500 = 510 mph.

Exercise Set 4.6 (continued)

9. An airplane goes 1236 miles in the same time that a boat goes 36 miles. If the airplane goes 400 mph faster than the boat, find the rate of each.

10. An airplane travels 1620 miles in the same time a train goes 180 miles. If the airplane goes 480 miles per hour faster than the train, find the rate of each.

11. A man and his wife go from San Francisco to Hawaii. The wife goes by plane in 5 hours and the man goes by boat in 100 hours. If the plane goes 285 mph faster than the boat, find the rate of each.

12. Some bank robbers leave Cleveland traveling at 60 miles per hour. Two hours later the cops leave Cleveland along the same road at 80 miles per hour. How far from Cleveland do the cops overtake the robbers?

13. After work a man jogs from his office to his car, which he has left parked 3 miles away. He then drives the remaining 10 miles home. Due to congestion on the freeway, he figures that he jogs as fast as he drives. If he spends 2 hours 20 minutes longer driving than jogging, how long does it take him to get home from the office?

14. A businessman takes a helicopter from his office to the airport and a plane from the airport to his business appointment. His trip was 500 miles altogether and took $2\frac{1}{2}$ hours. If the helicopter averaged 50 mph and the airplane averaged 300 mph, how far is his office from the airport?

MIXING SOLUTIONS

Example 3. We have a 60% acid solution which we dilute by adding a 20% acid solution to obtain a 35% solution. If we start with 10 liters of the 60% solution, how many liters of the 20% solution should we add?

Solution. Let x be the number of liters we should add. Before adding anything, we have 0.60(10) liters of acid in the solution. After adding x liters of the 20% solution, we have $0.60(10) + 0.20x$ liters of acid. We want to have $0.35(10 + x)$ liters of acid. Thus we have the equation

$$0.60(10) + 0.20x = 0.35(10 + x)$$

$$6 + 0.2x = 3.5 + 0.35x$$
$$2.5 = 0.15x$$
$$x = 2.5/0.15 = 16.67.$$

We must add 16.67 liters of the 20% solution.

Exercise Set 4.6 (continued)

15. We have 30 liters of a 40% acid solution and we add a 10% acid solution to obtain a 30% solution. How many liters of the 10% solution should we add?
16. We have 40 liters of a 45% acid solution and we add a 15% acid solution to obtain a 30% solution. How many liters of the 15% solution should we add?
17. How much pure alcohol should be added to 15 liters of a 40% solution to obtain a solution which is at least 50% alcohol?
18. How much water should be added to 6 liters of a 15% salt solution to obtain a solution which is 10% salt?
19. We have 40 liters of a 60% acid solution and we dilute it by adding a 10% acid solution. How many liters should we add to get a solution which is at most 20% acid?
20. A quart of 20% alcohol solution is mixed with four quarts of distilled water. What percent alcohol is the resulting solution?

DIET PROBLEMS

Example 4. We prepare an animal feed by mixing two foodstuffs—a less expensive foodstuff and a more expensive one. The amount of the less expensive foodstuff we mix will vary (depending on the growth rate of the animal), but we mix the constant amount of 4 units of the more expensive foodstuff. The resulting mixture, however, must contain at least 20 grams of protein. If each unit of the less expensive foodstuff contains 3 grams of protein and each unit of the more expensive foodstuff contains 2 grams of protein, how many units of the less expensive foodstuff must we mix to satisfy the protein requirement?

Solution. Let x denote the number of units of the less expensive foodstuff in the mixture. We have $3x$ grams of protein from the less expensive foodstuff and $2(4) = 8$ grams of protein from the more expensive foodstuff. So we have $3x + 8$ grams of protein in the mixture, and the protein requirement is expressed by the inequality

$$3x + 8 \geq 20$$
$$3x \geq 12$$
$$x \geq 4.$$

We must mix at least 4 units of the less expensive foodstuff.

#27 $4(.05)+x(.04)\leq .60$
$.2+.04x\leq .60$
$.04x\leq .4$
$x\leq 10$ - P2 02

Exercise Set 4.6 (continued)

21. We prepare an animal feed by mixing a variable amount of a less expensive foodstuff with 3 units of a more expensive foodstuff. Each unit of the less expensive foodstuff contains 4 grams of fats, and each unit of the more expensive foodstuff contains 5 grams of fats. How many units of the less expensive foodstuff must we mix to satisfy the requirement that the mixture contain at least 43 grams of fats?

22. We prepare an animal feed by mixing a variable amount of a less expensive foodstuff with 4 units of a more expensive foodstuff. Each unit of the less expensive foodstuff contains 3 grams of carbohydrates, and each unit of the more expensive foodstuff contains 5 grams of carbohydrates. How many units of the less expensive foodstuff must we mix to satisfy the requirement that the mixture contain at least 36 grams of carbohydrates?

23. A food company comes out with a drink that consists of a mixture of orange juice and pineapple juice. Each can contains 5 fluid ounces of pineapple juice and must contain at least 30 milligrams of vitamin C. How many fluid ounces of orange juice should be in the mixture to satisfy this requirement if each fluid ounce of orange juice contains 2 mg of vitamin C and each fluid ounce of pineapple juice contains 3 mg of vitamin C?

24. A food company comes out with a drink that consists of a mixture of orange juice and pineapple juice. Each can contains 6 fluid ounces of orange juice and must contain at least 21 milligrams of vitamin D. How many fluid ounces of pineapple juice should be in the mixture to satisfy this requirement if each fluid ounce of orange juice contains 1 mg of vitamin D and each fluid ounce of pineapple juice contains 3 mg of vitamin D?

25. We prepare an animal feed by mixing two basic foodstuffs, one costing \$4 per unit with 8 units of another costing \$5 per unit. How much of the less expensive foodstuff must we mix if the cost of the mixture is at most \$60?

26. We prepare an animal feed by mixing two basic foodstuffs, 5 units of one costing \$4 per unit with another costing \$5 per unit. How much of the more expensive foodstuff must we mix if the cost of the mixture is at most \$40?

27. A food company comes out with a drink that consists of a mixture of orange juice and pineapple juice. Each can contains 4 fluid ounces of pineapple juice and must cost at most \$0.60. How many fluid ounces of orange juice must be in a can if each fluid ounce of orange juice costs \$0.04 and each fluid ounce of pineapple juice costs \$0.05?

28. A food company comes out with a drink that consists of a mixture of orange juice and pineapple juice. Each can contains 4 fluid ounces of orange juice and must cost at most \$0.54. How many fluid ounces of pineapple juice must be in a can if each fluid ounce of orange juice costs \$0.03 and each fluid ounce of pineapple juice costs \$0.03?

BUSINESS AND MANAGEMENT

Example 5. A businessman has a report to be typed and two secretaries. He knows that one secretary alone could type it in 4 hours and the other one could type it in 3 hours. How long will it take them to do it together?

Solution. Let t be the time it takes them when they work together. Now the first secretary can type 1/4 of the report in one hour and the second secretary can type 1/3 of the report in one hour. Together they can type (1/4) + (1/3) of the report, which is $1/t$ of the report, in one hour. So

$$\frac{1}{4} + \frac{1}{3} = \frac{1}{t}$$

$$\frac{7}{12} = \frac{1}{t}$$

$$t = 12/7.$$

It takes 12/7 hours.

Exercise Set 4.6 (continued)

29. A businesswoman has a report to be typed and two secretaries. She knows that one secretary alone could type it in 5 hours and the other one alone could type it in 6 hours. How long will it take them to do it together?

30. The payroll checks for a large company are printed out by a computer in 20 hours. A computer salesman insists that the checks could be printed out by his faster computer in 16 hours. How long would it take both computers together?

31. It takes 2 hours for an electronic sorter to sort a certain quantity of cards. When a second sorter is used, the two together take 45 minutes to sort the same quantity of cards. How long would it take the second sorter alone to do the job?

32. A dealer determines that her price for a hi-fi is a 10% discount from the catalogue price for the hi-fi. She wants to give a sale so that her sale price is a 15% discount from the catalogue price. What percent discount of the dealer's usual price is her sale price? *Hint:* It's not 5%.

33. A man wishes to divide his fortune among his three sons and his wife. He gives his first son half his fortune plus $1000, his second son half of what remains plus $1000, and his third son half of what then remains plus $1000. He finds he then has $1000 left for his wife. How much was his fortune?

34. A total of $6000 is invested, part at 4% compounded annually and part at 5% compounded annually. If the annual interest is $275, how much is invested at each rate?

TOLERANCE

Example 6. An engineer must construct a square that has perimeter 36.56 ± 0.02 centimeters. That is, the perimeter P must be between 36.56 + 0.02 centimeters and 36.56 − 0.02 centimeters; we say that there is a *tolerance* of 0.02 centimeters.

a. Express this tolerance as an inequality in P using an absolute value.

b. What tolerance on the side x of the square is required to obtain the tolerance of 0.02 centimeters on the perimeter P?

Solution. a. The difference between P and 36.56 centimeters must be at most $\pm$ 0.02 centimeters. This condition can be expressed by the inequalities

$$-0.02 < P - 36.56 < 0.02.$$

Using an absolute value, this condition is expressed as

$$|P - 36.56| < 0.02.$$

b. Since $P = 4x$, this inequality is expressed by

$$|4x - 36.56| < 0.02.$$

We solve this inequality for x:

$$-0.02 < 4x - 36.56 < 0.02$$

$$36.54 < 4x < 36.58$$

$$9.135 < x < 9.145.$$

The side x must be between 9.135 centimeters and 9.145 centimeters. That is, we want the side x to be $36.56/4 = 9.14$ centimeters, and we must have

$$|x - 9.14| < 0.005.$$

Thus, the tolerance on x is 0.005 centimeters.

Exercise Set 4.6 (continued)

35. An engineer must construct a square that has a perimeter P of 8.24 ± 0.04 centimeters.
    a. Express this tolerance as an inequality in P using an absolute value.
    b. What tolerance on the side x of the square is required to obtain the tolerance of 0.04 centimeters on P?
36. An engineer must construct a square that has a perimeter P of 5.48 ± 0.01 centimeters.
    a. Express this tolerance as an inequality in P using an absolute value.
    b. What tolerance on the side x of the square is required to obtain the tolerance of 0.01 centimeters on P?
37. An architect must construct a square that has a perimeter P of 14.28 ± 0.06 meters.
    a. Express this tolerance as an inequality in P using an absolute value.
    b. What tolerance on the side x of the square is required to obtain the tolerance of 0.06 meters on P?
38. An architect must construct a square that has a perimeter P of 9.32 ± 0.03 meters.
    a. Express this tolerance as an inequality in P using an absolute value.
    b. What tolerance on the side x of the square is required to obtain the tolerance of 0.03 meters on P?

39. An engineer must construct a circle that has a circumference C of 28.15 ± 0.02 centimeters.
    a. Express this tolerance as an inequality in C using an absolute value.
    b. What tolerance on the radius r of the circle is required to obtain the tolerance of 0.02 centimeters on C? Use 3.14 as an approximation to π.

40. An engineer must construct a circle that has a circumference C of 18.17 ± 0.03 centimeters.
    a. Express this tolerance as an inequality in C using an absolute value.
    b. What tolerance on the radius r of the circle is required to obtain the tolerance of 0.03 centimeters on C? Use 3.14 as an approximation to π.

HISTORICAL EXERCISES

Throughout the history of algebra, word problems have played a central role. Many of these problems have an obvious application to commerce, but many others are intended as puzzles or "mind teasers." We give a sample of some typical problems. These problems are adopted from actual ancient books on algebra.

Exercise Set 4.6 (continued)

The oldest known algebra book dates from about 1650 B.C. and comes from ancient Egypt. Known now as the Rhind Papyrus, it's a copy of a mathematical textbook and handbook originally copied by the Scribe A'hmose. His copy contains 85 problems and solutions, most of the problems involving the solution of a linear equation. Exercises 41 and 42 are from the Rhind Papyrus.

41. (Problem 28) A quantity and its 2/3 are added together and from the sum 1/3 of the sum is subtracted and 10 remains. What is this quantity?

42. (Problem 65) Example of dividing 100 loaves among 10 men, including a boatman, a foreman, and a doorkeeper, who receive double portions. What is the share of each?

Exercises 43–46 are from the *Greek Anthology*, a collection of problems assembled about A.D. 500 by Metrodorus for Greek students.

43. How many apples are needed if four persons of six receive one third, one eighth, one fourth, and one fifth, respectively, of the total number, while the fifth receives ten apples, and one apple remains left for the sixth person?

44. Brickmaker, I am in a hurry to build this house. Today is cloudless, and I do not require many more bricks, for I have all I want but three hundred. Thou alone in one day couldst make as many, but thy son in one day could make two hundred, and thy son-in-law could make two hundred and fifty. Working all together, in how many days can you make these?

45. I am a brazen lion; my spouts are my two eyes, my mouth, and the flat of my right foot. My right eye fills a jar in two days, my left eye in three, and my foot in four. My mouth is capable of filling it in six hours. Tell me how long all four together will take to fill it.

The following problem from the *Greek Anthology* tells about all that is known about the personal life of the Greek algebraist Diophantus. How old did he live to be?

46. God granted him to be a boy for the sixth part of his life, and adding a twelfth part to this, He clothed his cheeks with down; He lit the lock of wedlock after a seventh part, and five years after his marriage He granted him a son. Alas! late-born wretched

child; after attaining the measure of half his father's life, chill Fate took him. After consoling his grief by the science of numbers for four years, he ended his life.

Hindu mathematics has a long history, and we give some examples of problems occurring there. Exercise 47 is from the *Triśatkâ* by Śrîdhara (A.D. 750); Exercise 48 is from the *Brahmasphuta-siddhânta* by Brahmagupta (628); and Exercise 49 is from the *Lîlâvatî* by Bhâskara (1150).

47. The third part of a necklace of pearls, broken in an amorous struggle, fell to the ground; its fifth part rested on the couch; the sixth part was saved by the wench; and the tenth part was taken by her lover: six pearls remained strung. Say, of how many pearls was the necklace composed?

48. In what time will four fountains, being let loose together, fill a cistern, which they would severally fill in a day, in half a day, in a quarter, and in a fifth part of a day?

49. A snake's hole is at the foot of a pillar which is 15 cubits high, and a peacock is perched on the top of the pillar. Seeing a snake gliding towards his hole at a distance of thrice the pillar's height, the peacock pounces obliquely on him. Say quickly at how many cubits from the snake's hole do they meet, both proceeding an equal distance?

Exercise 50 is due to Tartaglia (1499–1557), who wrote a two-volume book on arithmetic and algebra.

50. If 100 lire of Modon money amounts to 115 lire in Venice, and if 180 lire in Venice comes to 150 in Corfu, and if 240 lire Corfu money is worth as much as 360 lire in Negroponte, what is the value in Modon coinage of 666 lire Negroponte money?

Exercises 51 and 52 are adopted from Chuquet's *Triparty en la science des nombres* (1484).

51. A merchant visited three fairs. At the first, he doubled his money and spent \$30, at the second he tripled his money and spent \$54, at the third he quadrupled his money and spent \$72, and then he had \$48 left. How much money had he at the start?

52. A carpenter agrees to work under the conditions that he is to be paid \$5.50 every day he works, while he must pay \$6.60 every day he does not work. At the end of 30 days, he finds he has paid out as much as he has received. How many days did he work?

The following exercise is adopted from Clavius' arithmetic of 1583.

53. A merchant bought 50,000 pounds of pepper in Portugal for 10,000 scudi, paying a tax of 500 scudi. He carried it to Italy at a cost of 300 scudi and there paid another duty of 200 scudi. The transportation from the coast to Florence cost 100 scudi, and he was obliged to pay an impost of 100 scudi to that city. Lastly, the government demanded a tax from each merchant of 1000 scudi. Now he is perplexed to know what price to charge per pound so that, after all these expenses, he may make a profit of one-tenth of a scudi a pound.

Exercises 54–56 are from the *Liber Abaci* of Leonard Fibonacci (1202).

54. A pit is 50 handbreadths deep. A lion at the bottom climbs 1/7 handbreadth each day and falls back each night by 1/9 handbreadth. How long does it take the lion to get out of the pit? (Be careful with this one—the answer Fibonacci gave in his book is incorrect!)

55. A certain man doing business in Lucca doubled his money there, but he spent 12 denarii. Then he went to Florence, where he also doubled his money and spent 12 denarii. Returning to Pisa, he again doubled his money and spent 12 denarii; and he found he had no money remaining. How much did he have at the beginning?

56. A man entered an orchard through seven gates, and there took a certain number of apples. When he left the orchard, he gave the first guard half the apples that he had and one apple more. To the second guard he gave half the remaining apples and one apple more. He did the same to each of the remaining five guards, and left the orchard with one apple. How many apples did he gather in the orchard?

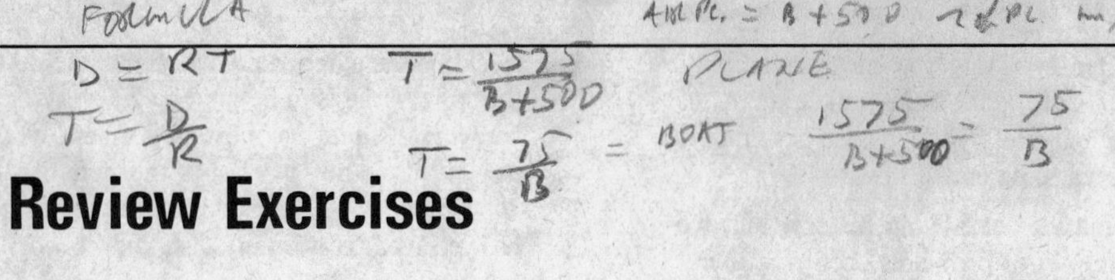

Review Exercises

In exercises 1–32, solve the given equation or inequality.

1. $x + 2 = -6$
2. $x - 3 = 1$
3. $3x \geq 5$
4. $4x \leq 14$
5. $4x + 5 = 7$
6. $6x - 2 = 8$
7. $5x + 3 < 18$
8. $4x - 3 < 15$
9. $2x + 2(x + 4) = 20$
10. $2x + 2(4x) = 28$
11. $6x - 4 > 3x + 8$
12. $2x + 1 > 5x - 8$
13. $\dfrac{2x + 3}{4} \leq 6$
14. $\dfrac{3x - 4}{5} \leq 1$
15. $(x + 2)^2 = (x - 3)^2$
16. $(x - 4)^2 = (x + 3)^2$
17. $(x + 1)^2 + 16 = (x + 3)^2$
18. $(x - 1)^2 + 16 = (x - 3)^2$
19. $(x + 2)(x - 3) = (x + 1)(x - 2)$
20. $(x + 1)(x - 2) = (x - 1)(x + 2)$
21. $\dfrac{x - 4}{x + 2} = \dfrac{x - 3}{x + 1}$
22. $\dfrac{x + 5}{x - 6} = \dfrac{x + 4}{x - 2}$
23. $\dfrac{1}{x - 1} + \dfrac{2}{x - 2} = \dfrac{2}{x^2 - 3x + 2}$
24. $\dfrac{1}{x + 1} + \dfrac{2}{x + 2} = \dfrac{3}{x^2 + 3x + 2}$
25. $-5 \leq 4x + 3 \leq 5$
26. $-3 \leq 2x - 3 \leq 7$
27. $-1 \leq \dfrac{2x - 1}{3} \leq 1$
28. $-2 \leq \dfrac{3x + 1}{4} \leq 2$
29. $0.3x + 0.4(5) \leq 2.6$
30. $0.5x + 0.3(6) \leq 3.8$
31. $0.4x + 0.7(10) = 0.25(x + 10)$
32. $0.6x + 0.3(12) = 0.35(x + 12)$

In exercises 33–40, solve the equation for the specified variable.

33. $F = ma$ for m
34. $\dfrac{P_1V_1}{T_1} = \dfrac{P_2V_2}{T_2}$ for T_1
35. $S = \pi r^2 + \pi rs$ for s
36. $P = A(1 - dt)$ for t
37. $4x + 3y + 24 = 0$ for y
38. $2x - 6y + 12 = 0$ for y
39. $2x + 2yy' = xy' + y$ for y'
40. $1 + y' = xy' + y$ for y'

In exercises 41–48, find the given absolute value.

41. $|-4|$
42. $|-2.2|$
43. $|3|$
44. $|3.6|$
45. $|5 - 8|$
46. $|3 - 9|$
47. $|4 - 2|$
48. $|5 - 2|$

In exercises 49–54, express the given interval as an absolute value (not including the end points).

49. From 2 to 8
50. From 4 to 16
51. From -3 to 5
52. From -1 to 6
53. From -4 to -1
54. From -6 to -2

In exercises 55–60, solve the given inequality.

55. $|x - 1| < 6$
56. $|x - 3| < 2$
57. $|3x - 5| \geq 2$
58. $|2x - 3| \geq 4$
59. $\left|\frac{x + 3}{3}\right| \leq 4$
60. $\left|\frac{x + 1}{5}\right| \leq 2$

61. One side of a rectangle is 6 meters longer than the other side. What is the length of each side if the perimeter is 28 meters?
62. An airplane travels 1575 meters in the same time that a boat goes 75 meters. If the airplane goes 500 meters per minute faster than the boat, find the rate of each.
63. We have a 70% acid solution which we dilute by adding a 30% acid solution to obtain a 45% solution. If we start with 20 liters of the 70% solution, how many liters of the 30% solution should we add?
64. We prepare an animal feed by mixing a less expensive foodstuff with a more expensive foodstuff. The amount of the less expensive foodstuff we mix will vary, but we mix the constant amount of 5 units of the more expensive foodstuff. The resulting mixture must contain at least 45 grams of protein. If each unit of the less expensive foodstuff contains 4 grams of protein and each unit of the more expensive foodstuff contains 5 grams of protein, how many units of the more expensive foodstuff must we mix to satisfy the protein requirement?
65. A man decides for breakfast to mix 3 ounces of a high-protein cereal with a variable amount of a more expensive but tastier cereal. If the high-protein cereal costs 4 cents an ounce and the tasty cereal costs 5 cents an ounce, how many ounces of the tasty cereal should he mix to keep the cost at most 52 cents per breakfast?
66. It takes $1\frac{1}{2}$ hours for an electronic sorter to sort a certain quantity of cards. When a second sorter is used, the two together take 45 minutes to sort the same quantity of cards. How long does it take the second sorter alone to do the job?
67. An engineer must construct a square that has a perimeter P of 64.16 ± 0.04 centimeters.
    a. Express this tolerance as an inequality in P using an absolute value.
    b. What tolerance on the side x of the square is required to obtain the tolerance of 0.04 centimeters on P?

Exercise 68 is from the Rhind Papyrus (1650 B.C.).

68. A quantity and its 1/5 added together become 21. What is the quantity?

chapter 5

Quadratic Equations and Inequalities

1. Solution by Factoring

2. Completing the Square

3. The Quadratic Formula

4. Further Quadratic Equations

5. Quadratic Inequalities

6. Equations Quadratic in Form

7. Squaring an Equation

8. Applications

In this chapter we discuss quadratic equations and inequalities in one variable and find techniques for solving them. In Chapter 4 we studied the technique of performing the operations of addition, subtraction, multiplication, and division to solve equations and inequalities. Although this technique can be used to solve linear equations and inequalities, it is not adequate to solve quadratic equations and inequalities. So we consider the further techniques of factoring, completing the square, and the quadratic formula.

We use factoring to solve quadratic equations in Section 5.1, and in Section 5.5 we use factoring to solve quadratic inequalities. We use completing the square to solve quadratic equations in Section 5.2. In Sections 5.3 and 5.4 we use the quadratic formula to solve quadratic equations.

Further topics in quadratic equations are equations quadratic in form (Section 5.6) and squaring an equation (Section 5.7). In Section 5.8 we apply quadratic equations and inequalities to geometry problems, rate problems, business and management considerations, projectiles, economics, and historical problems.

section 1 • Solution by Factoring

In this section we see how factoring can be used to solve quadratic equations. This technique depends on the Zero Factor Property and factoring a quadratic expression. At this point you may wish to review Section 2.4.

ZERO FACTOR PROPERTY

Recall from Section 4.2 that a linear or first-degree equation has a first-degree term and a constant term. A quadratic or second-degree equation has a second-degree term, a first-degree term, and a constant term.

DEFINITION OF A QUADRATIC OR SECOND-DEGREE EQUATION IN ONE VARIABLE. A *quadratic equation* or *second-degree equation* in one variable x is an equation equivalent to one of the form

$$ax^2 + bx + c = 0,$$

where a, b, and c are constants and $a \neq 0$.

Such an equation is called *quadratic* or *second-degree* because it involves a squared term but no higher degree terms. The term ax^2 is called the *quadratic term* and cannot be missing from the equation. The term bx is called the *linear term* and may or may not be missing from the equation, depending on whether or not $b = 0$. The term c is called the *constant term* and may or may not be missing from the equation, depending on whether or not $c = 0$. An equation is quadratic if it is *equivalent* to one of the appropriate form, but we may have to perform operations on the equation to put it in this form. Also, the variable need not be called x but may be denoted by some other letter.

ZERO FACTOR PROPERTY. If $ab = 0$, then $a = 0$ or $b = 0$; that is, if a product of two numbers is 0, then at least one of the numbers must be 0.

To prove the Zero Factor Property, suppose $ab = 0$. If $a \neq 0$, we can divide both sides of the equation by a to obtain

$$ab/a = 0/a \quad \text{or} \quad b = 0.$$

We can use the Zero Factor Property to solve a quadratic equation when it is written as a product of two factors equal to 0.

Example 1. Solve $(x + 3)(2x - 1) = 0$.

Solution. By the Zero Factor Property,

$$x + 3 = 0 \quad \text{or} \quad 2x - 1 = 0$$
$$x = -3 \quad \text{or} \quad x = 1/2.$$

We check both -3 and $1/2$ in the original equation:

$$[(-3) + 3][2(-3) - 1] = 0[-7] = 0$$
$$[(1/2) + 3][2(1/2) - 1] = [7/2]0 = 0$$

Therefore, the solutions are -3 and $1/2$.

Exercise Set 5.1

In exercises 1–8, use the Zero Factor Property to solve each given equation.

1. $(x - 3)(x - 8) = 0$
2. $(x + 4)(x - 2) = 0$
3. $(x + 5)(x + 6) = 0$
4. $(y + 1)(y + 7) = 0$
5. $(2x - 1)(3x - 2) = 0$
6. $(4x + 3)(x - 6) = 0$
7. $(5x + 4)(2x + 3) = 0$
8. $(x - 5)(3x + 1) = 0$

SOLUTION BY FACTORING

If we can factor a quadratic equation so that it possesses the form of a product of two factors equal to 0, we can then use the Zero Factor Property to solve the equation.

Example 2. Solve $x^2 - 4x - 21 = 0$.

Solution. We factor this equation:

$$(x + 3)(x - 7) = 0.$$

By the Zero Factor Property,

$$x + 3 = 0 \quad \text{or} \quad x - 7 = 0$$
$$x = -3 \quad \text{or} \quad x = 7.$$

Both -3 and 7 satisfy the equation, since

$$(-3)^2 - 4(-3) - 21 = 0 \quad \text{and} \quad (7)^2 - 4(7) - 21 = 0.$$

So the solutions are -3 and 7.

Example 3. Solve $6x^2 - 11x - 10 = 0$.

Solution. We factor this equation:

$$(3x + 2)(2x - 5) = 0.$$

By the Zero Factor Property,

$$3x + 2 = 0 \quad \text{or} \quad 2x - 5 = 0$$
$$x = -2/3 \quad \text{or} \quad x = 5/2.$$

We check both $-2/3$ and $5/2$ in the original equation:

$$6(-2/3)^2 - 11(-2/3) - 10 = (8/3) + (22/3) - 10 = 0$$
$$6(5/2)^2 - 11(5/2) - 10 = (75/2) - (55/2) - 10 = 0.$$

Therefore, the solutions are $-2/3$ and $5/2$.

Exercise Set 5.1 (continued)

In exercises 9–20, solve each given equation by factoring.

9. $x^2 + 3x - 4 = 0$
10. $x^2 + 5x + 6 = 0$
11. $u^2 - 7u + 12 = 0$
12. $y^2 + 2y - 24 = 0$
13. $x^2 + 13x - 30 = 0$
14. $y^2 + 2y - 35 = 0$
15. $2x^2 - 5x - 3 = 0$
16. $4x^2 + 8x + 3 = 0$
17. $3x^2 + 14x + 8 = 0$
18. $4x^2 + 21x - 18 = 0$
19. $25x^2 - 20x + 4 = 0$
20. $12y^2 - 16y - 3 = 0$

SIMPLIFYING EQUATIONS

Frequently, to solve a quadratic equation we must first put the equation in the form of a quadratic equation. That is, we must write the equation in the form of decreasing powers of the variable on one side of the equation and 0 on the other side of the equation.

Example 4. Solve $x(x + 5) = 84$.

Solution. First we write the equation in the form of a quadratic equation.

$$x(x + 5) = 84$$
$$x^2 + 5x = 84$$
$$x^2 + 5x - 84 = 0$$

Now the equation possesses the form of a quadratic equation, and we can factor it:

$$(x - 7)(x + 12) = 0$$
$$x - 7 = 0 \quad \text{or} \quad x + 12 = 0$$
$$x = 7 \quad \text{or} \quad x = -12.$$

We check 7 and -12 in the original equation:

$$7(7 + 5) = 84 \quad \text{and} \quad -12(-12 + 5) = 84.$$

Therefore, the solutions are 7 and -12.

Example 5. Solve $(x - 4)(2x - 4) = 48$.

Solution. First we write the equation in the form of a quadratic equation.

$$(x - 4)(2x - 4) = 48$$
$$2x^2 - 12x + 16 = 48$$
$$2x^2 - 12x - 32 = 0$$

Notice that every term of this quadratic equation is divisible by 2. We can simplify the work by dividing both sides of the equation by 2:

$$x^2 - 6x - 16 = 0.$$

Now we factor the equation.

$$(x - 8)(x + 2) = 0$$
$$x - 8 = 0 \quad \text{or} \quad x + 2 = 0$$
$$x = 8 \quad \text{or} \quad x = -2$$

We check 8 and -2 in the original equation:

$$[8 - 4][2(8) - 4] = (4)(12) = 48$$
$$[(-2) - 4][2(-2) - 4] = (-6)(-8) = 48.$$

Therefore, the solutions are 8 and -2.

Exercise Set 5.1 (continued)

In exercises 21–32, solve each given equation.

21. $x(x + 8) = 48$
22. $x(x + 5) = 104$
23. $x(x - 6) = 27$
24. $x(x - 3) = 18$
25. $x^2 + (x + 2)^2 = 100$
26. $x^2 + (x - 1)^2 = 25$
27. $(x - 4)(2x - 4) = 70$
28. $(x - 4)(2x - 4) = 16$
29. $(x - 1)(2x + 1) = 20$
30. $(x - 1)(2x + 1) = 90$
31. $(x - 2)(2x + 3) = 39$
32. $(x - 2)(2x + 3) = 72$

FRACTIONAL EQUATIONS

Example 6. Solve $\dfrac{x}{2x + 1} + \dfrac{1}{x + 2} = 2$.

Solution. This equation is not in the form of a quadratic equation. To clear it of fractions, we multiply both sides of the equation by $(2x + 1)(x + 2)$.

$$x(x + 2) + 1 \cdot (2x + 1) = 2(2x + 1)(x + 2)$$
$$x^2 + 2x + 2x + 1 = 4x^2 + 10x + 4$$
$$0 = 3x^2 + 6x + 3$$

Since each term of this equation is divisible by 3, we divide both sides of the equation by 3.

$$0 = x^2 + 2x + 1$$
$$0 = (x + 1)^2$$
$$0 = x + 1$$
$$x = -1$$

We check -1 in the original equation:

$$\frac{-1}{2(-1) + 1} + \frac{1}{-1 + 2} = 2.$$

So the solution is -1.

Exercise Set 5.1 (continued)

In exercises 33–40, solve each given equation.

33. $\frac{x - 1}{x - 2} = \frac{6}{x}$
34. $\frac{x - 3}{x + 3} = \frac{2}{x}$
35. $\frac{3x - 1}{x + 3} = \frac{x - 1}{x + 2}$
36. $\frac{x + 2}{x - 2} = \frac{2x + 1}{x - 1}$
37. $\frac{1}{t + 1} + \frac{2}{t + 2} = 2$
38. $\frac{2}{x + 3} - \frac{5}{x - 3} = 3$
39. $\frac{x}{x + 1} - \frac{3}{x - 3} = 2$
40. $\frac{s}{s + 4} - \frac{5}{s + 2} = 2$

section 2 • Completing the Square

In Section 5.1 we saw how to solve some quadratic equations by factoring. However, the factoring techniques we have studied do not help with quadratic equations that don't factor; for example,

$$x^2 + 1 = 0, \quad x^2 + x + 1 = 0, \quad x^2 + 2x + 2 = 0, \quad x^2 - 5 = 0.$$

In this section we study a method (called completing the square) which can be applied to solve any quadratic equation.

EQUATIONS OF THE FORM $x^2 = c$

An equation of the form $x^2 = c$ is quadratic because we can write it in the form

$$x^2 - c = 0.$$

Written in this form (with the linear term missing) we can factor it as a difference of squares:

$$(x - \sqrt{c})(x + \sqrt{c}) = 0.$$

By the Zero Factor Property,

$$x - \sqrt{c} = 0 \quad \text{or} \quad x + \sqrt{c} = 0$$

$$x = \sqrt{c} \quad \text{or} \quad x = -\sqrt{c}.$$

Both $\sqrt{c}$ and $-\sqrt{c}$ satisfy this equation because

$$(\sqrt{c})^2 = c \quad \text{and} \quad (-\sqrt{c})^2 = c.$$

We summarize what we have shown.

SOLUTION OF $x^2 = c$. The solutions of the quadratic equation $x^2 = c$ are $\sqrt{c}$ and $-\sqrt{c}$.

We frequently use a special notation for this situation. To write both $\sqrt{c}$ and $-\sqrt{c}$ together, we write $\pm\sqrt{c}$. This symbol means that we have two numbers, both the positive and negative square roots of c.

Example 1. Solve $x^2 = 5$.

Solution. The solutions are $\sqrt{5}$ and $-\sqrt{5}$, which we can write as $\pm\sqrt{5}$. Using the table of square roots in Appendix C, we find that approximate solutions to the equation are ± 2.236.

Example 2. Solve $x^2 + 1 = 0$.

Solution. We can write

$$x^2 = -1$$

$$x = \pm\sqrt{-1} = \pm i$$

The solutions are i and $-i$.

Notice in Examples 1 and 2 that neither equation factors, but each equation has two numbers for solutions. The solutions may be real numbers or complex numbers.

Exercise Set 5.2

In exercises 1–8, solve each given equation.

1. $x^2 - 2 = 0$
2. $x^2 - 3 = 0$
3. $x^2 - 12 = 0$
4. $x^2 - 8 = 0$
5. $x^2 + 4 = 0$
6. $x^2 + 25 = 0$
7. $x^2 + 8 = 0$
8. $x^2 + 12 = 0$

COMPLETING THE SQUARE

To prepare for solving quadratic equations, we show how to complete the square on the expression $x^2 + kx$. That is, if we have an expression of the form $x^2 + kx$, we want to add a constant so that the sum factors as the square of a binomial. We add a constant so that

$$x^2 + kx + \text{constant} = (x + p)^2 = x^2 + 2px + p^2.$$

We see that we must have $2p = k$, and so $p = k/2$. Therefore, the constant we add is

$$p^2 = k^2/4.$$

COMPLETING THE SQUARE ON $x^2 + kx$. To complete the square on $x^2 + kx$,

$$\text{add} \quad \left(\frac{k}{2}\right)^2 = \frac{k^2}{4}$$

$$\text{to obtain} \quad x^2 + kx + \frac{k^2}{4} = \left(x + \frac{k}{2}\right)^2.$$

Completing the square has a geometric interpretation that explains its name. This interpretation is given in Figure 5.1, where we interpret multiplication as an area. The area of the large square is $(x + (k/2))^2$. This area consists of four smaller areas, and the one that completes the square is $k^2/4$.

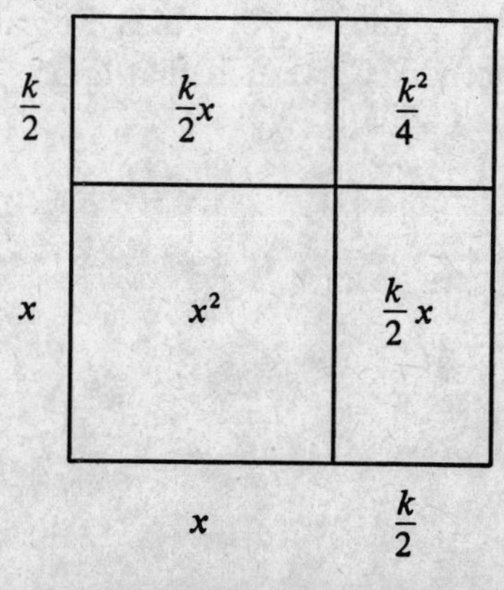

Figure 5.1

Example 3. Complete the square on:

a. $x^2 + 4x$; b. $x^2 + x$; c. $x^2 + \frac{2}{3}x$; d. $x^2 + \frac{3}{2}x$.

Solution.

a. Since $k = 4$, we add $k^2/4 = 4$ to obtain

$$x^2 + 4x + 4 = (x + 2)^2.$$

b. Since $k = 1$, we add $k^2/4 = 1/4$ to obtain

$$x^2 + x + \left(\frac{1}{2}\right)^2 = x^2 + x + \frac{1}{4} = \left(x + \frac{1}{2}\right)^2.$$

c. Since $k = 2/3$, we add $k^2/4 = 1/9$ to obtain

$$x^2 + \frac{2}{3}x + \left(\frac{1}{3}\right)^2 = x^2 + \frac{2}{3}x + \frac{1}{9} = \left(x + \frac{1}{3}\right)^2.$$

d. Since $k = 3/2$, we add $k^2/4 = 9/16$ to obtain

$$x^2 + \frac{3}{2}x + \left(\frac{3}{4}\right)^2 = x^2 + \frac{3}{2}x + \frac{9}{16} = \left(x + \frac{3}{4}\right)^2.$$

Exercise Set 5.2 (continued)

In exercises 9–22, complete the square on the given expression and factor the result as a binomial square.

9. $x^2 + 2x$
10. $x^2 + 4x$
11. $x^2 - 2x$
12. $x^2 - 4x$
13. $y^2 - 8y$
14. $x^2 - 10x$
15. $x^2 + 6x$
16. $z^2 + 6z$
17. $x^2 - x$
18. $x^2 - 3x$
19. $x^2 - \frac{5}{3}x$
20. $x^2 - \frac{5}{2}x$
21. $x^2 + \frac{4}{3}x$
22. $x^2 - \frac{1}{2}x$

SOLVING QUADRATIC EQUATIONS

Now we put completing the square together with factoring a difference of squares to solve quadratic equations.

Example 4. Solve $x^2 + 4x - 21 = 0$.

Solution. First we write

$$x^2 + 4x = 21.$$

To complete the square on $x^2 + 4x$ we add 4 to both sides of the equation.

$$x^2 + 4x + 4 = 21 + 4$$
$$(x + 2)^2 = 25$$

Solving an equation of this form, we find

$$x + 2 = \pm 5$$

$$x + 2 = 5 \quad \text{or} \quad x + 2 = -5$$

$$x = 3 \quad \text{or} \quad x = -7$$

The solutions are 3 and -7. (Note that this equation could have been solved by factoring.)

Example 5. Solve $x^2 + x + 1 = 0$.

Solution. This equation cannot be solved by factoring. First we write

$$x^2 + x = -1$$

To complete the square on $x^2 + x$, we add 1/4 to both sides of the equation.

$$x^2 + x + \frac{1}{4} = -1 + \frac{1}{4}$$

$$\left(x + \frac{1}{2}\right)^2 = -\frac{3}{4}$$

Solving an equation of this form, we find

$$x + \frac{1}{2} = \pm\sqrt{-\frac{3}{4}} = \pm\frac{\sqrt{3}}{2}i$$

$$x = -\frac{1}{2} \pm \frac{\sqrt{3}}{2}i$$

The solutions are $(-1 + \sqrt{3}i)/2$ and $(-1 - \sqrt{3}i)/2$.

Exercise Set 5.2 (continued)

In exercises 23–38, use the method of completing the square to solve the given quadratic equation.

23. $x^2 + 2x - 2 = 0$
24. $x^2 + 4x - 3 = 0$
25. $x^2 - 2x - 24 = 0$
26. $u^2 + 8u + 12 = 0$
27. $x^2 + 2x + 2 = 0$
28. $x^2 - 4x + 8 = 0$
29. $y^2 - 8y + 15 = 0$
30. $x^2 - 10x + 21 = 0$
31. $x^2 + 6x + 3 = 0$
32. $z^2 + 6z - 2 = 0$
33. $x^2 - 8x + 14 = 0$
34. $x^2 - 6x + 33 = 0$
35. $t^2 + t - 1 = 0$
36. $u^2 + 3u + 1 = 0$
37. $x^2 - x + 1 = 0$
38. $x^2 - 3x + 3 = 0$

FURTHER QUADRATIC EQUATIONS

So far in this section we have considered only quadratic equations in which the coefficient of the x^2 term is 1. If this coefficient is not 1, we divide both sides of the equation by the coefficient to make it 1. Examples 6 and 7 illustrate this.

Example 6. Solve $3x^2 + 2x - 2 = 0$.

Solution. Divide both sides of the equation by 3 to obtain

$$x^2 + \frac{2}{3}x - \frac{2}{3} = 0$$

$$x^2 + \frac{2}{3}x = \frac{2}{3}.$$

We complete the square by adding 1/9 to both sides of the equation.

$$x^2 + \frac{2}{3}x + \frac{1}{9} = \frac{2}{3} + \frac{1}{9}$$

$$\left(x + \frac{1}{3}\right)^2 = \frac{7}{9}$$

$$x + \frac{1}{3} = \pm\sqrt{\frac{7}{9}} = \pm\frac{\sqrt{7}}{3}$$

$$x = -\frac{1}{3} \pm \frac{\sqrt{7}}{3}$$

The solutions are $(-1 + \sqrt{7})/3$ and $(-1 - \sqrt{7})/3$.

Example 7. Solve $2x^2 + 3x + 3 = 0$.

Solution. Divide both sides of the equation by 2 to obtain

$$x^2 + \frac{3}{2}x + \frac{3}{2} = 0$$

$$x^2 + \frac{3}{2}x = -\frac{3}{2}.$$

To complete the square we add 9/16 to both sides of the equation.

$$x^2 + \frac{3}{2}x + \frac{9}{16} = -\frac{3}{2} + \frac{9}{16}$$

$$\left(x + \frac{3}{4}\right)^2 = -\frac{15}{16}$$

$$x + \frac{3}{4} = \pm\sqrt{-\frac{15}{16}} = \pm\frac{\sqrt{15}}{4}i$$

$$x = -\frac{3}{4} \pm \frac{\sqrt{15}}{4}i$$

The solutions are $(-3 + \sqrt{15}i)/4$ and $(-3 - \sqrt{15}i)/4$.

Exercise Set 5.2 (continued)

In exercises 39–48, use the method of completing the square to solve each given quadratic equation.

39. $4x^2 + 8x + 5 = 0$
40. $2s^2 + s + 4 = 0$
41. $3x^2 - 5x - 2 = 0$
42. $2x^2 - 5x + 3 = 0$
43. $4u^2 - 4u - 1 = 0$
44. $4x^2 - 6x - 1 = 0$
45. $4x^2 + 12x + 9 = 0$
46. $9y^2 - 6y + 1 = 0$
47. $3z^2 - 2z - 3 = 0$
48. $3x^2 + 4x - 1 = 0$

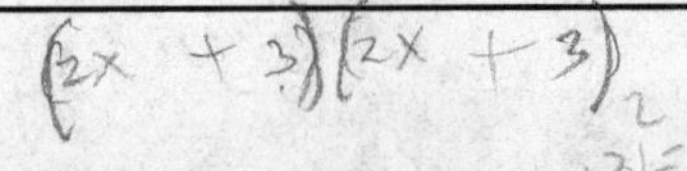

section 3 • The Quadratic Formula

In this section we derive the quadratic formula by applying the method of completing the square to a general quadratic equation. The quadratic formula can be used to solve any quadratic equation. We begin by defining the discriminant of a quadratic equation.

THE DISCRIMINANT

We will see that the discriminant of a quadratic equation occurs in the quadratic formula.

DEFINITION OF THE DISCRIMINANT OF A QUADRATIC EQUATION. The *discriminant* of the quadratic equation

$$ax^2 + bx + c = 0,$$

where a, b, c are real numbers, is the number

$$D = b^2 - 4ac.$$

Example 1. Find the discriminant of $x^2 + 2x - 4 = 0$.

Solution. Since $a = 1$, $b = 2$, and $c = -4$,

$$D = 2^2 - 4(1)(-4) = 20.$$

Example 2. Find the discriminant of $4x^2 - 4x - 1 = 0$.

Solution. Since $a = 4$, $b = -4$, and $c = -1$,

$$D = (-4)^2 - 4(4)(-1) = 32.$$

Example 3. Find the discriminant of $x^2 - 6x + 13 = 0$.

Solution. Since $a = 1$, $b = -6$, and $c = 13$,

$$D = (-6)^2 - 4(1)(13) = -16.$$

Exercise Set 5.3

In exercises 1–20, find the discriminant of the given quadratic equation. (Save your results for use later in this section.)

1. $x^2 + 4x + 2 = 0$
2. $x^2 + 6x - 1 = 0$
3. $x^2 - 2x + 2 = 0$
4. $t^2 + 4t + 5 = 0$
5. $x^2 + x - 1 = 0$
6. $x^2 - x - 1 = 0$
7. $u^2 - u + 1 = 0$
8. $x^2 + x + 1 = 0$
9. $2x^2 + 3x - 2 = 0$
10. $2x^2 - 3x + 2 = 0$
11. $5x^2 - 2x + 1 = 0$
12. $5x^2 - 5x + 1 = 0$
13. $x^2 + 5x - 4 = 0$
14. $24x^2 + 2x - 15 = 0$
15. $5x^2 - 6x + 2 = 0$
16. $y^2 + 10y + 2 = 0$
17. $x^2 + 3x + 3 = 0$
18. $2x^2 + 19x + 35 = 0$
19. $4x^2 - 2x + 3 = 0$
20. $6x^2 - 9x - 2 = 0$

THE QUADRATIC FORMULA

Consider the general quadratic equation

$$ax^2 + bx + c = 0,$$

where a, b, and c are real number constants and $a \neq 0$. We apply the method of completing the square (Section 5.3) to this quadratic equation. First divide both sides of the equation by a to obtain

$$x^2 + \frac{b}{a}x + \frac{c}{a} = 0$$

$$x^2 + \frac{b}{a}x = -\frac{c}{a}.$$

To complete the square we add $(b/2a)^2 = b^2/4a^2$ to both sides of the equation and obtain

$$x^2 + \frac{b}{a}x + \left(\frac{b}{2a}\right)^2 = \frac{b^2}{4a^2} - \frac{c}{a}$$

$$\left(x + \frac{b}{2a}\right)^2 = \frac{b^2 - 4ac}{4a^2} = \frac{D}{4a^2}.$$

where $D = b^2 - 4ac$ is the discriminant. Solving an equation of this form, we find

$$x + \frac{b}{2a} = \pm\sqrt{\frac{D}{4a^2}} = \pm\frac{\sqrt{D}}{2a}$$

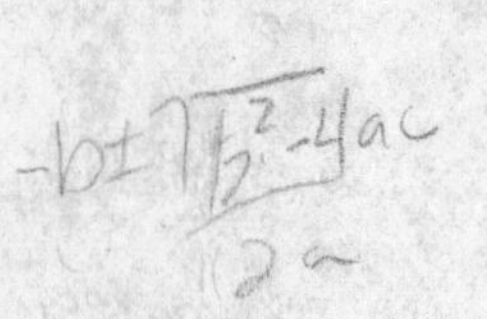

$$x = -\frac{b}{2a} \pm \frac{\sqrt{D}}{2a}$$

THE QUADRATIC FORMULA. If a quadratic equation is in the form

$$ax^2 + bx + c = 0$$

with a, b, and c numbers with $a \neq 0$, the solution is

$$\frac{-b + \sqrt{D}}{2a} \quad \text{and} \quad \frac{-b - \sqrt{D}}{2a}$$

where $D = b^2 - 4ac$ is the discriminant.

The quadratic formula shows that we can find the solution to a quadratic equation simply by substituting the constants a, b, c into the formula. The quadratic formula can be used to solve any quadratic equation, but the equation must be written in the *form* $ax^2 + bx + c = 0$.

Example 1 (continued). Solve $x^2 + 2x - 4 = 0$.

Solution. We have calculated that the discriminant $D = 20$. By the quadratic formula,

$$x = \frac{-2 \pm \sqrt{20}}{2} = \frac{-2 \pm 2\sqrt{5}}{2} = -1 \pm \sqrt{5}.$$

The solutions are $-1 + \sqrt{5}$ and $-1 - \sqrt{5}$.

Example 2 (continued). Solve $4x^2 - 4x - 1 = 0$.

Solution. We have seen that the discriminant $D = 32$. By the quadratic formula,

$$x = \frac{-(-4) \pm \sqrt{32}}{2(4)} = \frac{4 \pm 4\sqrt{2}}{2(4)} = \frac{1 \pm \sqrt{2}}{2}.$$

The solutions are $(1 + \sqrt{2})/2$ and $(1 - \sqrt{2})/2$.

Example 3 (continued). Solve $x^2 - 6x + 13 = 0$.

Solution. We have seen that the discriminant $D = -16$. By the quadratic formula,

$$x = \frac{6 \pm \sqrt{-16}}{2} = \frac{6 \pm 4i}{2} = 3 \pm 2i.$$

The solutions are $3 + 2i$ and $3 - 2i$.

Exercise Set 5.3 (continued)

Use the quadratic formula to solve the quadratic equations given in exercises 1–20.

SIMPLIFYING EQUATIONS

To apply the quadratic formula it is important that the equation be written in the form of a quadratic equation so we can correctly identify the constants a, b, and c. Frequently we must simplify an equation to write it in the proper form.

Example 4. Solve $x^2 + (x + 2)^2 = (x + 1)^2$.

Solution. First we bring the equation to the form of a quadratic equation.

$$x^2 + x^2 + 4x + 4 = x^2 + 2x + 1$$
$$x^2 + 2x + 3 = 0$$

Discriminant is $2^2 - 4(1)(3) = -8$. By the quadratic formula,

$$x = \frac{-2 \pm \sqrt{-8}}{2(1)} = \frac{-2 \pm 2\sqrt{2}i}{2} = -1 \pm \sqrt{2}i.$$

The solutions are $(-1 + \sqrt{2}i)/2$ and $(-1 - \sqrt{2}i)/2$.

Example 5. Solve $\dfrac{x}{x - 1} + \dfrac{1}{x + 1} = 4$.

Solution. We clear of fractions by multiplying by $(x - 1)(x + 1) = x^2 - 1$.

$$x(x + 1) + 1(x - 1) = 4(x^2 - 1)$$
$$x^2 + x + x - 1 = 4x^2 - 4$$
$$0 = 3x^2 - 2x - 3$$

Discriminant is $(-2)^2 - 4(3)(-3) = 40$. By the quadratic formula,

$$x = \frac{2 \pm \sqrt{40}}{6} = \frac{2 \pm 2\sqrt{10}}{6} = \frac{1 \pm \sqrt{10}}{3}.$$

The solutions are $(1 + \sqrt{10})/3$ and $(1 - \sqrt{10})/3$.

Exercise Set 5.3 (continued)

In exercises 21–40, use the quadratic formula to solve each given equation.

21. $x^2 + (x - 1)^2 = (x + 2)^2$
22. $x^2 + (x - 2)^2 = (x + 1)^2$
23. $x^2 + (x + 2)^2 = (x + 3)^2$
24. $x^2 + (x + 3)^2 = (x + 4)^2$
25. $(x + 2)(2x - 1) = (x + 1)(x - 4)$
26. $(x + 1)(2x - 3) = (x - 2)(x - 3)$
27. $(x + 1)(3x + 2) = x(2x + 7)$
28. $(x - 1)(3x - 2) = x(2x - 7)$
29. $\dfrac{x}{x + 2} = \dfrac{2}{x}$
30. $\dfrac{t - 4}{t - 3} = \dfrac{4}{t + 1}$
31. $\dfrac{2x + 1}{x - 1} = \dfrac{3x}{x + 5}$
32. $\dfrac{3x - 1}{x + 2} = \dfrac{2x}{x - 3}$
33. $\dfrac{1}{y - 1} + \dfrac{1}{y + 1} = 4$
34. $2 + \dfrac{1}{x + 1} = \dfrac{x^2}{x^2 - 1}$
35. $\dfrac{1}{x} + \dfrac{1}{x - 1} = \dfrac{6}{5}$
36. $\dfrac{2}{x} + \dfrac{1}{x - 1} = \dfrac{6}{5}$
37. $\dfrac{x}{x - 2} + \dfrac{x}{x + 2} = \dfrac{8}{x^2 - 4}$
38. $\dfrac{x}{x - 1} + \dfrac{x}{x + 1} = 1$
39. $\dfrac{x}{x + 1} + \dfrac{1}{x^2 - x - 2} = 2$
40. $\dfrac{x}{x - 3} + \dfrac{x}{x + 2} = \dfrac{2}{x^2 - x - 6}$

section 4 • Further Quadratic Equations

In this section we consider further the topic of quadratic equations. First we use the discriminant to classify the possible solutions to a quadratic equation. Then we consider solving additional quadratic equations.

CHARACTER OF SOLUTIONS

It is important to discuss the types of solutions we obtain from a quadratic equation. How many solutions are there? When are they real numbers and when are they complex numbers? We now answer these questions.

The solutions to the quadratic equation

$$ax^2 + bx + c = 0$$

are

$$u = \frac{-b + \sqrt{D}}{2a} \quad \text{and} \quad v = \frac{-b - \sqrt{D}}{2a},$$

where $D = b^2 - 4ac$ is the discriminant. If we form the sum and product of the solutions,

$$u + v = \frac{-b + \sqrt{D}}{2a} + \frac{-b - \sqrt{D}}{2a} = \frac{-2b}{2a} = -\frac{b}{a}$$

$$uv = \left(\frac{-b + \sqrt{D}}{2a}\right)\left(\frac{-b - \sqrt{D}}{2a}\right) = \frac{(-b)^2 - D}{4a^2} = \frac{b^2 - (b^2 - 4ac)}{4a^2} = \frac{4ac}{4a^2} = \frac{c}{a}$$

SUM AND PRODUCT OF THE SOLUTIONS OF A QUADRATIC EQUATION. If u and v are the solutions to the quadratic equation $ax^2 + bx + c = 0$, then

$$u + v = -b/a \quad \text{and} \quad uv = c/a.$$

The significance of the sum and product of the solutions of a quadratic equation is that *every* quadratic equation factors over the complex numbers. Specifically, we calculate

$$a(x - u)(x - v) = a[x^2 - (u + v)x + uv] = a\left[x^2 - \left(-\frac{b}{a}\right)x + \frac{c}{a}\right]$$
$$= ax^2 + bx + c$$

FACTORIZATION OF A QUADRATIC EQUATION. If u and v are the solutions to the quadratic equation $ax^2 + bx + c = 0$, then we have the factorization

$$ax^2 + bx + c = a(x - u)(x - v).$$

Example 1. The numbers 6 and -4 are solutions of the quadratic equation

$$x^2 - 2x - 24 = 0.$$

The sum of the solutions is 2, the product is 24, and we have the factorization

$$x^2 - 2x - 24 = (x - 6)(x + 4).$$

Now we classify the types of solutions we can obtain to a quadratic equation. We consider three cases: (1) The discriminant is positive; (2) The discriminant is zero; and (3) The discriminant is negative.

CHARACTER OF SOLUTIONS—POSITIVE DISCRIMINANT. If the discriminant D is positive, then $\sqrt{D}$ is a real number, so we obtain two different real number solutions to the quadratic equation.

CHARACTER OF SOLUTIONS—ZERO DISCRIMINANT. If the discriminant $D = 0$, then we obtain one real number solution, $-b/2a$. Also, we have the factorization

$$ax^2 + bx + c = a\left(x + \frac{b}{2a}\right)^2.$$

So the equation must factor as the square of a binomial if the discriminant is 0. Because we have a square, we call this solution a *repeated solution.*

CHARACTER OF SOLUTIONS—NEGATIVE DISCRIMINANT. If the discriminant D is negative, then $\sqrt{D}$ is imaginary, so we obtain two complex number solutions to the quadratic equation. These two complex numbers are conjugates of each other.

In conclusion, we can say that a quadratic equation always has at most two solutions. If the discriminant is positive, the solutions are two distinct real numbers. If the discriminant is 0, there is a real number repeated solution. If the discriminant is negative, the two solutions are complex numbers that are conjugates of each other.

Exercise Set 5.4

In exercises 1–8, find the sum and product of the solutions to each given quadratic equation without solving the equation. Also, specify whether the solution is two real numbers, a repeated solution, or two complex numbers.

1. $x^2 - 7x + 12 = 0$
2. $x^2 - 5x + 10 = 0$
3. $x^2 + 4x - 12 = 0$
4. $x^2 + 6x - 8 = 0$
5. $3x^2 + x - 1 = 0$
6. $2x^2 - 5x + 4 = 0$
7. $4x^2 + 5x + 5 = 0$
8. $6x^2 - 7x + 9 = 0$

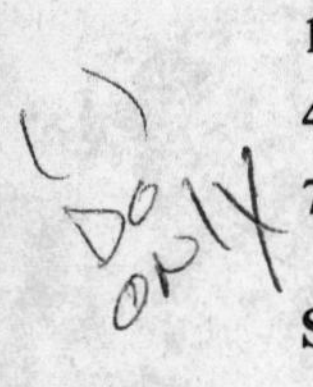

SOLVING QUADRATIC EQUATIONS

We have three methods for solving a quadratic equation: (1) Factoring; (2) Completing the square; and (3) The quadratic formula. Each method has its advantages and disadvantages:

1. Factoring is a quick method to solve an equation if you see the factors quickly. However, factoring does not apply to every equation.
2. Completing the square can be applied to solve any quadratic equation, but this method can be time-consuming.
3. The quadratic formula can be applied to solve any quadratic equation and is less time-consuming, but you must remember the formula and apply it correctly.

Example 2. Solve $\dfrac{600}{x} = 10x + 40$.

Solution. We write the equation in the form of a quadratic equation:

$$600 = 10x^2 + 40x$$
$$0 = 10x^2 + 40x - 600$$
$$0 = x^2 + 4x - 60$$

This equation is easy to factor:

$$0 = (x + 10)(x - 6)$$
$$x = -10 \quad \text{or} \quad x = 6$$

The solutions are -10 and 6.

In many applications of quadratic equations, complex number solutions do not possess an interpretation. In such applications we are only interested in finding real number solutions.

Example 3. Find any real number solutions of

$$x^2 + (x + 30)^2 = (x + 10)^2.$$

Solution. First we simplify the equation:

$$x^2 + x^2 + 60x + 900 = x^2 + 20x + 100$$
$$x^2 + 40x + 800 = 0$$

In this case the coefficients are relatively large numbers and it would take a long time to consider all the possible factors. However, by the quadratic formula;

$$\text{discriminant is } 40^2 - 4(1)(800) = -1600.$$

Because the discriminant is negative, this equation does not possess real number solutions.

Exercise Set 5.4 (continued)

In exercises 9–32, find any real number solutions of the given equation; specify if the solution is repeated.

9. $\dfrac{15}{x} = 24x + 2$

10. $\dfrac{18}{x} = 12x - 6$

11. $\dfrac{150}{x} = 4x + 50$

12. $\dfrac{2400}{x} = 5x - 20$

13. $\frac{400}{x} = 80 - 4x$

14. $\frac{180}{x} = 60 - 5x$

15. $\frac{10}{x} = 6 - x$

16. $\frac{8}{x} = 6 - 2x$

17. $x(300 - 2x) = 12{,}000$

18. $x(400 - 2x) = 40{,}000$

19. $x(300 + 2x) = 12{,}000$

20. $x(400 + 2x) = 40{,}000$

21. $x(300 - 2x) = 11{,}200$

22. $x(400 - 2x) = 19{,}800$

23. $x^2 + (x + 20)^2 = (x + 10)^2$

24. $x^2 + (x + 11)^2 = (x + 10)^2$

25. $x^2 + (x - 20)^2 = (x - 10)^2$

26. $x^2 + (x - 11)^2 = (x - 10)^2$

27. $x^2 + (x + 4)^2 = 26$

28. $x^2 + (x - 4)^2 = 10$

29. $4x^2 + 4\left(x + \frac{1}{2}\right)^2 = 85$

30. $9x^2 + 9\left(x + \frac{1}{3}\right)^2 = 221$

31. $\frac{90}{x} + \frac{1}{2} = \frac{90}{x + 15}$

32. $\frac{10}{x} + \frac{20}{x + 10} = \frac{5}{6}$

Decimal Coefficients (Calculator)

In many applications, quadratic equations occur in which the coefficients are not integers. We now consider solving quadratic equations in which the coefficients are decimals. Of the methods we have discussed, the quadratic formula is probably the easiest to apply to such equations.

CALCULATOR NOTE: The quadratic formula is well-suited for use with an electronic calculator. The following procedure is suggested for solving $ax^2 + bx + c = 0$:

1. Calculate the discriminant $D = b^2 - 4ac$.
2. Approximate $\sqrt{D}$ to the desired degree of accuracy.
3. Calculate $(\sqrt{D} - b)/2a$ and $(-\sqrt{D} - b)/2a$. We can find the first number, for example, by entering

$$D \; \sqrt{\;} \; - \; b \div 2 \div a =$$

Example 4. Find any real number solutions of

$$4.9x^2 - 760x + 2000 = 0$$

to one decimal place accuracy.

Solution. We use the quadratic formula with a calculator; discriminant is $D = 538{,}400$.

$$(\sqrt{D} - b)/2a = 152.424$$
$$(-\sqrt{D} - b)/2a = 2.67781$$

To one decimal place accuracy, the solutions are 152.4 and 2.7.

Note: In this example we could also multiply both sides of the equation by 10 and obtain an equivalent equation with integer coefficients. You may prefer to solve the equation this way.

Exercise Set 5.4 (continued)

In exercises 33–38, find any real number solutions of the given equation. Round off each solution to one decimal place.

33. $3.1x^2 - 2.1x - 1.1 = 0$
34. $2.2x^2 + 3.3x - 4.4 = 0$
35. $2000 = 600x - 4.9x^2$
36. $2000 = 500x - 4.9x^2$
37. $x(2x - 5.7) = 3.6x + 9.8$
38. $x(x + 4.5) = 7.2x + 4.5$

In exercises 39–44, find any real number solutions of the given equation. Round off each solution to two decimal places.

39. $x^2 + 7.62x + 5.43 = 0$
40. $x^2 - 3.21x - 12.16 = 0$
41. $x(x + 36.52) = 75.12$
42. $x(x - 14.58) = 25.44$
43. $x(2x - 3.14) = 2.72$
44. $x(3x + 6.28) = 2.72$

section 5 • Quadratic Inequalities

In this section, we consider quadratic inequalities. The techniques used to solve quadratic inequalities are factoring and completing the square.

DEFINITION OF A QUADRATIC OR SECOND-DEGREE INEQUALITY IN ONE VARIABLE. A *quadratic inequality* or *second-degree inequality* in one variable x is any inequality equivalent to one of the form

$$ax^2 + bx + c > 0 \quad \text{or} \quad ax^2 + bx + c \geq 0$$

$$\text{or} \quad ax^2 + bx + c < 0 \quad \text{or} \quad ax^2 + bx + c \leq 0,$$

where a, b, c are real number constants and $a \neq 0$.

INEQUALITIES THAT FACTOR

First we give some examples of quadratic inequalities that can be solved by factoring. The basic observation is that the product of two positive factors is positive, the product of a positive and negative factor is negative, and the product of two negative factors is positive.

Example 1. Solve $x^2 - x - 6 > 0$.

Solution. This quadratic inequality factors as $(x + 2)(x - 3) > 0$. In order for the product of the two factors to be positive, both factors must be positive or both factors must be negative. We consider two cases:

Case **1**. If both factors are positive, then $x + 2 > 0$ and $x - 3 > 0$. But then $x > -2$ and $x > 3$. The solution for this case is $x > 3$.

Case **2**. If both factors are negative, then $x + 2 < 0$ and $x - 3 < 0$. But then $x < -2$ and $x < 3$. The solution for this case is $x < -2$.

Putting these csses together, the solution for the quadratic inequality is the set of numbers x such that $x > 3$ or $x < -2$. The graph of the solution is shown in Figure 5.2.

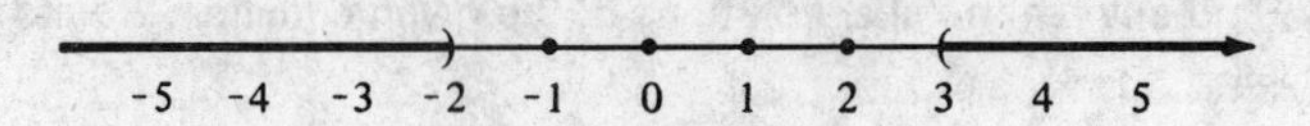

Figure 5.2

We can organize this inequality in the form of a chart:

$x + 2$	$x - 3$	$(x + 2)(x - 3)$
+	+	+
+	−	−
−	+	−
−	−	+

The first line corresponds to Case 1, and the fourth line corresponds to Case 2. The second and third lines do not satisfy the inequality.

Example 2. Solve $x^2 + 4x \leq -3$.

Solution. We write this inequality as $x^2 + 4x + 3 \leq 0$ and factor as $(x + 1)(x + 3) \leq 0$. To make the product of the factors negative, one of the two factors must be positive and the other must be negative. We consider two cases:

Case **1**. $x + 1$ is positive and $x + 3$ is negative. So $x + 1 \geq 0$ and $x + 3 \leq 0$, which implies $x \geq -1$ and $x \leq -3$. Since it is impossible for x to be both larger than -1 and smaller than -3, this case cannot occur.

Case **2**. $x + 1$ is negative and $x + 3$ is positive. So $x + 1 \leq 0$ and $x + 3 \geq 0$, which implies $x \leq -1$ and $x \geq -3$; in other words, $-3 \leq x \leq -1$. Therefore, the solution of the inequality is the set of numbers x such that $-3 \leq x \leq -1$; the graph is shown in Figure 5.3.

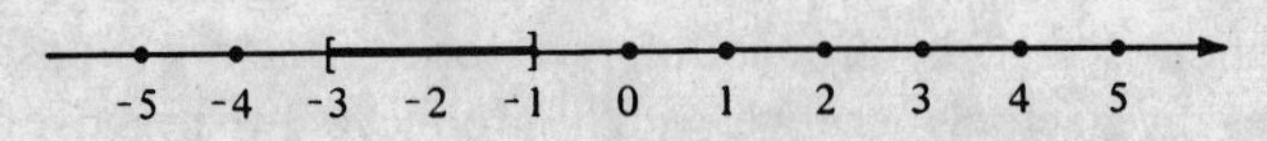

Figure 5.3

Here is a chart for this inequality.

$x + 1$	$x + 3$	$(x + 1)(x + 3)$
+	+	+
+	−	−
−	+	−
−	−	+

Case 1 corresponds to the second line; Case 2 corresponds to the third line.

Exercise Set 5.5

In exercises 1–12, solve each given inequality.

1. $(x - 1)(x - 4) \geq 0$
2. $(x - 5)(x - 6) \geq 0$
3. $(x + 2)(x - 1) < 0$
4. $(y - 6)(y + 1) \leq 0$
5. $(x + 4)(x + 2) \geq 0$
6. $x(x - 3) > 0$
7. $2x^2 + 7x + 3 < 0$
8. $4x^2 + 5x - 6 < 0$
9. $x^2 - 2x + 1 > 0$
10. $u^2 - 2u + 1 < 0$
11. $4x^2 - 3x - 1 < 0$
12. $6x^2 + x - 1 \geq 0$

INEQUALITIES THAT DON'T FACTOR

We give two examples of inequalities that cannot be solved by factoring alone but that yield to completing the square (Section 5.2).

Example 3. Solve $3x^2 + 6x \leq 6$.

Solution. Dividing both sides of the inequality by 3, we obtain

$$x^2 + 2x \leq 2$$
$$x^2 + 2x - 2 \leq 0$$

Completing the square on $x^2 + 2x$ by adding 1, we have

$$x^2 + 2x + 1 - 1 - 2 \leq 0$$
$$(x + 1)^2 - 3 \leq 0.$$

We factor as a difference of squares:

$$[(x + 1) - \sqrt{3}][(x + 1) + \sqrt{3}] \leq 0.$$

Now we have a negative product of two factors, and so we proceed as in the previous examples to examine positive and negative factors. We obtain as solution the set of numbers x such that $-1 - \sqrt{3} \leq x \leq -1 + \sqrt{3}$. Figure 5.4 depicts the graph.

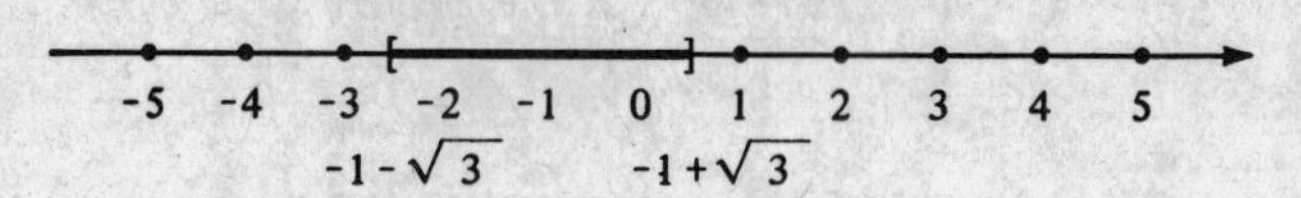

Figure 5.4

Example 4. Solve $x^2 + 4x + 5 < 0$.

Solution. To complete the square on $x^2 + 4x$, add 4:

$$x^2 + 4x + 4 - 4 + 5 < 0$$
$$(x + 2)^2 + 1 < 0$$

Now $(x + 2)^2 \geq 0$ for *all* values of x. Therefore, $(x + 2)^2 + 1 \geq 1$; and it is impossible for the inequality to be fulfilled. Thus the inequality has no solution.

Exercise Set 5.5 (continued)

In exercises 13–20, solve the given inequality.

13. $x^2 + 2x - 1 > 0$
14. $x^2 + 4x + 1 \geq 0$
15. $y^2 + 6y + 3 \leq 0$
16. $x^2 - 6x - 2 < 0$
17. $x^2 + 2x + 2 > 0$
18. $x^2 + 2x + 2 \leq 0$
19. $4x^2 + 4x + 3 < 0$
20. $4u^2 + 4u + 3 \geq 0$

SIMPLIFYING INEQUALITIES

Frequently we must simplify an inequality before solving it.

Example 5. Solve $(x + 1)(2x - 3) \geq 3$

Solution. Simplifying this inequality, we have

$$(x + 1)(2x - 3) \geq 3$$
$$2x^2 - x - 3 \geq 3$$
$$2x^2 - x - 6 \geq 0$$
$$(x - 2)(2x + 3) \geq 0$$

The product of these factors is positive if both factors are positive (giving $x \geq 2$) or if both factors are negative (giving $x \leq -3/2$). Therefore, the solution is the set of numbers x such that $x \geq 2$ or $x \leq -3/2$.

Exercise Set 5.5 (continued)

In exercises 21–32, solve the given inequality.

21. $2x^2 > x$
22. $2z^2 \leq 3 - z$
23. $x^2 < 2(x - 1)$
24. $8x^2 \leq 10x + 3$
25. $16t^2 + 8t + 1 \geq 0$
26. $16x^2 + 8x + 1 \leq 0$
27. $10x^2 \geq x + 2$
28. $18u^2 < 9u + 3$
29. $(x - 1)(60x + 10) \geq 50$
30. $(x - 3)(120x - 40) \geq 440$
31. $(x - 4)(10x + 40) \geq 200$
32. $(x - 4)(2x - 10) \geq 60$

section 6 • Equations Quadratic in Form

Some equations are said to be quadratic in form because they are quadratic equations when appropriately viewed. We will look at two types of equations that one quadratic in form: equations in one variable that become quadratic when an appropriate substitution is made, and equations in more than one variable which we wish to solve for one variable in terms of other variables. For equations quadratic in form, we can apply the techniques of quadratic equations—factoring, completing the square, and the quadratic formula.

EOUATIONS IN ONE VARIABLE

An equation in one variable is quadratic in form if we can make a substitution, say u, in terms of the variable so that the resulting equation in u is quadratic.

Example 1. Solve $x^4 - 5x^2 + 6 = 0$.

Solution. We can write the equation as

$$(x^2)^2 - 5(x^2) + 6 = 0.$$

If we let $u = x^2$, the equation becomes

$$u^2 - 5u + 6 = 0$$

and so the equation is quadratic in form. Factoring, we have

$$(u - 2)(u - 3) = 0$$

$$u = 2 \quad \text{or} \quad u = 3$$

$$x^2 = 2 \quad \text{or} \quad x^2 = 3$$

$$x = \pm\sqrt{2} \quad \text{or} \quad x = \pm\sqrt{3}$$

The solutions are $\sqrt{2}$, $-\sqrt{2}$, $\sqrt{3}$, and $-\sqrt{3}$.

Example 2. Solve $4x + 11\sqrt{x} - 3 = 0$.

Solution. We can write this equation as

$$4(\sqrt{x})^2 + 11(\sqrt{x}) - 3 = 0.$$

If we let $u = \sqrt{x}$ we obtain

$$4u^2 + 11u - 3 = 0;$$

so the equation is quadratic in form. Factoring we have

$$(4u - 1)(u + 3) = 0$$

$$u = 1/4 \quad \text{or} \quad u = -3$$

$$\sqrt{x} = 1/4 \quad \text{or} \quad \sqrt{x} = -3$$

Notice, however, that $\sqrt{x}$ must be positive, so we cannot have $\sqrt{x} = -3$. Therefore, we have $x = (1/4)^2 = 1/16$. The solution is 1/16.

Exercise Set 5.6

In exercises 1–20, solve the given equation.

1. $x^4 - 7x^2 + 12 = 0$
2. $x^4 - 5x^2 + 4 = 0$
3. $4x^4 - 9x^2 + 2 = 0$
4. $3y^4 - 28y^2 + 9 = 0$
5. $x^4 + 16 = 10x^2$
6. $x^4 + 4 = 4x^2$
7. $x^4 + 2304 = 100x^2$
8. $4x^2 + 144 = 25x^2$
9. $x - 7\sqrt{x} + 12 = 0$
10. $x - 8\sqrt{x} + 15 = 0$
11. $3t - 11\sqrt{t} - 4 = 0$
12. $4x + 4\sqrt{x} + 1 = 0$
13. $x^{-2} - 7x^{-1} + 12 = 0$
14. $z^{-2} - 10z^{-1} + 9 = 0$
15. $\frac{9}{x^2} + \frac{6}{x} + 1 = 0$
16. $\frac{2}{x^2} + \frac{3}{x} - 2 = 0$
17. $x^{2/3} - 13x^{1/3} + 36 = 0$
18. $y^{2/3} - 10y^{1/3} + 9 = 0$
19. $(x - 1) - 5\sqrt{x - 1} + 4 = 0$
20. $(x + 1) - 13\sqrt{x + 1} + 36 = 0$

SOLVING FOR A SPECIFIED VARIABLE

If we have an equation in more than one variable, we frequently wish to solve for one of the variables in terms of the others. The equation is quadratic in form if it is quadratic in the variable we want to solve for.

Example 3. Solve $x^2 + y^2 = 1$ for y.

Solution. We want to solve for y in terms of x, so we consider x to be a constant in the equation. We can write the equation

$$y^2 = 1 - x^2$$
$$y = \pm\sqrt{1 - x^2}$$

We have $y = \sqrt{1 - x^2}$ or $y = -\sqrt{1 - x^2}$, both a positive and a negative square root. We have to be careful here because $\sqrt{1 - x^2}$ is a real number only for those values of x that satisfy $1 - x^2 \geq 0$; these values of x are $-1 \leq x \leq 1$.

Example 4. Solve $x^2 + xy + y^2 = 1$ for y.

Solution. We want to solve for y in terms of x, so we consider x as a constant. Writing the equation as a quadratic equation in y, we have

$$y^2 + xy + (x^2 - 1) = 0.$$

We solve this using the quadratic formula with $a = 1$, $b = x$, and $c = x^2 - 1$. The discriminant is

$$D = x^2 - 4(1)(x^2 - 1) = 4 - 3x^2$$

$$y = \frac{-x \pm \sqrt{4 - 3x^2}}{2}.$$

This gives y in terms of x. Notice that we have real number values only when $4 - 3x^2 \geq 0$; that is, $-2\sqrt{3}/3 \leq x \leq 2\sqrt{3}/3$.

Exercise Set 5.6 (continued)

In exercises 21–40, solve the equation for the specified variable.

21. $x^2 + y^2 = 1$ for x
22. $x^2 + xy + y^2 = 1$ for x
23. $x^2 + y^2 = a^2$ for y
24. $y^2 = x$ for y
25. $x^2 - y^2 = 1$ for y
26. $x^2 - y^2 = a^2$ for y
27. $x^2 + 4y^2 = 4$ for y
28. $x^2 + 4y^2 = 4$ for x
29. $9x^2 + 4y^2 = 36$ for y
30. $4x^2 + 25y^2 = 100$ for y
31. $a^2x^2 + b^2y^2 = a^2b^2$ for y
32. $4x^2 - y^2 = 4$ for y
33. $4x^2 - 9y^2 = 36$ for y
34. $25x^2 - 4y^2 = 100$ for y
35. $x^2 + y^2 + 3x + 6y + 1 = 0$ for y
36. $x^2 + y^2 + 3x + 6y + 1 = 0$ for x
37. $x^2 + 2ax - a^2 = 0$ for x
38. $a^2x^2 + abx - 2b^2 = 0$ for x
39. $x^2 - (a + b)x - (a + b + c)c = 0$ for x
40. $a^2x^2 + abx + b^2 = 0$ for x

section 7 • Squaring an Equation

Sometimes an equation contains square roots. To remove the square roots from the equation, we can square both sides of the equation. However, this operation may enlarge the solution set of the equation; that is, we may introduce what we will call an *extraneous solution* to the equation.

THE OPERATION OF SQUARING

Squaring both sides of an equation is essentially a matter of multiplying it twice. Suppose we have an equation $a = b$. If we multiply the equation by a, we obtain $a^2 = ab$. If we multiply the same equation by b we obtain $ab = b^2$. Since a^2 and b^2 are both equal to ab, they are equal to each other. So we have $a^2 = b^2$.

SQUARING AN EQUATION. If $a = b$, then $a^2 = b^2$.

An important thing to remember in dealing with equations involving square roots is that we assume that the expression under the square root sign denotes a *positive* real number and $\sqrt{a}$ denotes the *positive* square root of a.

Example 1. Solve $\sqrt{3 - x} = 2$.

Solution. Squaring both sides of the equation, we have

$$3 - x = 4$$

$$-1 = x$$

We check -1 in the original equation: $\sqrt{3 - (-1)} = 2$. The solution is -1.

Example 2. Solve $\sqrt{x + 6} = \sqrt{3x + 2}$.

Solution. Squaring both sides of the equation, we have

$$x + 6 = 3x + 2$$

$$4 = 2x$$

$$2 = x$$

We check 2 in the original equation: $\sqrt{2 + 6} = \sqrt{3 \cdot 2 + 2}$. Therefore, the solution is 2.

Exercise Set 5.7

In exercises 1–10, solve each given equation.

1. $\sqrt{x} = 1$
2. $\sqrt{x} = 6$
3. $\sqrt{x - 3} = 2$
4. $\sqrt{x + 1} = 5$
5. $\sqrt{2x} = 4$
6. $\sqrt{3x} = 3$
7. $\sqrt{x + 4} = \sqrt{2x - 5}$
8. $\sqrt{3y + 6} = \sqrt{7y - 6}$
9. $\sqrt{2z - 1} = \sqrt{1 - 2z}$
10. $\sqrt{4x - 3} = \sqrt{4x + 3}$

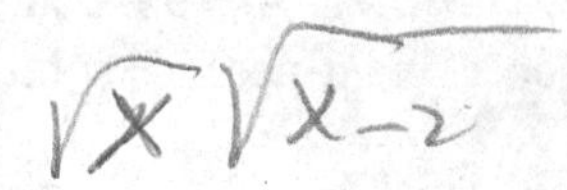

EXTRANEOUS SOLUTIONS

When we square both sides of an equation, the solution set of the resulting equation may be larger than the solution set of the original equation. That is, there may be numbers that satisfy the squared equation but do not satisfy the original equation. Such a number is called an *extraneous solution.*

Example 3. Solve $\sqrt{3 - x} = -2$.

Solution. Squaring both sides of the equation, we have

$$3 - x = 4$$
$$-1 = x$$

We check -1 in the original equation: $\sqrt{3 - (-1)} = \sqrt{4} = 2 \neq -2$. So -1 is not a solution to the equation, and the equation has no solution.

Note: Actually, we can see just by inspecting the equation that there is no solution because on the left-hand side we have a square root, which must be positive, and on the right-hand side we have a negative number; so the two sides cannot be equal. However, after squaring the equation, we get a *possible* solution, $x = -1$. This solution does not check back in the equation, so it is an extraneous solution.

Example 4. Solve $\sqrt{2x + 7} = x - 4$.

Solution. Squaring both sides of the equation, we have

$$2x + 7 = x^2 - 8x + 16$$
$$0 = x^2 - 10x + 9$$

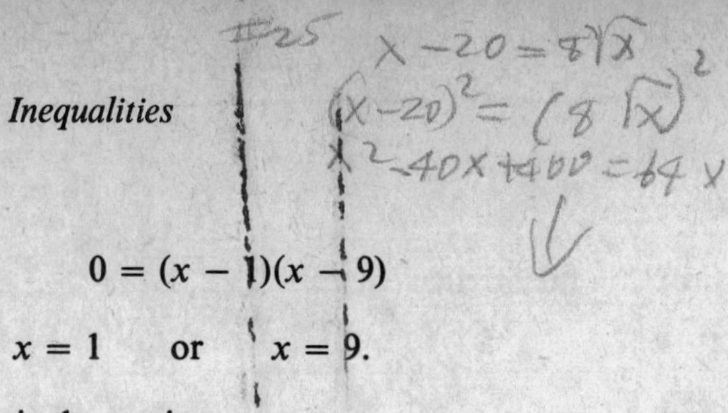

$$0 = (x - 1)(x - 9)$$

$$x = 1 \quad \text{or} \quad x = 9.$$

We check 1 and 9 in the original equation:

$$\sqrt{2(1) + 7} \neq 1 - 4 \quad \text{and} \quad \sqrt{2(9) + 7} = 9 - 4.$$

Thus, 9 satisfies the equation but 1 does not. The solution is 9, and 1 is an extraneous solution.

Note: The possibility of obtaining an extraneous solution shows the importance of checking a possible solution in the original equation.

Exercise Set 5.7 (continued)

In exercises 11–28, solve each given equation.

11. $\sqrt{5x - 6} = 8$
12. $\sqrt{5x - 6} = -8$
13. $\sqrt{3x + 1} = -2$
14. $\sqrt{3x + 1} = 2$
15. $\sqrt{x - 2} = x - 2$
16. $\sqrt{x + 6} = x + 6$
17. $\sqrt{x + 3} = x - 3$
18. $\sqrt{2y + 3} = 2 + 3y$
19. $\sqrt{2x} = x - 4$
20. $\sqrt{3x} = x - 6$
21. $x - 3 = 2\sqrt{x}$
22. $x - 4 = 3\sqrt{x}$
23. $x - 1 = \sqrt{x/2}$
24. $x - 2 = \sqrt{x/3}$
25. $x - 20 = 8\sqrt{x}$
26. $x - 4 = 3\sqrt{x}$
27. $\frac{1}{9}x - 2 = \sqrt{\frac{x}{2}}$
28. $x - 4 = \sqrt{\frac{x}{3}}$

SEPARATING THE RADICALS

If an equation contains two square roots, we first write the equation so that the square roots occur on opposite sides of the equation. This separation of the radicals simplifies subsequent work. If an equation contains more than two square roots, we cannot separate the radicals (see exercises 39 and 40).

Example 5. Solve $\sqrt{5x + 6} + \sqrt{5x - 3} = 1$.

Solution. First we subtract $\sqrt{5x - 3}$ from both sides of the equation to obtain

$$\sqrt{5x + 6} = 1 - \sqrt{5x - 3}.$$

Now we have separated the radicals, and we square both sides of the equation. Be careful to square out correctly the right-hand side, which is a *binomial* expression. We obtain

$$5x + 6 = 1 - 2\sqrt{5x - 3} + 5x - 3$$

$$8 = -2\sqrt{5x - 3}$$

$$4 = -\sqrt{5x - 3}$$

At this point we realize that there is no solution because the left-hand side is positive and the right-hand side is negative. Let's continue, however, by squaring out the equation again. We obtain

$$16 = 5x - 3$$

$$19 = 5x$$

$$19/5 = x$$

We check 19/5 in the original equation:

$$\sqrt{5\left(\frac{19}{5}\right) + 6} + \sqrt{5\left(\frac{19}{5}\right) - 3} = \sqrt{25} + \sqrt{16} = 5 + 4 \neq 1.$$

Therefore, 19/5 is an extraneous solution, and the equation has no solution.

Exercise Set 5.7 (continued)

In exercises 29–40, solve each given equation.

29. $\sqrt{x + 4} = 1 + \sqrt{x - 5}$
30. $\sqrt{y + 5} = 2 - \sqrt{y - 4}$
31. $\sqrt{x + 1} - \sqrt{x - 1} = 2$
32. $\sqrt{2x + 1} + \sqrt{2x - 1} = 1$
33. $\sqrt{x + 13} = \sqrt{x + 1} + 2$
34. $\sqrt{x + 4} = \sqrt{x - 1} + 1$
35. $\sqrt{x + 2} + \sqrt{2x - 3} = 3$
36. $\sqrt{3x + 4} + \sqrt{2x + 3} = 2$
37. $\sqrt{u + 3} - \sqrt{2u - 1} = 1$
38. $\sqrt{x + 3} - \sqrt{3x - 2} = 1$
39. $\sqrt{t + 3} - \sqrt{t} = \sqrt{2t - 1}$
40. $\sqrt{x - 3} + \sqrt{x} = \sqrt{2x + 1}$

section 8 • Applications

We examine applications of quadratic equations and inequalities in one variable to geometry problems, rate problems, business and management considerations, projectiles, economics, and historical problems.

GEOMETRY PROBLEMS

Example 1. One side of a rectangle is 5 meters longer than the other side. If the area of the rectangle is 84 square meters, what is the length of each side?

Solution. If x denotes the length of the shorter side, then $x + 5$ denotes the length of

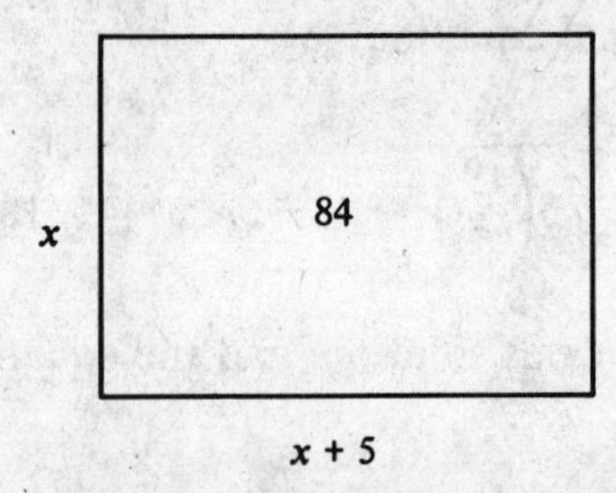

Figure 5.5

the longer side. The area is $x(x + 5)$ on the one hand and 84 on the other (see Figure 5.5). So

$$x(x + 5) = 84$$
$$x^2 + 5x - 84 = 0$$
$$(x - 7)(x + 12) = 0$$
$$x = 7 \quad \text{or} \quad x = -12$$

Since the side of a rectangle cannot be negative, we do not use the solution $x = -12$. Therefore the shorter side has length 7 meters and the longer side has length $7 + 5 = 12$ meters.

Exercise Set 5.8

1. One side of a rectangle is 8 meters longer than the other side. If the area of the rectangle is 48 square meters, find the length of each siue.

2. One side of a rectangle is 5 inches longer than the other side. If the area of the rectangle is 104 square inches, what is the length of the sides?

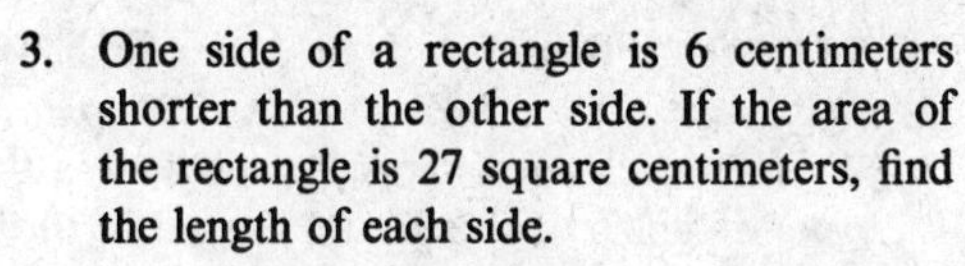

3. One side of a rectangle is 6 centimeters shorter than the other side. If the area of the rectangle is 27 square centimeters, find the length of each side.
4. The diagonal of a square is 10 feet. What is the length of a side?
5. One side of a rectangle is 2 inches longer than the other side. If the diagonal is 10 inches, what is the length of each side?
6. One side of a rectangle is 3 inches longer than the other side. If the diagonal is 4 inches longer than the shorter side, what is the length of each side?

RATE PROBLEMS

In rate problems we use the formula

$$\text{rate} = \frac{\text{distance}}{\text{time}}.$$

Example 2. A riverboat can go 10 miles per hour in still water. When it makes a trip both upstream and downstream between two towns 24 miles apart, the total time is 5 hours. What is the rate of the river?

Solution. Let r denote the rate of the river. The rate of the boat downstream is $10 + r$, and the rate of the boat upstream is $10 - r$. Therefore, the time downstream is $24/(10 + r)$ and the time upstream is $24/(10 - r)$. We obtain an equation from the total time for the round trip:

$$\frac{24}{10 + r} + \frac{24}{10 - r} = 5$$

Multiplying both sides of this equation by $(10 + r)(10 - r)$, we obtain

$$24(10 - r) + 24(10 + r) = 5(10 - r)(10 + r)$$

$$240 - 24r + 240 + 24r = 500 - 5r^2$$

$$5r^2 - 20 = 0$$

$$r^2 - 4 = 0$$

$$(r - 2)(r + 2) = 0$$

$$r = 2 \quad \text{or} \quad r = -2$$

Since the rate cannot be negative, we find that the rate of the stream is 2 miles per hour.

Exercise Set 5.8 (continued)

7. A riverboat can go 9 mph in still water. When it makes a trip both upstream and downstream between two towns 36 miles apart, the total time is 9 hours. What is the rate of the river?
8. A riverboat can go 6 mph in still water. When it makes a pleasure trip both upstream and downstream between two towns 24 miles apart, it takes 8 hours and 6 minutes. What is the rate of the river?

9. If a motorist driving between two towns 90 miles apart were to increase his speed by 15 miles per hour, he would save half an hour. How fast is he going?

10. A man drove 10 miles to a train station and then took a train 20 miles to work. His total time traveling was 50 minutes. If the rate of the train was 10 mph faster than that of the car, find the rate of each.

11. An airplane flying north at a rate of 600 mph passed over Memphis at 2:00 p.m. Ten minutes later, a second plane flew east over Memphis at a rate of 600 mph at the same altitude. At what time will the planes be 500 miles apart?

12. Two hikers leave from the same point at the same time, one heading south at a certain rate and the other east but 1/3 mph faster. After three hours, they are $\sqrt{221}$ miles apart. How fast is each going?

BUSINESS AND MANAGEMENT

Example 3. Suppose we wish to design a page for a book. We want to have 2-inch margins on each side and at the top and bottom. We want the height to be twice the width, and we want the area of the printed matter to be 48 square inches. What should be the dimensions of the page?

Solution. Let x be the width of the page and $2x$ the height of the page. So the printed matter has dimensions $x - 4$ and $2x - 4$. The area of the printed matter is $(x - 4)(2x - 4)$ on the one hand and 48 on the other (see Figure 5.6). So we have the equation

$$(x - 4)(2x - 4) = 48$$
$$2x^2 - 12x + 16 = 48$$
$$x^2 - 6x - 16 = 0$$
$$(x - 8)(x + 2) = 0$$
$$x = 8 \quad \text{or} \quad x = -2$$

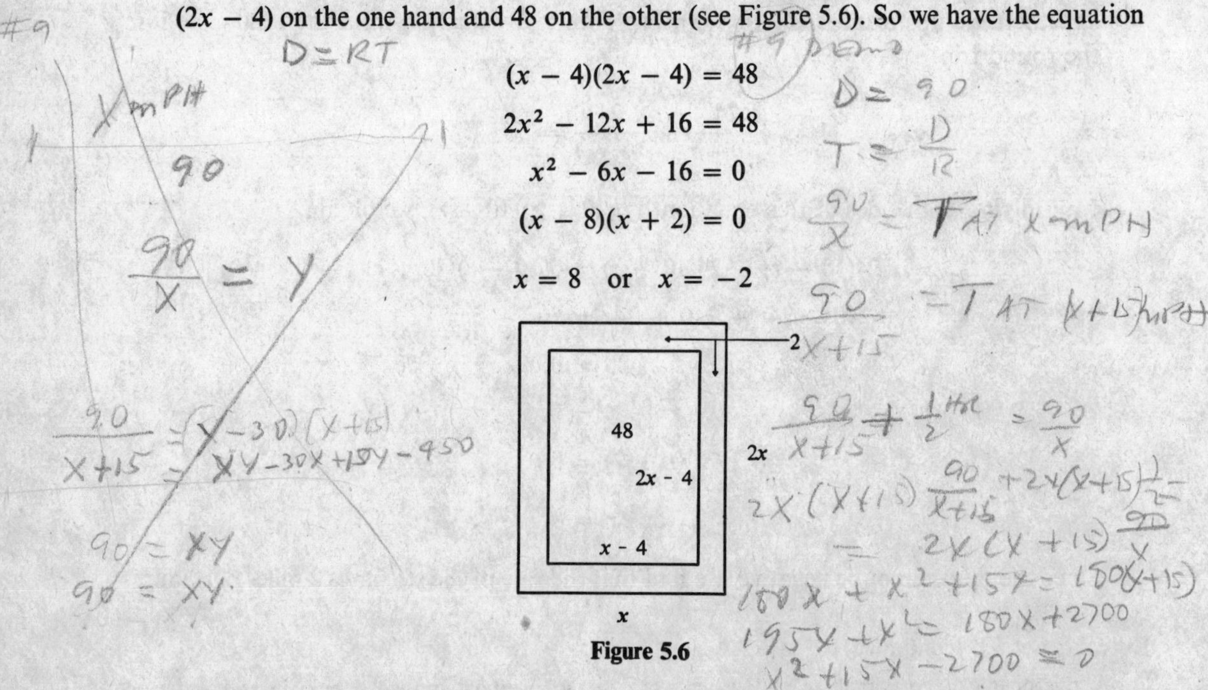

Figure 5.6

Since the width cannot be negative, the width is 8 inches. Thus the height is $2(8) = 16$ inches.

13. A page for a book has a height that is twice its width and margins of 2 inches at top and bottom and on each side. If the area of the printed matter is 70 square inches, what is the width of the page?

14. A page for a book has height 3 inches more than its width, and the margins are 3/4 inch at top and bottom and 1 inch on each side. If the area of the printed matter is 36 square inches, what are the dimensions of the page?

15. A box without a top is to be made from a square piece of cardboard by cutting a 4-centimeter square from each corner and folding up the sides. If the box is to hold 121 cubic centimeters what should be the length of a side of the cardboard?

16. A square garden plot 30 meters on a side is to be surrounded by a walk of uniform width. If the area of the walk is to be 700 square meters, what is the width of the walk?

17. Two electronic sorters are used together to sort a quantity of cards. One of the sorters working alone can sort all of the cards in an hour less than the other sorter can. If it takes 1 hour 12 minutes when they sort together, how long does it take each to do the job alone?

18. Three electronic sorters are used together to sort a quantity of cards. Alone, two particular sorters can sort all of the cards in equal lengths of time, but the third sorter can sort all of the cards in an hour less than these other two. If it takes one hour, 12 minutes when all three sort together, how long does it take each to do the job alone?

In exercises 19–22, a farmer wishes to fence in a rectangular field. However, the farmer uses a straight river for one side of the field and fencing for the other three sides. (Variations on this type of problem occur frequently in calculus.)

19. Find the dimensions of the field if the farmer has 300 meters of fencing and wants to enclose an area of 11,200 square meters.

20. Find the dimensions of the field if the farmer has 400 meters of fencing and wants to enclose an area of 19,800 square meters.

21. Find the dimensions of the field if the farmer has 300 meters of fencing and wants to enclose an area of 12,000 square meters.

22. Find the dimensions of the field if the farmer has 400 meters of fencing and wants to enclose an area of 40,000 square meters.

PROJECTILES

An object is thrown or fired upward. Let t denote the time elapsed (in seconds), let h denote the height of the object (in meters), and let v denote the vertical component of the initial velocity (in meters per second). Then

$$h = vt - 4.9t^2.$$

Example 4. Some rifles used for big game fire a 200-grain bullet at an initial velocity of 760 meters per second. If such a bullet is fired straight upward, how long does it take for the bullet to reach a height of 2000 meters?

Solution. We have $v = 760$ and $h = 2000$. So we must solve

$$2000 = 760t - 4.9t^2$$

$$4.9t^2 - 760t + 2000 = 0$$

discriminant is $(-760)^2 - 4(4.9)(2000) = 538{,}400$

$$t = \frac{-(-760) \pm \sqrt{538400}}{2(4.9)} = \frac{760 \pm 733.8}{9.8}$$

$$t = 152.4 \quad \text{or} \quad t = 2.7$$

There are two times when the bullet is at a height of 2000 meters: after 2.7 seconds (going up) and 152.4 seconds (coming down).

Exercise Set 5.8 (continued)

23. Some rifles used for big game fire a 300-grain bullet at an initial velocity of 600 meters per second. If such a bullet is fired straight upward, how long does it take for the bullet to reach a height of 2000 meters?
24. Some rifles used for big game fire a 500-grain bullet at an initial velocity of 500 meters per second. If such a bullet is fired straight upward, how long does it take for the bullet to reach a height of 2000 meters?

Exercises 25–28 relate to the first experiments in liquid propellant rockets performed by R. H. Goddard.

25. In the first experiment in 1926, the rocket was fired with an initial vertical velocity of 15.3 meters per second. How long did it take the rocket to reach a height of 15 meters?
26. In an experiment in 1930, the rocket was fired with an initial vertical velocity of 108.4 meters per second. How long did it take the rocket to reach a height of 500 meters?
27. In an experiment in 1935, the rocket was fired with an initial vertical velocity of 212.3 meters per second. How long did it take the rocket to reach a height of 2000 meters?
28. In the 1935 experiment the rocket was fired with an initial velocity of 212.3 meters per second. How long did it take the rocket to reach a height of 2500 meters?

ECONOMICS: SUPPLY EQUATIONS

Sometimes a company uses a *supply equation* to obtain pricing information about a product. Let p denote the price per unit of the product and let q_s denote the quantity supplied by the company (supply quantity). The supply equation shows the relation between p and q_s and usually has the form

$$q_s = ap + b,$$

where a and b are constants depending on the particular product.

Example 5. Suppose it costs a potato farmer \$4 per bushel to grow potatoes. The supply equation is $q_s = 8p + 90$. What price should the farmer charge per bushel to realize a profit of at least \$315?

Solution. The profit per bushel is $p - 4$. So the total profit from selling q_s bushels is $(p - 4)q_s = (p - 4)(8p + 90)$. We want to determine p so that

$$(p - 4)(8p + 90) \geq 315.$$

This inequality is equivalent to

$$8p^2 + 58p - 675 \geq 0$$

$$(4p - 25)(2p + 27) \geq 0.$$

$$p \leq -27/2 = -13.5 \quad \text{or} \quad p \geq 25/4 = 6.25$$

Since we discard negative prices, a price of at least \$6.25 per bushel will realize a profit of at least \$315.

Exercise Set 5.8 (continued)

29. It costs a dairy \$1 a gallon to make ice cream, and the supply equation is determined to be $q_s = 60p + 10$. What price will realize a profit of at least \$50?

30. It costs a cigarette manufacturer \$3 to make a carton of cigarettes, and the supply equation is $q_s = 120p - 40$. What price will realize a profit of at least \$440?

31. It costs a toy manufacturer \$4 per toy to make a particular toy. If the supply equation is $q_s = 10p + 40$, what price will realize a profit of at least \$200?

32. It costs \$4 per book to publish a book, and the supply equation is $q_s = 2p - 10$. What price will realize a profit of at least \$60?

ECONOMICS: EQUILIBRIUM MARKET PRICE

Sometimes a company uses an *equilibrium market price* to obtain pricing information about a product. Let p denote the market price of the product, let q_d denote the quantity of the product sold on the market (demand quantity), and let q_s denote the quantity of the product supplied by the company (supply quantity). The demand equation relates p and q_d, and the supply equation relates p and q_s. These equations usually have the forms

$$q_d = k/p \quad \text{and} \quad q_s = ap + b,$$

where k, a, and b are constants depending on the particular product. The equilibrium market price results from solving the equation $q_d = q_s$ for p.

Example 6. Suppose the demand equation for wine is $q_d = 600/p$, where p is the price per quart. The supply equation is $q_s = 10p + 40$. What is the equilibrium market price?

Solution. We set $q_d = q_s$,

$$600/p = 10p + 40.$$

$$600 = 10p^2 + 40p$$

$$p^2 + 4p - 60 = 0$$

$$(p - 6)(p + 10) = 0$$

$$p = 6 \quad \text{or} \quad p = -10$$

We discard the negative solution, and the equilibrium market price is $6 per quart.

Exercise Set 5.8 (continued)

33. Determine the equilibrium market price for a product whose demand equation is $q_d = 15/p$ and whose supply equation is $q_s = 24p + 2$.

34. Determine the equilibrium market price for a product whose demand equation is $q_d = 18/p$ and whose supply equation is $q_s = 12p - 6$.

35. Suppose the demand equation for a product is $q_d = 150/p$ and the supply equation is $q_s = 4p + 50$. What is the equilibrium market price? Now a tax of $.50 per unit is imposed on the product. What is the new equilibrium market price? *Hint:* After the tax, the demand equation remains the same but the supply equation is changed because p is replaced by $p - \frac{1}{2}$.

HISTORICAL EXERCISES

Exercise Set 5.8 (continued)

The first mathematical work published in the New World was the *Sumario Compendioso* of Juan Diez, published in the city of Mexico in 1556. Exercises 37 and 38 are adopted from this book.

37. A man takes a ship and asks the master what the fare is. The master replies, "The fare is the number of pesos which, when squared and added to this number, gives 1260." Find the fare.

38. A man travelling on a road asks another man how many leagues it is to a certain place. The other man replies, "There are so many leagues that, dividing the square of the number by 5, the quotient is 80." Find the number of leagues.

Exercises 39 and 40 are adopted from the *Lîlâvatî* by Bhâskaa II, a Hindu problem book from about 1150.

39. Arjuna, irritated in fight, shot a quiver of arrows to slay Karna. With half of his arrows he parried those of his antagonist; with four times the square root of the quiver-full he killed his horses; with three he demolished the umbrella; three more the standard; three more the bow; and with one he cut off the head of his foe. How many were the arrows which Arjuna let fly?

40. The square root of half the number of bees in a swarm has flown out upon a jessamine bush; 8/9 of the swarm has remained behind; one female bee flies about a male that is buzzing within a lotus flower into which he was allured in the night by its sweet odor but in which he is now imprisoned. Tell me, most enchanting lady, the number of bees.

Review Exercises

In exercises 1–40, solve each given equation. Some of these equations are linear and some are quadratic.

1. $(x - 3)(x + 2) = 0$
2. $(x + 4)(x - 5) = 0$
3. $(4x + 3)(2x - 1) = 0$
4. $(3x - 2)(3x + 2) = 0$

5. $x^2 = 12$
6. $x^2 = 18$
7. $x^2 + 7x + 10 = 0$
8. $x^2 + 7x + 6 = 0$
9. $x^2 + 3x - 18 = 0$
10. $x^2 + x - 20 = 0$
11. $4x^2 + 36 = 0$
12. $3x^2 + 12 = 0$
13. $x^2 - 2x + 10 = 0$
14. $x^2 - 4x + 8 = 0$
15. $x^2 + 2x - 1 = 0$
16. $x^2 - 2x - 2 = 0$
17. $(2x + 1)^2 = (2x + 3)^2$
18. $(5x - 1)^2 = (5x + 4)^2$
19. $(x + 1)^2 = (2x + 1)^2$
20. $(x - 1)^2 = (5x - 1)^2$
21. $x(2x - 3) = 54$
22. $x(3x + 1) = 10$
23. $(x + 3)(x - 5) = 9$
24. $(x + 2)(2x + 3) = 1$
25. $x^2 + (x + 1)^2 = (x + 2)^2$
26. $x^2 + (x + 1)^2 = (x + 3)^2$
27. $x^2 + (x + 2)^2 = 36$
28. $x^2 + (x + 1)^2 = 25$
29. $(x + 3)^2 + (x + 4)^2 = x(x + 4)$
30. $(x - 9)(x + 1) = 2x(x - 1)$
31. $\frac{1}{x} = x + 1$
32. $\frac{3}{x} = x + 2$
33. $\frac{x - 3}{x + 2} = \frac{x + 1}{x - 4}$
34. $\frac{x - 3}{x + 4} = \frac{x - 5}{x + 6}$
35. $\frac{1}{x^2} + \frac{1}{x} = 2$
36. $\frac{1}{x^2} + \frac{1}{x} = 6$
37. $\frac{x}{x + 2} = \frac{1}{x}$
38. $\frac{x}{x + 6} = \frac{3}{x}$
39. $\frac{x}{x - 1} + \frac{x}{x + 1} = \frac{2}{x^2 - 1}$
40. $\frac{x}{x - 1} + \frac{x}{x + 1} = \frac{8}{x^2 - 1}$

In exercises 41–48, solve each given inequality.

41. $(x - 1)(x + 4) < 0$
42. $(x - 2)(x + 3) > 0$
43. $x^2 - 4 \geq 0$
44. $x^2 - 16 \leq 0$
45. $2x^2 - 5x - 3 > 0$
46. $3x^2 + 5x - 2 < 0$
47. $x^2 + x + 1 \geq 0$
48. $x^2 + 5x + 25 < 0$

In exercises 49–56, solve each given equation which is quadratic in form.

49. $x^4 + 25 = 10x^2$
50. $x^4 + 9 = 10x^2$
51. $x + 2\sqrt{x} - 8 = 0$
52. $x + 4\sqrt{x} - 5 = 0$
53. $x^2 + y^2 = 9$ for y
54. $x^2 + y^2 = 4$ for y
55. $4x^2 + 9y^2 = 36$ for y
56. $9x^2 + y^2 = 9$ for y

In exercises 57–64, solve each given equation.

57. $\sqrt{4x + 5} = \sqrt{3x + 10}$
58. $\sqrt{x + 3} = \sqrt{3x + 7}$
59. $\sqrt{3x + 1} = x - 3$
60. $\sqrt{3x + 6} = x - 4$

61. $\sqrt{4x+1} - \sqrt{2x} = 1$

62. $\sqrt{2x+1} - \sqrt{x-3} = 2$

63. $\sqrt{x+1} + \sqrt{x-1} = 1$

64. $\sqrt{x+1} + \sqrt{x-1} = 2$

65. One side of a rectangle is 7 meters longer than the shorter side. Find the length of each side if the area is 60 square meters.

66. One side of a rectangle is 3 meters longer than the shorter side. Find the length of each side if the diagonal is $\sqrt{29}$ meters.

67. A riverboat can go 12 kilometers per hour in still water. When it makes a trip both upstream and downstream between two towns 45 kilometers apart, the total time is 8 hours. What is the rate of the river?

68. We design a page for a book with 3-centimeter margins on each side and at the top and bottom. We want the height of the page to be twice its width, and we want the area of the printed matter to be 108 square centimeters. What are the dimensions of the page?

69. A farmer wishes to fence in a rectangular field using a straight river for one side of the field and fencing for the other three sides. Find the dimensions of the field if the farmer has 500 meters of fencing and wants to enclose an area of 30,000 square meters.

70. A number x is the *mean proportional* between a and b if $a/x = x/b$. We wish to divide a line segment of length 1 into two parts so that one part is the mean proportional between the whole segment and the other part. How long should each part be? *Note:* This proportion, known as the *golden section*, was famous throughout antiquity.

chapter 6

Polynomial Equations

1. Polynomials

2. Operations on Polynomials

3. Synthetic Division

4. Remainder Theorem

5. Integer Roots

In Chapter 4 we considered linear or first-degree equations; these equations can be put in the form

$$ax + b = 0 \qquad \text{with } a \neq 0.$$

In Chapter 5 we considered quadratic or second-degree equations; these equations can be put in the form

$$ax^2 + bx + c = 0 \qquad \text{with } a \neq 0.$$

It is natural to consider third-degree or fourth-degree equations; these equations would be put in the form

$$ax^3 + bx^2 + cx + d = 0 \qquad \text{with } a \neq 0;$$

$$ax^4 + bx^3 + cx^2 + dx + e = 0 \qquad \text{with } a \neq 0.$$

Equations of degree n (where n is a positive integer) are called *polynomial equations.*

In Section 6.1 we prepare for the topics of this chapter by defining polynomials. We show how to add, subtract, multiply, and divide polynomials in Section 6.2. Synthetic division is a special type of polynomial division which we examine in Section 6.3. We apply synthetic division in Section 6.4 by using the Remainder Theorem and the Factor Theorem. In Section 6.5 we examine a technique for finding integer solutions of a polynomial equation.

section 1 • Polynomials

In this section we specify what it means for an algebraic expression to be a polynomial and we show how to substitute numbers for the variable in a polynomial.

DEFINITION OF POLYNOMIALS

A polynomial in a variable x is an algebraic expression which can be written in decreasing powers of x. The exponent of the highest power of x that occurs is called the *degree* of the polynomial. For example (assuming $a \neq 0$), polynomials possess the form:

first degree	$ax + b$
second degree	$ax^2 + bx + c$
third degree	$ax^3 + bx^2 + cx + d$
fourth degree	$ax^4 + bx^3 + cx^2 + dx + e$

But we need a more systematic way to write the constants in a polynomial, because otherwise we will run out of letters of the alphabet. So we write a general polynomial using *subscripts* to correspond to the degree of the term. For example, a general third-degree polynomial would have the form $a_3x^3 + a_2x^2 + a_1x + a_0$. A general polynomial has the same form but may have higher degree.

DEFINITION OF A POLYNOMIAL. A *polynomial* in one variable x is an algebraic expression which can be written in the form

$$a_nx^n + a_{n-1}x^{n-1} + \cdots + a_1x + a_0,$$

where $a_0, a_1, \ldots, a_{n-1}, a_n$ are constants, n is a non-negative integer, and $a_n \neq 0$.

A polynomial in one variable x is an algebraic expression that can be written in decreasing powers of x. The expression does not have to be *actually* written in decreasing powers of x, so long as it *can* be written that way. The powers of x must be integral powers; a polynomial has no fractional powers. There may be $x^0 = 1$, but there are no negative powers of x. So we may have terms involving x^0, x^1, x^2, x^3, etc., but not x^{-1}, x^{-2}, x^{-3}, etc.

Example 1. Expressions which are polynomials.

a. $3x^2 + 2x - 1$

b. $x^3 + 7$

c. $x^4 - 2x^2 + 1$

d. $x^7 + x^6 - x^5 + 13x^3 - 12$

e. $x^3 + x^4 - x^2 + x$ is a polynomial although it's not written in the specified *form* of a polynomial. It is, however, *equal to* the expression $x^4 + x^3 - x^2 + x$, which has the proper form.

Example 2. Expressions which are not polynomials.

a. $\frac{1}{x}$, since x occurs as x^{-1}.

b. $\sqrt{x}$, since x occurs as $x^{1/2}$.

c. $\frac{x+2}{x-3}$; this is a quotient of two first-degree polynomials.

d. $\sqrt{x^{10} + 2}$; this is the square root of a polynomial.

DEFINITION OF DEGREE AND COEFFICIENTS OF A POLYNOMIAL. In the polynomial $a_n x^n + a_{n-1}x^{n-1} + \cdots + a_1 x + a_0$, the a's are called *coefficients* and a_i is called the *coefficient of the* ith*-degree term.* The degree of the term of largest degree is called the *degree of the polynomial.* The coefficient of the term of largest degree is called the *leading coefficient*; and the coefficient of x^0, namely a_0, is called the *constant term.*

Example 3. Degree and coefficients.

a. For the polynomial $8x^3 + 3x^2 - 2x - 1$, the degree is 3, the leading coefficient is 8, and the constant term is -1.

b. For the polynomial $x^{100} + x^{17}$, the degree is 100, the leading coefficient is 1, and the constant term is 0.

c. For the polynomial $x^6 + x^4 - x^2 - 1$, the degree is 6, the leading coefficient is 1, and the constant term is -1.

Finally, it's possible that a polynomial is not written in the form of a polynomial because the terms involving the same power of x have not been collected together.

Example 4. Collecting terms in a polynomial.

a. $3x^2 + 2x - 1 + x^2 - x - 2$

$$= (3x^2 + x^2) + (2x - x) + (-1 - 2)$$

$$= 4x^2 + x - 3$$

b. $x^5 + 6x - 2 + x^2 - x + 2x^3 - 1$

$$= x^5 + 2x^3 + x^2 + (6x - x) + (-2 - 1)$$

$$= x^5 + 2x^3 + x^2 + 5x - 3$$

c. $x^3 + x^2 + x + 1 + 3x^3 - x^2 - 2x + 2$
$$= (x^3 + 3x^3) + (x^2 - x^2) + (x - 2x) + (1 + 2)$$
$$= 4x^3 - x + 3$$

Exercise Set 6.1

In exercises 1–10, specify the degree, leading coefficient, and constant term of each given polynomial.

1. $3x^2 - x + 7$
2. $x^5 + 6x - 2$
3. $x^{17} + x^4 - 3x^2 + 1$
4. $-3x^8 - 6x^4 - 7x + 2$
5. $12x^{87} + 13x^{26} - 6x^2$
6. $12x^4 + x - x^2 - 2$
7. $x^2 + x^6 - 7x^8$
8. $x^4 - x^{12} + 4 - x^{11}$
9. $x^2 + 3x^3 - x^6 + 7x$
10. $1 + x^{99}$

In exercises 11–20, collect together terms of the same power in each given polynomial.

11. $x^2 + 3x - 2 + 2x^2 - 3x + 1$
12. $x^3 - 3x + 2 + 3x^2 - 6x - 8$
13. $x^2 + 5x - 6 + 7x^2 - 7x - 7$
14. $x^3 + 2x + 2 - x^3 - 6x^2 + 6x - 4$
15. $x^4 + 3x^2 - x + 2x^4 - x - 3 - x^3$
16. $x^4 + 2x^3 + 3x - x^3 - 4x - 3$
17. $x^3 + x^4 - x^2 + 3x^3 - x - 2x^2$
18. $x^2 + x^3 + x^4 - 1 - 3x^2 + x^4$
19. $x^5 + x^3 + x + x^5 + x + 1$
20. $x^5 - x^4 + x^5 - x^4 + x^3$

SUBSTITUTING NUMBERS

To avoid writing the long expression

$$a_nx^n + a_{n-1}x^{n-1} + \cdots + a_1x + a_0$$

all the time, we will frequently use the notation $P(x)$ or $Q(x)$ to denote polynomials. The x in the parentheses refers to the variable x in the polynomial. An advantage of this notation is that, when we substitute a number, say a, for x in the polynomial $P(x)$, we write $P(a)$.

Example 5. If $P(x) = x^3 + 4x^2 - x - 4$, we substitute the numbers 0, 1, 2, and -1 for x:

$$P(0) = 0^3 + 4(0)^2 - 0 - 4 = -4$$
$$P(1) = 1^3 + 4(1)^2 - 1 - 4 = 0$$
$$P(2) = 2^3 + 4(2)^2 - 2 - 4 = 18$$
$$P(-1) = (-1)^3 + 4(-1)^2 - (-1) - 4 = 0$$

DEFINITION OF ROOT. A number a is a *root* of a polynomial $P(x)$ if $P(a) = 0$.

In Example 5, of the four numbers we substituted for x, we see that 1 and -1 are roots of $P(x)$ and that 0 and 2 are not roots of $P(x)$.

Example 6. If $P(x) = 2x^3 - x^2 - 8x + 4$, we substitute the numbers 0, 1, -1, and 0.5 for x:

$$P(0) = 2(0)^3 - (0)^2 - 8(0) + 4 = 4$$
$$P(1) = 2(1)^3 - (1)^2 - 8(1) + 4 = -3$$
$$P(-1) = 2(-1)^3 - (-1)^2 - 8(-1) + 4 = 9$$
$$P(0.5) = 2(0.5)^3 - (0.5)^2 - 8(0.5) + 4 = 0$$

We see that 0.5 is a root of $P(x)$ and that 0, 1, and -1 are not roots of $P(x)$.

Example 7. If $Q(x) = x^4 - 1$, we substitute the numbers 1, -1, $\sqrt{2}$, and i for x:

$$Q(1) = (1)^4 - 1 = 0$$
$$Q(-1) = (-1)^4 - 1 = 0$$
$$Q(\sqrt{2}) = (\sqrt{2})^4 - 1 = 3$$
$$Q(i) = i^4 - 1 = 0$$

We see that 1, -1, and i are roots of $Q(x)$ and that $\sqrt{2}$ is not a root of $Q(x)$.

Exercise Set 6.1 (continued)

In exercises 21–32, substitute the given numbers for x in each given polynomial and specify whether each number is or is not a root.

21. $P(x) = x^3 + 4x^2 - x - 4$; $-2, -3, -4, 4$
22. $P(x) = 2x^3 - x^2 - 8x + 4$; $2, -2, 3, -3$
23. $Q(x) = x^4 - 1$; $0, 2, -2, -i$
24. $Q(x) = x^4 + 1$; $1, -1, 0, i$
25. $P(x) = x^3 - 2x^2 - x + 2$; $1, -1, 2, -2$
26. $P(x) = x^3 - 6x^2 + x - 6$; $1, -1, 2, -2$
27. $P(x) = x^3 + 3x^2 - x + 12$; $1, -1, 3, -3$
28. $P(x) = x^3 - 10x - 75$; $1, -1, 3, -3$
29. $Q(x) = x^4 - x^2 - 2$; $1, -1, \sqrt{2}, -\sqrt{2}$
30. $Q(x) = x^6 + 1$; $1, -1, \sqrt{2}, -\sqrt{2}$
31. $P(x) = x^{100} - 1$; $0, 1, -1, i$
32. $P(x) = x^{100} + 1$; $0, 1, -1, i$

---

section 2 • Operations on Polynomials

In this section we examine the four operations—addition, subtraction, multiplication, and division—with respect to polynomials.

ADDITION AND SUBTRACTION

Example 1. If $P(x) = 3x^3 - 2x^2 + x - 1$ and $Q(x) = x^2 - 3x + 2$, find $P(x) + Q(x)$.

Solution. Just by placing a plus sign between the polynomials, we have $P(x) + Q(x) = 3x^3 - 2x^2 + x - 1 + x^2 - 3x + 2$. The only thing left is to write this as a polynomial, and so we collect together the terms having the same powers of x. We obtain

$$\begin{aligned} P(x) + Q(x) &= 3x^3 - 2x^2 + x^2 + x - 3x - 1 + 2 \\ &= 3x^3 - x^2 - 2x + 1 \end{aligned}$$

This is written as a polynomial and is the sum of $P(x)$ and $Q(x)$.

It is sometimes convenient to do the addition by writing powers of x of the same degree under each other.

$$\begin{array}{r} 3x^3 - 2x^2 + x - 1 \\ + \qquad\quad x^2 - 3x + 2 \\ \hline 3x^3 - x^2 - 2x + 1 \end{array}$$

Example 2. If $P(x) = x^4 - x^3 + 3x - 2$ and $Q(x) = x^4 + 2x^2 + 2x$, find $P(x) - Q(x)$.

Solution. If we perform the subtraction and collect together the terms having the same powers of x, we have

$$\begin{aligned} P(x) - Q(x) &= (x^4 - x^3 + 3x - 2) - (x^4 + 2x^2 + 2x) \\ &= (x^4 - x^4) - x^3 - 2x^2 + (3x - 2x) - 2 \\ &= -x^3 - 2x^2 + x - 2 \end{aligned}$$

We can also perform subtraction by using the following scheme:

$$\begin{array}{r} x^4 - x^3 + 0x^2 + 3x - 2 \\ -\,x^4 + 0x^3 + 2x^2 + 2x \quad \\ \hline - x^3 - 2x^2 + x - 2 \end{array}$$

Exercise Set 6.2

In exercises 1–12, find $P(x) + Q(x)$ and $P(x) - Q(x)$. Write each result as a polynomial.

1. $P(x) = 2x^3 + 3x^2 - 2x + 1$; $Q(x) = x^3 - x^2 + 3x - 3$
2. $P(x) = 2x^3 - x^2 + 4x - 2$; $Q(x) = x^3 + 2x^2 + 3x + 4$
3. $P(x) = 3x^3 - 3x^2 + 4$; $Q(x) = x^3 + 3x - 2$
4. $P(x) = 3x^3 + 2x - 6$; $Q(x) = x^3 - x^2 + 4$
5. $P(x) = 3x^3 - 3x + 4$; $Q(x) = 2x^3 - x^2 + x - 2$
6. $P(x) = 4x^3 + 3x^2 + 2x + 1$; $Q(x) = 2x^3 - x^2 + x$
7. $P(x) = x^4 + x^2 - 2$; $Q(x) = x^3 + x^2 + x + 1$

8. $P(x) = x^4 - 1$; $Q(x) = x^3 - x^2 + x - 1$
9. $P(x) = x^4 + x^3 + x^2 + x + 1$; $Q(x) = x^4 - x^3 + x^2 - x + 1$
10. $P(x) = x^4 - x^3 + x - 2$; $Q(x) = x^4 + x^2 + 1$
11. $P(x) = x^5 + 1$; $Q(x) = x^5 - 1$
12. $P(x) = x^6 + x$; $Q(x) = x^6 - x$

MULTIPLICATION

Example 3. If $P(x) = 2x^2 + x - 3$ and $Q(x) = 3x^2 - x + 1$, find $P(x) \cdot Q(x)$.

Solution. We use the basic rule for multiplying any algebraic expressions (Section 2.1): We form all possible monomial products, paying attention to positive and negative numbers, and add them together.

$$\begin{aligned} P(x) \cdot Q(x) &= (2x^2 + x - 3)(3x^2 - x + 1) \\ &= 2x^2(3x^2) + x(3x^2) - 3(3x^2) + 2x^2(-x) + x(-x) - 3(-x) \\ &\quad + 2x^2(1) + x(1) - 3(1) \\ &= 6x^4 + 3x^3 - 9x^2 - 2x^3 - x^2 + 3x + 2x^2 + x - 3 \\ &= 6x^4 + x^3 - 8x^2 + 4x - 3. \end{aligned}$$

When we are multiplying polynomials, it is particularly convenient to use the following scheme:

$$\begin{array}{rrrrr} & & 2x^2 & + x & - 3 \\ & & 3x^2 & - x & + 1 \\ \hline 6x^4 & + 3x^3 & - 9x^2 & & \\ & - 2x^3 & - x^2 & + 3x & \\ & & 2x^2 & + x & - 3 \\ \hline 6x^4 & + x^3 & - 8x^2 & + 4x & - 3 \end{array}$$

The scheme simply provides a convenient way of forming all possible monomial products.

Exercise Set 6.2 (continued)

In exercises 13–24, find $P(x) \cdot Q(x)$. Write the product as a polynomial.

13. $P(x) = x^2 + 4x - 3$; $Q(x) = x + 2$
14. $P(x) = x^2 - x + 6$; $Q(x) = 2x + 1$
15. $P(x) = x - 3$; $Q(x) = 3x^2 - 7$
16. $P(x) = 2x + 7$; $Q(x) = x^2 + 5x - 1$
17. $P(x) = x^2 + 2x - 1$; $Q(x) = x^2 + x + 1$
18. $P(x) = 7x^2 + 5$; $Q(x) = x^2 - 3x - 2$
19. $P(x) = 2x^2 - x + 4$; $Q(x) = x^2 - 4x$
20. $P(x) = x^2 + 3x - 4$; $Q(x) = 3x^2 - 2x + 1$

21. $P(x) = x^3 + 1$; $Q(x) = x^3 - 1$
22. $P(x) = 2x^3 + x^2 - 5x + 4$; $Q(x) = x^2 - 4$
23. $P(x) = x^3 + 6x^2 - 2x + 1$; $Q(x) = x^2 + 2x + 2$
24. $P(x) = 3x^4 + x^2 + 7$; $Q(x) = x - 1$

DIVISION

Example 4. Divide $6x^4 + 4x^3 + 3x^2 + 9x - 4$ by $2x^2 + 2x - 1$.

Solution. We use the familiar scheme for long division:

$$\begin{array}{r|l}
 & 3x^2 - x + 4 \\
2x^2 + 2x - 1 & 6x^4 + 4x^3 + 3x^2 + 9x - 4 \\
 & \underline{6x^4 + 6x^3 - 3x^2} \\
 & -2x^3 + 6x^2 + 9x \\
 & \underline{-2x^3 - 2x^2 + x} \\
 & 8x^2 + 8x - 4 \\
 & \underline{8x^2 + 8x - 4}
\end{array}$$

The quotient of $6x^4 + 4x^3 + 3x^2 + 9x - 4$ by $2x^2 + 2x - 1$ is $3x^2 - x + 4$.

What happens when we divide two polynomials? If we divide $A(x)$ by $B(x)$, we obtain a quotient polynomial $Q(x)$ and a remainder polynomial $R(x)$(although the remainder may be zero). Furthermore, we can keep dividing $A(x)$ by $B(x)$ until the degree of $R(x)$ is *less than* the degree of $B(x)$. We write this as

$$\frac{A(x)}{B(x)} = Q(x) + \frac{R(x)}{B(x)} \quad \text{and} \quad \text{degree } R(x) < \text{degree } B(x).$$

If we multiply both sides of this equation by $B(x)$, we can write it in the form

$$A(x) = Q(x)\,B(x) + R(x) \quad \text{and} \quad \text{degree } R(x) < \text{degree } B(x).$$

Exercise Set 6.2 (continued)

In exercises 25–36, divide the first polynomial by the second polynomial; specify the quotient and remainder.

25. $3x^4 + 5x^3 + 7x^2 + 3x - 2$; $x^2 + x + 2$
26. $2x^4 - 7x^3 + 5x^2 - 9x - 18$; $x^2 - x - 3$
27. $6x^4 - 2x^3 - 33x^2 + 29x - 6$; $2x^2 + 4x - 3$
28. $9x^4 - 6x^3 + 3x^2 - 12x - 30$; $3x^2 - 2x - 5$
29. $3x^4 - 4x^3 - 5x^2 + 5x + 3$; $x^2 - 2x - 1$
30. $4x^4 + x^3 - 10x^2 - x$; $x^2 - 1$
31. $x^6 - 1$; $x^2 - 1$
32. $x^8 - 1$; $x^2 - 1$

33. $x^4 + 5x^3 + 3x^2 - 5x + 2;\ x + 3$
34. $x^4 - 2x^3 + 2x^2 - 5x + 6;\ x - 2$
35. $x^4 - 1;\ x - 1$
36. $x^5 - 1;\ x - 1$

---

section 3 • Synthetic Division

In this section we consider a special method for dividing polynomials, *synthetic division.* Synthetic division does not apply to all cases of division of polynomials. It applies only to the case in which the divisor is a polynomial of the form $x + c$ or $x - c$. Yet this is an important case, and synthetic division provides a fast way of doing this type of division.

METHOD OF SYNTHETIC DIVISION

We use the example of dividing $3x^4 + 4x^3 - 2x + 4$ by $x + 2$. We do this first by long division.

$$
\begin{array}{r|l}
 & 3x^3 - 2x^2 + 4x - 10 \\ \hline
x + 2 & 3x^4 + 4x^3 \qquad\quad - 2x + 4 \\
 & \underline{3x^4 + 6x^3} \\
 & \quad - 2x^3 + 0x^2 \\
 & \quad \underline{- 2x^3 - 4x^2} \\
 & \qquad\qquad 4x^2 - 2x \\
 & \qquad\qquad \underline{4x^2 + 8x} \\
 & \qquad\qquad\quad - 10x + 4 \\
 & \qquad\qquad\quad \underline{- 10x - 20} \\
 & \qquad\qquad\qquad\qquad 24
\end{array}
$$

The quotient is $3x^3 - 2x^2 + 4x - 10$ and the remainder is 24.

Now synthetic division is just another way of *organizing* long division. The first thing to notice is that the powers of x are in some sense placeholders—it's not really necessary to write them down. Suppose we omit the powers of x and simply write the coefficients.

$$
\begin{array}{rr|rrrrr}
 & & 3 & -2 & 4 & -10 & \\
\hline
1 & 2 & 3 & 4 & 0 & -2 & 4 \\
 & & 3 & 6 & & & \\
\cline{3-4}
 & & & -2 & 0 & & \\
 & & & -2 & -4 & & \\
\cline{4-5}
 & & & & 4 & -2 & \\
 & & & & 4 & 8 & \\
\cline{5-6}
 & & & & & -10 & 4 \\
 & & & & & -10 & -20 \\
\cline{6-7}
 & & & & & & 24
\end{array}
$$

The next thing to notice is that there is repetition among the numbers we have written down. If we leave out the repetitions, we obtain

$$
\begin{array}{rr|rrrrr}
 & & 3 & -2 & 4 & -10 & \\
\hline
1 & 2 & 3 & 4 & 0 & -2 & 4 \\
 & & & 6 & & & \\
 & & & -2 & -4 & & \\
 & & & & 4 & 8 & \\
 & & & & & -10 & -20 \\
 & & & & & & 24
\end{array}
$$

We still have repetitions from the numbers on the top line, the answer line. To avoid these repetitions, we reorganize the work so the answer line is the bottom line.

$$
\begin{array}{rr|rrrrr}
1 & 2 & 3 & 4 & 0 & -2 & 4 \\
\cline{1-2}
 & & & 6 & -4 & 8 & -20 \\
\hline
 & & 3 & -2 & 4 & -10 & \multicolumn{1}{|r}{24}
\end{array}
$$

Finally, the 1 in the divisor plays no part, so we omit that. Also, it's easier to add than subtract, and we can add by changing the 2 in the divisor to -2. Now we have the form for synthetic division of these polynomials:

$$
\begin{array}{r|rrrrr}
-2 & 3 & 4 & 0 & -2 & 4 \\
\cline{1-1}
 & & -6 & 4 & -8 & 20 \\
\hline
 & 3 & -2 & 4 & -10 & \multicolumn{1}{|r}{24}
\end{array}
$$

Now we outline the technique of synthetic division. We want to divide $3x^4 + 4x^3 - 2x + 4$ by $x + 2$. On the first line, we change the sign of the constant in the divisor and write the coefficients of the dividend.

$$
\begin{array}{r|rrrrr}
-2 & 3 & 4 & 0 & -2 & 4 \\
\cline{1-1}
\end{array}
$$

The answer—the quotient and remainder—will appear on the third line. To obtain this arrangement, we bring down the first number to the third line. Then we multiply that number by -2 and enter the product in the second line, one space to the right. We obtain the next entry on the third line by *adding*.

$$\begin{array}{r|rrrr|r}
-2 & 3 & 4 & 0 & -2 & 4 \\
 & & -6 & & & \\ \hline
 & 3 & -2 & & &
\end{array}$$

We proceed in this way, working to the right. Next we multiply the entry on the third line by -2 and enter the answer on the second line, one space to the right. Then we add again to obtain the next entry on the third line.

$$\begin{array}{r|rrrr|r}
-2 & 3 & 4 & 0 & -2 & 4 \\
 & & -6 & 4 & & \\ \hline
 & 3 & -2 & 4 & &
\end{array}$$

Now we simply continue in this way, finally reading the answer off the third line. The number furthest to the right is the remainder, and the rest are the coefficients of the various powers of x.

$$\begin{array}{r|rrrr|r}
-2 & 3 & 4 & 0 & -2 & 4 \\
 & & -6 & 4 & -8 & 20 \\ \hline
 & 3 & -2 & 4 & -10 & 24
\end{array}$$

The quotient is $3x^3 - 2x^2 + 4x - 10$, and the remainder is 24.

Example 1. Divide $6x^4 + x^3 - 3x + 4$ by $x + 1$.

Solution.

$$\begin{array}{r|rrrr|r}
-1 & 6 & 1 & 0 & -3 & 4 \\
 & & -6 & 5 & -5 & 8 \\ \hline
 & 6 & -5 & 5 & -8 & 12
\end{array}$$

The quotient is $6x^3 - 5x^2 + 5x - 8$, and the remainder is 12.

Example 2. Divide $x^6 + 3x^5 - 2x^4 + x^2 - 3x + 2$ by $x - 2$.

Solution.

$$\begin{array}{r|rrrrrr|r}
2 & 1 & 3 & -2 & 0 & 1 & -3 & 2 \\
 & & 2 & 10 & 16 & 32 & 66 & 126 \\ \hline
 & 1 & 5 & 8 & 16 & 33 & 63 & 128
\end{array}$$

The quotient is $x^5 + 5x^4 + 8x^3 + 16x^2 + 33x + 63$, and the remainder is 128.

Exercise Set 6.3

In exercises 1–6, use synthetic division to divide each given polynomial by $x - 1$; specify the quotient polynomial and the remainder.

1. $4x^3 - 4x^2 - 5x + 3$
2. $3x^3 + 5x^2 + 5x + 2$
3. $x^3 - 9x^2 - 25x + 33$
4. $x^3 + x - 2$
5. $3x^4 + 3x^2 + 4$
6. $2x^4 + 7x^3 + 6x^2 + 7x - 6$

In exercises 7–12, use synthetic division to divide each given polynomial by $x + 1$; specify the quotient polynomial and the remainder.

7. $x^3 - 2x^2 - x + 2$
8. $x^3 - 6x^2 + x - 6$
9. $x^3 + 2x^2 + 10x - 20$
10. $x^3 + 2x^2 + 10x + 20$
11. $2x^5 + 3x^3 - 16x^2 - 24$
12. $x^6 + x^4 + 5x^2 - 4$

In exercises 13–18, use synthetic division to divide each given polynomial by $x - 2$; specify the quotient polynomial and the remainder.

13. $x^3 - 2x^2 - x + 2$
14. $x^3 - 6x^2 + x - 6$
15. $x^3 + x - 2$
16. $3x^3 + 5x^2 + 5x + 2$
17. $x^4 - 2x^3 - 7x^2 + 8x - 12$
18. $x^4 - x^3 + x^2 - x + 12$

In exercises 19–24, use synthetic division to divide each given polynomial by $x + 2$; specify the quotient polynomial and the remainder.

19. $x^3 + x - 2$
20. $3x^3 + 5x^2 + 5x + 2$
21. $3x^4 + 3x^2 + 4$
22. $2x^4 + 7x^3 + 6x^2 + 7x - 6$
23. $2x^5 + 3x^3 - 16x^2 - 24$
24. $x^6 + x^4 + 5x^2 - 4$

In exercises 25–28, use synthetic division to divide each given polynomial by $x - 0.5$; specify the quotient polynomial and the remainder.

25. $2x^3 - 5x^2 - 4x + 3$
26. $2x^3 - 7x^2 + 9$
27. $4x^3 - 4x^2 - 5x + 3$
28. $2x^4 + 7x^3 + 6x^2 + 7x - 6$

In exercises 29–32, use synthetic division to divide each given polynomial by $x + 0.5$; specify the quotient polynomial and the remainder.

29. $2x^3 - 5x^2 - 4x + 3$
30. $2x^3 - 7x^2 + 9$
31. $8x^3 + 1$
32. $32x^5 + 1$

section 4 • Remainder Theorem

In this section we examine the Remainder Theorem and the Factor Theorem. The Remainder Theorem can be used to find the value of a polynomial when a number is substituted for the variable, and the Factor Theorem can be used to test for factors of a polynomial.

THE REMAINDER THEOREM

If we divide a polynomial $P(x)$ by a polynomial of the form $x - a$, we obtain a quotient polynomial $Q(x)$ and a remainder. Because the divisor $x - a$ is a first-degree polynomial, the remainder must be a number R. We can write (see Section 6.2)

$$P(x) = (x - a)Q(x) + R.$$

If we substitute the number a for x, we find

$$P(a) = (a - a)Q(a) + R = 0 \cdot Q(a) + R = R.$$

We have proven the Remainder Theorem.

REMAINDER THEOREM. Suppose $P(x)$ is a polynomial and a is a number. If $P(x) = (x - a)Q(x) + R$, then $P(a) = R$; that is, $P(a)$ equals the remainder when we divide $P(x)$ by $x - a$.

We can use the Remainder Theorem to find the value of a polynomial when a number a is substituted for x. Usually it is easier to use synthetic division to divide by $x - a$ than to substitute a directly in the polynomial.

Example 1. If $P(x) = x^3 - 3x^2 + 2x + 2$, find $P(-1)$.

Solution. Using synthetic division, we divide $P(x)$ by $x - (-1)$.

$$\begin{array}{r|rrrr} -1 & 1 & -3 & 2 & 2 \\ & & -1 & 4 & -6 \\ \hline & 1 & -4 & 6 & \vert -4 \end{array}$$

Since the remainder is -4, we have $P(-1) = -4$. We can check this by direct substitution:

$$P(-1) = (-1)^3 - 3(-1)^2 + 2(-1) + 2 = -1 - 3 - 2 + 2 = -4.$$

Example 2. If $P(x) = 3x^4 + x^3 - 2x^2 + 5x - 1$, find $P(3)$.

Solution. Using synthetic division, we divide $P(x)$ by $x - 3$.

$$\begin{array}{r|rrrrr} 3 & 3 & 1 & -2 & 5 & -1 \\ & & 9 & 30 & 84 & 267 \\ \hline & 3 & 10 & 28 & 89 & \vert\ 266 \end{array}$$

The remainder term is 266, so $P(3) = 266$. We check this by direct substitution:

$$P(3) = 3 \cdot 3^4 + 3^3 - 2 \cdot 3^2 + 5 \cdot 3 - 1 = 243 + 27 - 18 + 15 - 1 = 266.$$

Exercise Set 6.4

In exercises 1–6, find $P(1)$ and $P(-1)$ by using the Remainder Theorem.

1. $x^3 - 2x^2 - x + 2$
2. $x^3 - 6x^2 + x - 6$
3. $2x^3 - 5x^2 - 4x + 3$
4. $2x^3 - 7x^2 + 9$
5. $3x^4 + 3x^2 + 4$
6. $2x^4 + 7x^3 + 6x^2 + 7x - 6$

In exercises 7–12, find $P(2)$ and $P(-2)$ by using the Remainder Theorem.

7. $x^3 - x^2 - 4x + 4$
8. $x^3 + 4x^2 + x - 6$
9. $x^3 + 2x^2 + 10x + 20$
10. $x^3 + 3x^2 - x + 12$
11. $3x^4 + 3x^2 + 4$
12. $2x^4 + 7x^3 + 6x^2 + 7x - 6$

In exercises 13–16, find $P(3)$ and $P(-3)$ by using the Remainder Theorem.

13. $x^3 + 3x^2 - x + 12$
14. $x^3 - 10x - 75$
15. $2x^3 - 5x^2 - 4x + 3$
16. $2x^3 - 7x^2 + 9$

In exercises 17–20, find $P(0.5)$ and $P(-0.5)$ by using the Remainder Theorem.

17. $2x^3 - 5x^2 - 4x + 3$
18. $2x^3 - 7x^2 + 9$
19. $x^3 + 2x^2 + 10x + 20$
20. $x^3 - 6x^2 + 6x - 5$

THE FACTOR THEOREM

We specify what it means for a polynomial to be a factor of another polynomial.

DEFINITION OF FACTOR. A polynomial $A(x)$ is a *factor* of a polynomial $P(x)$ if $P(x) = A(x)B(x)$ for some polynomial $B(x)$.

Example 3. Factors.

a. $x^2 + 1$ is a factor of $x^3 - x^2 + x - 1$ because

$$x^3 - x^2 + x - 1 = (x^2 + 1)(x - 1);$$

$x - 1$ is also a factor.

b. $x + 2$ is a factor of $x^3 + x^2 - 8x - 12$ because

$$x^3 + x^2 - 8x - 12 = (x + 2)(x^2 - x - 6);$$

$x^2 - x - 6$ is also a factor.

The Factor Theorem relates roots of a polynomial with factors of the polynomial.

FACTOR THEOREM. A number a is a root of a polynomial $P(x)$ if and only if $x - a$ is a factor of $P(x)$.

Recall from Section 6.1 that a number a is a root of $P(x)$ if and only if $P(a) = 0$. By the Remainder Theorem, $P(a) = 0$ if and only if the remainder when $P(x)$ is divided by $x - a$ is 0; that is, if and only if $x - a$ is a factor of $P(x)$.

Example 4. If $P(x) = x^3 - 15x - 4$ we find $P(4)$ by using synthetic division:

$$\begin{array}{r|rrrr} 4 & 1 & 0 & -15 & -4 \\ & & 4 & 16 & 4 \\ \hline & 1 & 4 & 1 & \vert\ 0 \end{array}$$

We have $P(4) = 0$, so 4 is a root of $P(x)$. Also, since the remainder is 0, $x - 4$ is a factor of $P(x)$. We see that

$$x^3 - 15x - 4 = (x - 4)(x^2 + 4x + 1).$$

Now we find $P(-1)$ by synthetic division:

$$\begin{array}{r|rrrr} -1 & 1 & 0 & -15 & -4 \\ & & -1 & 1 & 14 \\ \hline & 1 & -1 & -14 & \vert\ 10 \end{array}$$

We have $P(-1) = 10$, so -1 is not a root of $P(x)$. Also, since the remainder is not 0, $x + 1$ is not a factor of $P(x)$.

Exercise Set 6.4 (continued)

In exercises 21–26, determine whether $A(x)$ is a factor of $P(x)$ by dividing $P(x)$ by $A(x)$ to see if the remainder is 0.

21. $P(x) = x^4 + x^3 - x - 1$; $A(x) = x^2 - 1$
22. $P(x) = x^4 - x^3 + 2x^2 - x + 1$; $A(x) = x^2 + 1$
23. $P(x) = x^4 + x^3 + x + 1$; $A(x) = x^2 - 1$
24. $P(x) = x^4 + x^3 + 2x^2 + x + 1$; $A(x) = x^2 + 1$
25. $P(x) = x^5 - 1$; $A(x) = x^2 + x + 1$
26. $P(x) = x^5 + 1$; $A(x) = x^2 + 2x + 1$

In exercises 27–38, determine whether the given number is a root of $P(x)$ by using synthetic division.

27. $P(x) = x^3 - 2x^2 - x + 2$; 1
28. $P(x) = x^3 - 6x^2 + x - 6$; 1
29. $P(x) = 2x^3 - 5x^2 - 4x + 3$; -1
30. $P(x) = 2x^3 - 7x^2 + 9$; -1
31. $P(x) = x^3 - 9x^2 - 25x + 33$; 1
32. $P(x) = x^3 + x - 2$; 1
33. $P(x) = 3x^4 + 3x^2 + 4$; -1
34. $P(x) = 2x^4 + 7x^3 + 6x^2 + 7x - 6$; -1
35. $P(x) = 3x^4 + 3x^2 + 4$; -2
36. $P(x) = 2x^4 + 7x^3 + 6x^2 + 7x - 6$; -2
37. $P(x) = 2x^5 + 3x^3 - 16x^2 - 24$; 2
38. $P(x) = x^6 + x^4 + 5x^2 - 4$; 2

In exercises 39–46, determine whether the given polynomial $x - a$ is a factor of $P(x)$; if it is, find the factorization.

39. $P(x) = x^3 - x^2 - 4x + 4;\ x - 1$
40. $P(x) = x^3 + 4x^2 + x - 6;\ x - 1$
41. $P(x) = x^3 + 2x^2 + 10x + 20;\ x + 1$
42. $P(x) = x^3 + 3x^2 - x + 12;\ x + 1$
43. $P(x) = x^3 - x^2 - 4x + 4;\ x - 2$
44. $P(x) = x^3 + 4x^2 + x - 6;\ x - 2$
45. $P(x) = x^3 - 6x^2 + 6x - 5;\ x - 1$
46. $P(x) = x^3 - 10x - 75;\ x - 5$
47. Show that, if n is a positive integer, then $x - 1$ is always a factor of $x^n - 1$.
48. Show that, if n is a positive even integer, then $x + 1$ is always a factor of $x^n - 1$.
49. Show that, if n is a positive odd integer, then $x + 1$ is always a factor of $x^n + 1$.

section 5 • Integer Roots

In this section we examine a theorem about the integer roots of a polynomial. Then we show how we can use this theorem to solve polynomial equations.

FACTORS OF INTEGERS

DEFINITION OF FACTOR. An integer a is a *factor* of an integer n if $n = ab$ for some integer b.

The concept of factor is defined when the numbers involved are all integers. Notice that ± 1 and $\pm n$ are always factors of n, and there may or may not be other factors. But for a non-zero integer, there is only a finite number of factors.

We can check to see if an integer a is a factor of an integer n by dividing n by a; if the result is an integer, then a is a factor of n; but if the result is not an integer, then a is not a factor of n. Of course, this check can be done quickly using an electronic calculator.

Example 1. Find all the factors of the following numbers:

a. The factors of 10 are ± 1, ± 2, ± 5, and ± 10.
b. The factors of 12 are ± 1, ± 2, ± 3, ± 4, ± 6, and ± 12.
c. The factors of -15 are ± 1, ± 3, ± 5, and ± 15.
d. The factors of 17 are ± 1 and ± 17.

Exercise Set 6.5

In exercises 1–12, find all the factors of each given integer.

1. 6
2. 20
3. -8
4. -51
5. 19
6. 24
7. -75
8. 33
9. 35
10. -21
11. -91
12. 1001

THEOREM ON INTEGER ROOTS

Suppose we have a polynomial $P(x)$ with integer coefficients; that is,

$$P(x) = a_n x^n + a_{n-1} x^{n-1} + \cdots + a_1 x + a_0$$

where $a_n, a_{n-1}, \ldots, a_1, a_0$ are integers. In looking for roots to $P(x)$, we naturally start tr[illegible]mbers that are integers. For example, we might try 1, 2, 3, -1, and so on. Because [illegible]e infinitely many integers, we could go on forever looking for integer roots. The [illegible]ng theorem reduces the integer possibilities to a finite number; that is, we need only [illegible] many integers as possibilities for roots.

[illegible] ON INTEGER ROOTS. If $P(x)$ is a polynomial with integer coefficients [illegible]nteger root of $P(x)$, then r is a factor of the constant term of $P(x)$.
[illegible]e this theorem, suppose we have

$$P(x) = a_n x^n + a_{n-1} x^{n-1} + \cdots + a_1 x + a_0$$

[illegible] is an integer. If r is a root, then

$$0 = P(r) = a_n r^n + a_{n-1} r^{n-1} + \cdots + a_1 r + a_0.$$

[illegible]oring r from each term except a_0, we find $a_0 = rt$, where t is the integer

$$t = -(a_n r^{n-1} + a_{n-1} r^{n-2} + \cdots + a_1),$$

thus showing that r is a factor of the constant term a_0.

Example 2. Find all the integer roots of $P(x) = x^3 - 3x^2 - 3x - 4$.

Solution. The theorem on integer roots says that we only need to try the factors of -4. These factors are ± 1, ± 2, and ± 4. So there are six possibilities; if there is an integer root it must be one of these six. We try the possibilities:

$$P(1) = -9 \qquad P(-1) = -5$$
$$P(2) = -14 \qquad P(-2) = -18$$
$$P(4) = 0 \qquad P(-4) = -104$$

We see that 4 is the only integer root.

Example 3. Find all the integer roots of $P(x) = 2x^3 - 3x^2 + 2x - 3$.

Solution. The theorem on integer roots says that we only need to try the factors of -3. These factors are ± 1 and ± 3. So there are four possibilities; if there is an integer root it has to be one of these four. We try the possibilities:

$$P(1) = -2 \qquad P(-1) = -10$$
$$P(3) = 30 \qquad P(-3) = -90$$

We see that $P(x)$ has no integer roots.

Note: The fact that a polynomial has no integer roots doesn't mean that it has no roots at all. For example, the polynomial $P(x)$ of Example 3 has roots 1.5, i, and $-i$, as you can verify by substituting these numbers for x. In fact, a theorem of advanced mathematics, called The Fundamental Theorem of Algebra, states that every polynomial of degree at least one possesses at least one root among the complex numbers. For example, $P(x)$ possesses two complex number roots, i and $-i$, and one rational number root, 1.5.

Exercise Set 6.5 (continued)

In exercises 13–34, find all the integer roots of the given polynomial.

13. $x^3 - 2x^2 - x + 2$
14. $x^3 - 6x^2 + x - 6$
15. $x^3 + 3x^2 - x + 12$
16. $x^3 - 10x - 75$
17. $x^3 + 2x^2 + 10x - 20$
18. $x^3 + 2x^2 + 10x + 20$
19. $x^3 - 9x^2 - 25x + 33$
20. $x^3 + x - 2$
21. $2x^3 - 5x^2 - 4x + 3$
22. $2x^3 - 7x^2 + 9$
23. $4x^3 - 4x^2 - 5x + 3$
24. $3x^3 + 5x^2 + 5x + 2$
25. $x^4 - 2x^3 - 7x^2 + 8x - 12$
26. $x^4 - x^3 + x^2 - x + 12$
27. $3x^4 + 3x^2 + 4$
28. $2x^4 + 7x^3 + 6x^2 + 7x - 6$
29. $2x^5 + 3x^3 - 16x^2 - 24$
30. $x^6 + x^4 + 5x^2 - 4$
31. $x^{100} - 1$
32. $x^{100} + 1$
33. $x^{99} + 1$
34. $x^{99} - 1$

SOLVING POLYNOMIAL EQUATIONS

A *polynomial equation* is an equation of the form $P(x) = 0$, where $P(x)$ is a polynomial. In some cases we can find the solution of a polynomial equation; we now give some examples in which we can do so.

Example 2 (continued). Solve $x^3 - 3x^2 - 3x - 4 = 0$.

Solution. We have seen in Example 2 that 4 is a solution of this equation. That is, 4 is a root of the polynomial. By the Factor Theorem, $x - 4$ is a factor of $x^3 - 3x^2 - 3x - 4$. We use synthetic division to find the factorization:

$$\begin{array}{r|rrrr} 4 & 1 & -3 & -3 & -4 \\ & & 4 & 4 & 4 \\ \hline & 1 & 1 & 1 & 0 \end{array}$$

We see that we have the factorization

$$x^3 - 3x^2 - 3x - 4 = (x - 4)(x^2 + x + 1).$$

To solve the quadratic equation

$$x^2 + x + 1 = 0,$$

we use the quadratic formula; discriminant is $1^2 - 4(1)(1) = -3$.

$$x = \frac{-1 \pm \sqrt{-3}}{2} = \frac{-1 \pm \sqrt{3}i}{2}$$

Therefore the solutions are 4, $(-1 + \sqrt{3}i)/2$, and $(-1 - \sqrt{3}i)/2$.

Example 4. Solve $x^3 - 15x - 4 = 0$.

Solution. First we look for solutions that are integers. The constant term of this polynomial is -4, and the factors of -4 are ± 1, ± 2, and ± 4. When we try these possibilities, we find that 4 is a solution. By the Factor Theorem, $x - 4$ is a factor of $x^3 - 15x - 4$. To find the factorization we use synthetic division:

$$\begin{array}{r|rrrr} 4 & 1 & 0 & -15 & -4 \\ & & 4 & 16 & 4 \\ \hline & 1 & 4 & 1 & 0 \end{array}$$

We have the factorization

$$x^3 - 15x - 4 = (x - 4)(x^2 + 4x + 1).$$

We use the quadratic formula to solve the equation

$$x^2 + 4x + 1 = 0$$

discriminant is $4^2 - 4(1)(1) = 12$.

$$x = \frac{-4 \pm \sqrt{12}}{2} = \frac{-4 \pm 2\sqrt{3}}{2} = -2 \pm \sqrt{3}$$

The solutions are 4, $-2 + \sqrt{3}$, and $-2 - \sqrt{3}$.

Exercise Set 6.5 (continued)

In exercises 35–42, solve the polynomial equation.

35. $x^3 - x^2 - 4x + 4 = 0$
36. $x^3 + 4x^2 + x - 6 = 0$
37. $x^3 + 2x^2 + 10x + 20 = 0$
38. $x^3 + 3x^2 - x + 12 = 0$
39. $x^3 - 6x^2 + 6x - 5 = 0$
40. $x^3 - 10x - 75 = 0$
41. $3x^3 - 2x^2 - 3x + 2 = 0$
42. $12x^3 - 23x^2 - 3x + 2 = 0$

Review Exercises

In exercises 1–6, find $P(x) + Q(x)$, $P(x) - Q(x)$, $P(x)Q(x)$, and $P(x)/Q(x)$.

1. $P(x) = x^3 + 1;\ Q(x) = x + 1$
2. $P(x) = x^3 - 1;\ Q(x) = x - 1$
3. $P(x) = x^3 - x^2 + x - 1;\ Q(x) = x^2 + 1$
4. $P(x) = x^3 - x^2 + x - 1;\ Q(x) = x - 1$
5. $P(x) = x^3 - x^2 - x + 1;\ Q(x) = x^2 - 1$
6. $P(x) = x^3 - x^2 - x + 1;\ Q(x) = x - 1$

In exercises 7–18, use syntnetic division to divide the first polynomial by the second polynomial. Specify the quotient polynomial and the remainder.

7. $x^2 + 7x - 2;\ x + 3$
8. $3x^2 - 5x + 7;\ x - 2$
9. $x^3 - 3x^2 + 3x - 1;\ x - 1$
10. $x^3 - 3x^2 + 3x - 1;\ x + 1$
11. $3x^3 + 12x - 4;\ x - 2$
12. $4x^3 + x^2 + 12;\ x - 2$
13. $x^3 + 2x^2 + 10x - 20;\ x + 2$
14. $x^3 + 2x^2 + 10x + 20;\ x + 2$
15. $3x^4 + x^3 - x^2 + 3x - 2;\ x - 1$
16. $4x^4 + 7x^2 - 2x - 5;\ x + 3$
17. $x^5 + x^4 - 3x^3 + 2x - 3;\ x + 2$
18. $2x^5 - 3x^4 + 4x^3 - x^2 - x - 2;\ x - 3$

In exercises 19–28, find $P(a)$ for each given polynomial $P(x)$ by using the Remainder Theorem.

19. $P(x) = x^3 + 3x^2 - 7x + 4$; find $P(1)$.
20. $P(x) = x^3 - 5x^2 + 4x - 6$; find $P(-2)$.
21. $P(x) = 3x^3 + 7x^2 - 2x + 5$; find $P(-3)$.
22. $P(x) = 4x^3 - 5x^2 + 6x - 4$; find $P(2)$.
23. $P(x) = x^4 + x^2 - 2$; find $P(-1)$.
24. $P(x) = x^4 - 3x^2 + 2x - 5$; find $P(-2)$.
25. $P(x) = 2x^4 + x^3 - 3x^2 - x - 2$; find $P(-3)$.
26. $P(x) = 4x^4 - 2x^3 + x^2 - x - 4$; find $P(2)$.
27. $P(x) = x^5 - x^4 - x^3 + x^2 - x + 3$; find $P(2)$.
28. $P(x) = x^5 - 2x^4 - 3x^3 + 4x^2 - 5x + 6$; find $P(2)$.

In exercises 29–34, the given integer a is a root of $P(x)$. Find a factorization of the form $P(x) = (x - a)Q(x)$.

29. $P(x) = x^3 - 2x^2 + x - 2;\ a = 2$
30. $P(x) = x^3 + 2x^2 + x + 2;\ a = -2$
31. $P(x) = x^3 - 3x^2 + 3x - 1;\ a = 1$
32. $P(x) = x^3 - 6x^2 + 12x - 8;\ a = 2$
33. $P(x) = x^5 + 1;\ a = -1$
34. $P(x) = x^5 - 1;\ a = 1$

In exercises 35–42, find all the integer roots of each given polynomial.

35. $x^3 + x^2 + 9x + 9$
36. $x^3 + 4x^2 + x + 4$
37. $x^3 + x^2 - 9x - 9$
38. $x^3 + 4x^2 - x - 4$

39. $2x^3 + x^2 - 5x + 2$

40. $2x^3 - x^2 - 13x - 6$

41. $2x^3 + 3x^2 + 3x + 1$

42. $2x^3 - 3x^2 + 3x - 1$

In exercises 43–48, solve each given polynomial equation.

43. $x^3 - 2x^2 - 5x + 6 = 0$

44. $x^3 - 7x - 6 = 0$

45. $x^3 + 3x^2 + x + 3 = 0$

46. $x^3 - 2x^2 + x - 2 = 0$

47. $2x^3 - 5x^2 + 4x - 1 = 0$

48. $2x^3 + x^2 - 4x - 3 = 0$

chapter 7

Functions and Graphs

1. Functions

2. The Cartesian Coordinate System

3. Graphing Functions

4. Distance and Slope

5. Straight Lines

6. Linear Inequalities

7. Circles

8. Applications

The two basic concepts in this chapter are functions and graphs. These concepts are related to each other—we will graph functions in the Cartesian coordinate system. The graph provides a visual representation of the function.

In Section 7.1 we define and give examples of functions. We construct the Cartesian coordinate system and graph ordered pairs in Section 7.2. In Section 7.3 we show how to graph functions.

We examine the concepts of distance and slope in the Cartesian coordinate system in Section 7.4. In Section 7.5 we learn to recognize equations whose graphs are straight lines, and in Section 7.7 we learn to recognize equations whose graphs are circles. We graph an important class of inequalities in Section 7.6.

In Section 7.8 we apply functions and graphs to geometry, business, diet problems, budget problems, grade and pitch considerations, and linear correlation.

section 1 • Functions

In this section we define functions and functional notation and we substitute numbers for the independent variable in a function.

DEFINITION OF A FUNCTION

A function is a special type of correspondence between variables. The function concept describes the situation in which one variable depends on another variable in a specified way. That is, a function is a rule that assigns a number to a variable for each value of the other variable.

DEFINITION OF A FUNCTION. A *function* is a correspondence (or rule) between two variables by which the number values of one specified variable (the *dependent variable*) are uniquely determined by the corresponding number values of the other variable (the *independent variable*).

Example 1. The equation $y = 2x$ expresses y as a function of x. The independent variable is x and the dependent variable is y. The rule for this function is that each value of x is doubled to obtain the corresponding value of y. If we call this function f, we can write

$$y = f(x) = 2x.$$

The notation $f(x)$ is called *functional notation*. The name of the function is f, and the equation $y = f(x)$ expresses the fact that the dependent variable y is a function of the independent variable x.

Example 2. The equation $y = x + 2$ expresses y as a function of x. The independent variable is x and the dependent variable is y. The rule for this function is that each value of x is added to 2 to obtain the corresponding value of y. If we call this function f, we can write the functional notation

$$y = f(x) = x + 2.$$

Example 3. The equation $y = x^2$ expresses y as a function of x. The independent variable is x and the dependent variable is y. The rule for this function is that each value of x is squared to obtain the corresponding value of y. If we call this function g, we can write the functional notation

$$y = g(x) = x^2.$$

Note: Many different letters are used to name various functions; the ones you are most likely to see are f, g, and h. The letter f is frequently used to name functions because it is the first letter of the word *function.*

Note: Many different letters are used to denote the independent and dependent variables of a function. In this book we will frequently use x and y.

Example 4. The equation $y = x(x + 1)$ expresses y as a function of x. The independent variable is x and the dependent variable is y. The rule for this function is that each value of x is multiplied by $x + 1$ to obtain the corresponding value of y. If we call this function f, we can write the functional notation

$$y = f(x) = x(x + 1).$$

Exercise Set 7.1

In exercises 1–12, the rule for a function is given; if the function has the given name, express the function using functional notation.

1. The function f triples each value of x to obtain the corresponding value of y.
2. The function f quadruples each value of x to obtain the corresponding value of y.
3. The function f adds 3 to each value of x to obtain the corresponding value of y.
4. The function f adds 4 to each value of x to obtain the corresponding value of y.
5. The function f subtracts 2 from each value of x to obtain the corresponding value of y.
6. The function f subtracts 1 from each value of x to obtain the corresponding value of y.
7. The function g cubes x to obtain the corresponding value of y.
8. The function g takes the fourth power of x to obtain the corresponding value of y.
9. The function g subtracts each value of x from 4 to obtain the corresponding value of y.
10. The function g subtracts each value of x from 2 to obtain the corresponding value of y.
11. The function f multiplies each value of x by $x + 3$ to obtain the corresponding value of y.
12. The function f multiplies each value of x by $x + 4$ to obtain the corresponding value of y.

SUBSTITUTING NUMBERS

A principal advantage of using the functional notation $f(x)$ is that, if a is a number, $f(a)$ denotes the value of the dependent variable when a is substituted for x. We give examples of this use of functional notation.

Example 1 (continued). Consider the function $f(x) = 2x$. We substitute the numbers 2, 1, 0, -1, -2 for x:

$$f(2) = 2(2) = 4$$
$$f(1) = 2(1) = 2$$
$$f(0) = 2(0) = 0$$
$$f(-1) = 2(-1) = -2$$
$$f(-2) = 2(-2) = -4$$

Example 2 (continued). Consider the function $f(x) = x + 2$. We substitute the numbers 2, 1, 0, -1, -2 for x:

$$f(2) = (2) + 2 = 4$$
$$f(1) = (1) + 2 = 3$$
$$f(0) = (0) + 2 = 2$$
$$f(-1) = (-1) + 2 = 1$$
$$f(-2) = (-2) + 2 = 0$$

Example 3. Consider the function $g(x) = x^2 + 2x$. We substitute the numbers 2, 1, 0, -1, -2 for x:

$$g(2) = (2)^2 + 2(2) = 8$$
$$g(1) = (1)^2 + 2(1) = 3$$
$$g(0) = (0)^2 + 2(0) = 0$$
$$g(-1) = (-1)^2 + 2(-1) = -1$$
$$g(-2) = (-2)^2 + 2(-2) = 0$$

Exercise Set 7.1 (continued)

In exercises 13–32, substitute the numbers 2, 1, 0, -1, -2 for x and express the result using functional notation.

13. $f(x) = 3x$
14. $f(x) = 4x$
15. $f(x) = -x$
16. $f(x) = -2x$
17. $g(x) = x^2$
18. $g(x) = -x^2$
19. $f(x) = 5 - x$
20. $f(x) = 6 - x$
21. $g(x) = x^3$
22. $g(x) = x^4$
23. $f(x) = 1/x$
24. $f(x) = 1/x^2$
25. $g(x) = x^2 + x$
26. $g(x) = x^2 - x$
27. $f(x) = 4x + 6$
28. $f(x) = 4x + 8$
29. $f(x) = x(x + 3)$
30. $f(x) = x(x + 4)$
31. $f(x) = x(4 - x)$
32. $f(x) = x(15 - x)$
33. If $f(x) = 0.5x - 32{,}000$, find $f(5000)$, $f(10{,}000)$, $f(20{,}000)$.

34. If $f(x) = 0.7x - 28{,}000$, find $f(5000)$, $f(10{,}000)$, $f(20{,}000)$.
35. If $f(x) = 2.4x - 46{,}000$, find $f(5000)$, $f(10{,}000)$, $f(20{,}000)$.
36. If $f(x) = 1.8x - 39{,}000$, find $f(5000)$, $f(10{,}000)$, $f(20{,}000)$.

section 2 • The Cartesian Coordinate System

An important theme in mathematics is the representation of algebraic concepts in geometric terms and, conversely, the representation of geometric concepts in algebraic terms. The basic way of accomplishing this is by using a *coordinate system*. Many coordinate systems have been invented, but the most famous is the Cartesian coordinate system. This coordinate system was invented by the French mathematician René Descartes (1596–1650); hence the name "Cartesian."

In this section we construct the Cartesian coordinate system and show how to plot ordered pairs. As preparation, we define ordered pairs of real numbers.

ORDERED PAIRS

In previous chapters we have concentrated on solving equations and inequalities in one variable. In this and subsequent chapters we turn our attention to equations and inequalities in two variables. Usually we call the variables x and y. To consider two variables, we define ordered pairs.

DEFINITION OF AN ORDERED PAIR. An *ordered pair* of real numbers is a pair of numbers of the form (x, y), where x and y denote real numbers. The number denoted by x is called the *first coordinate* of the ordered pair, and the number denoted by y is called the *second coordinate* of the ordered pair. Two ordered pairs are *equal* only if their first coordinates are equal and their second coordinates are equal.

For example, $(2, 1)$ is an ordered pair with first coordinate 2 and second coordinate 1. The ordered pair $(1, 2)$ has first coordinate 1 and second coordinate 2. Therefore, $(2, 1)$ is not equal to $(1, 2)$. Also, the ordered pair $(1, -1)$ has first coordinate 1 and second coordinate -1; the ordered pair $(1, -1)$ is not equal to the ordered pair $(1, 2)$ because the second coordinates are not equal. Any real number can be the first or second coordinate of an ordered pair. However, a specific ordered pair may or may not satisfy a given equation or inequality.

Example 1. Consider the equation $x + 2y = 4$.

a. $(-2, 3)$ satisfies this equation since $\quad (-2) + 2(3) = 4.$

b. $(3, -2)$ does not satisfy this equation since $(3) + 2(-2) \neq 4$.
c. $(6, -1)$ satisfies this equation since $(6) + 2(-1) = 4$.
d. $(-1, -1)$ does not satisfy this equation since $(-1) + 2(-1) \neq 4$.

Example 2. Consider the inequality $3x + 2y \geq 6$.
a. $(2, 1)$ satisfies this inequality since $3(2) + 2(1) \geq 6$.
b. $(0, 0)$ does not satisfy this inequality since $3(0) + 2(0) < 6$.
c. $(0, 3)$ satisfies this inequality since $3(0) + 2(3) \geq 6$.
d. $(-2, 0)$ does not satisfy this inequality since $3(-2) + 2(0) < 6$.

DEFINITION OF THE SOLUTION SET OF AN EQUATION OR INEQUALITY IN TWO VARIABLES. The *solution set* of an equation or inequality in two variables x and y is the set of ordered pairs (x, y) that satisfy the equation or inequality when the first coordinate is substituted for x and the second coordinate is substituted for y.

Usually (but not always), the solution set of an equation or inequality in two variables is an infinite set of ordered pairs.

We also define *equivalence* of two equations or inequalities in two variables. This definition is analogous to the definition of equivalence for equations and inequalities in one variable (Section 4.1).

DEFINITION OF EQUIVALENT EQUATIONS OR INEQUALITIES. Two equations or inequalities in two variables are *equivalent* if they have the same solution sets.

Again, just as with equations and inequalities in one variable, we can transform equations and inequalities in two variables by performing operations. We are interested in the operations of addition, subtraction, multiplication, and division by non-zero real numbers because these operations lead to equivalent equations and inequalities.

OPERATIONS APPLIED TO EQUATIONS AND INEQUALITIES. For equations and inequalities in two variables we have:

Addition and Subtraction: For any expression c,

$a = b$ is equivalent to $a + c = b + c$ and $a - c = b - c$

$a > b$ is equivalent to $a + c > b + c$ and $a - c > b - c$

$a < b$ is equivalent to $a + c < b + c$ and $a - c < b - c$

Multiplication and Division: For any non-zero real number c,

$a = b$ is equivalent to $ac = bc$ and $a/c = b/c$

If c is a positive real number, then

$a > b$ is equivalent to $ac > bc$ and $a/c > b/c$

$a < b$ is equivalent to $ac < bc$ and $a/c < b/c$

If c is a negative real number, then

$a > b$ is equivalent to $ac < bc$ and $a/c < b/c$

$a < b$ is equivalent to $ac > bc$ and $a/c > b/c$

We can say that, if c is a positive real number, then multiplying or dividing both sides of an inequality by c *preserves* the inequality. However, if c is a negative real number, then multiplying or dividing both sides of an inequality by c *reverses* the inequality.

Example 3. Find the ordered pairs in the solution set of $x + 2y = 4$ corresponding to $x = 1, 2, 3$, and 4.

Solution. We solve for y in terms of x:

$$2y = 4 - x$$

$$y = (4 - x)/2$$

This equation is equivalent to the given equation, and it expresses y as a function of x. Now

$$\text{when } x = 1, y = (4 - 1)/2 = 3/2$$

$$\text{when } x = 2, y = (4 - 2)/2 = 1$$

$$\text{when } x = 3, y = (4 - 3)/2 = 1/2$$

$$\text{when } x = 4, y = (4 - 4)/2 = 0$$

Therefore, the ordered pairs in the solution set are $(1, 3/2)$, $(2, 1)$, $(3, 1/2)$, and $(4, 0)$.

Example 4. Find the ordered pairs in the solution set of $y = x^2 - 2x$ corresponding to $x = -1, 0$, and 1.

Solution. We already have y expressed as a function of x. Now

$$\text{when } x = -1, y = (-1)^2 - 2(-1) = 3$$

$$\text{when } x = 0, y = (0)^2 - 2(0) = 0$$

$$\text{when } x = 1, y = (1)^2 - 2(1) = -1$$

Therefore, the ordered pairs in the solution set are $(-1, 3)$, $(0, 0)$, and $(1, -1)$.

Exercise Set 7.2

In exercises 1–8, specify which of the given ordered pairs belong to the solution set of the given equation or inequality.

1. $y = 2x$; (1, 2); (2, 1); (2, 2); (2, 4); (0, 0).
2. $y = 3x$; (1, 3); (3, 1); (2, 6); (2, 4); (0, 0).
3. $y = x - 3$; (0, 0); (3, 0); (0, 3); (−3, 0); (0, −3).
4. $y = 3 - x$; (0, 0); (3, 0); (0, 3); (−3, 0); (0, −3).
5. $x + y \leq 4$; (0, 0); (2, 2); (3, 2); (−1, −1); (−2, 2).
6. $x - y \leq 4$; (0, 0); (2, 2); (3, 2); (−2, 2); (2, −2).
7. $y < 2x + 5$; (0, 0); (2, 1); (−2, 1); (1, −2); (0, −1).
8. $y \geq 2x - 3$; (0, 0); (2, 1); (−2, −2); (1, −2); (0, −2).

In exercises 9–20, a function f is given. Find the ordered pairs $(x, f(x))$ when $x = -1, 0, 1, 2, 3$, and 4.

9. $f(x) = -x$
10. $f(x) = -3x$
11. $f(x) = x^2 - 4$
12. $f(x) = -4x^2 + 8x - 3$
13. $f(x) = x^2 - 1$
14. $f(x) = x^2 + 1$
15. $f(x) = -\frac{3}{2}x + 3$
16. $f(x) = -\frac{5}{2}x + 5$
17. $f(x) = x^2 - 2x$
18. $f(x) = x^2 + 2x$
19. $f(x) = 4/x$
20. $f(x) = 6/x$

CONSTRUCTION OF THE CARTESIAN COORDINATE SYSTEM

Now we describe the construction of the Cartesian coordinate system. We construct this coordinate system in such a way that points in a plane correspond with ordered pairs of real numbers. (Recall that real number lines are reviewed in Section 1.1.)

Draw two intersecting real number lines in a plane. The point at which the real number lines intersect must correspond to the number 0 on each line. This point is called the *origin* of the coordinate system. To obtain the Cartesian coordinate system, we draw the lines perpendicular to each other. Otherwise we get an *oblique* coordinate system.

We now have two perpendicular real number lines which intersect at the origin (Figure 7.1). These lines are called *axes*. The names of the axes depend upon the variables we are

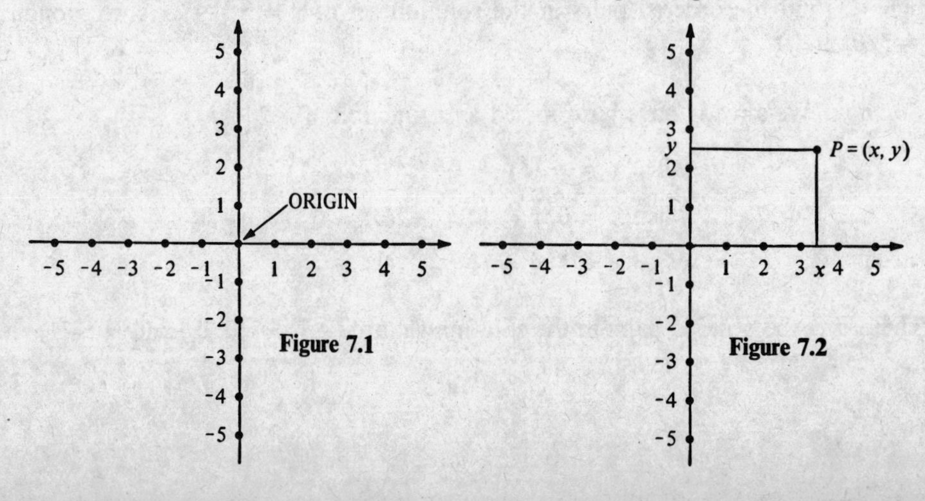

Figure 7.1

Figure 7.2

going to graph. If the variables are x and y, then the horizontal axis will be the x-axis and the vertical axis will be the y-axis. The horizontal axis is directed to the right, and the vertical axis is directed upward.

We want ordered pairs to correspond to points in the plane. We make the first coordinate correspond to the horizontal axis and the second coordinate correspond to the vertical axis. So what we do is this: To find the point corresponding to the ordered pair (x, y), we find the point on the x-axis corresponding to x and draw a line perpendicular to the x-axis through this point. We also find the point on the y-axis corresponding to y and draw a line perpendicular to the y-axis through this point. These two perpendicular lines intersect in a point P; we make this point correspond to (x, y) (Figure 7.2).

Conversely, if we're given a point P in the plane, how do we find the ordered pair (x, y) corresponding to it? We drop a perpendicular line to the x-axis and find the number corresponding to the foot of the perpendicular. This number is the first coordinate of the ordered pair. We also drop a perpendicular line to the y-axis and find the number corresponding to the foot of the perpendicular. This number is the second coordinate of the ordered pair.

In this way, we have established a one-to-one correspondence between points in a plane and ordered pairs of real numbers. This is the Cartesian coordinate system.

The Cartesian coordinate system naturally divides the plane into four regions called *quadrants*. The quadrants are labelled first, second, third, and fourth, as depicted in Figure 7.3.

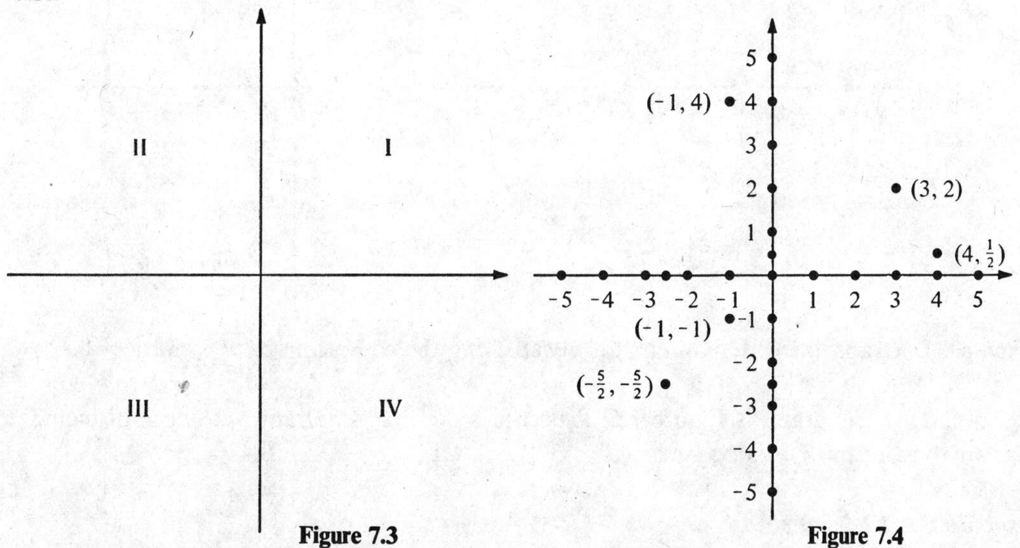

Figure 7.3 **Figure 7.4**

The point in the Cartesian coordinate system corresponding to a given ordered pair is called the *graph* of the ordered pair.

Example 5. Graph the ordered pairs $(3, 2)$, $(-1, 4)$, $(-1, -1)$, $(4, 1/2)$, and $(-5/2, -5/2)$.

Solution. The graph is Figure 7.4.

Example 6. Graph the ordered pairs (x, y) satisfying the equation $x + y = 4$ for $x = -1$, 0, 1, and 2.

Solution. Expressing y as a function of x, we have

$$y = 4 - x.$$

$$\text{when } x = -1, y = 4 - (-1) = 5$$

$$\text{when } x = 0, y = 4 - (0) = 4$$

$$\text{when } x = 1, y = 4 - (1) = 3$$

$$\text{when } x = 2, y = 4 - (2) = 2$$

The ordered pairs are $(-1, 5)$, $(0, 4)$, $(1, 3)$, and $(2, 2)$, and the graph is Figure 7.5.

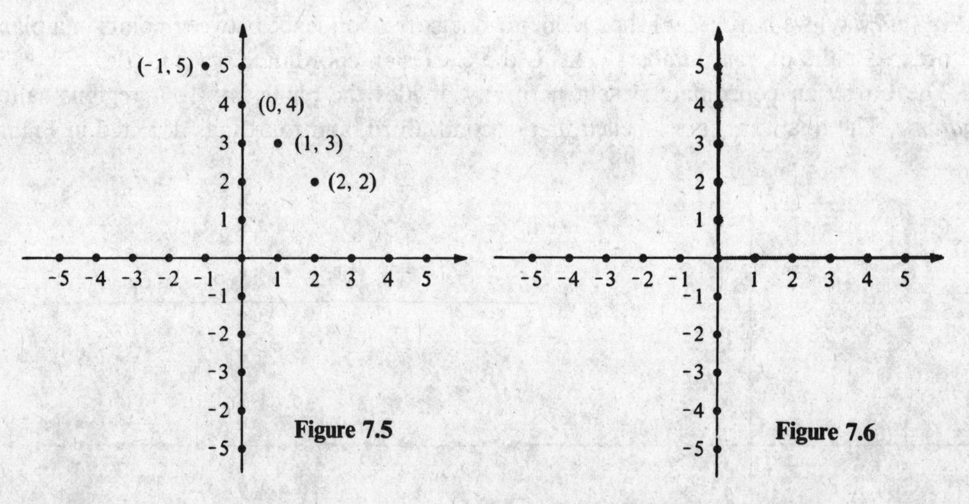

Figure 7.5

Figure 7.6

Example 7. Graph the ordered pairs (x, y) satisfying the conditions $x > 0$ and $y > 0$.

Solution. The graph is Figure 7.6. It occupies the first quadrant but does not include any part of the x-axis or y-axis.

Exercise Set 7.2 (continued)

In exercises 21–40, graph each given set of ordered pairs.

21. $(2, 5), (-4, 3), (-4, -4), (0, 5), (5, -2)$.
22. $(0, 0), (-3, 3), (3, -3), (3, 3), (-3, -3)$.
23. $(-1, 0), (0, 5/2), (3/2, 0), (0, -4)$.
24. $(1/2, -1/2), (2/3, 4), (-3/2, 0), (-7/8, 7/8), (5, -4/5)$.
25. The set of (x, y) such that $y = x - 3$ and $x = -2, -1, 0, 1, 2$.

26. The set of (x, y) such that $y = 3 - x$ and $x = -2, -1, 0, 1, 2$.
27. The set of (x, y) such that $2x + y = 2$ and $x = -1, 0, 1$.
28. The set of (x, y) such that $3x - 2y = 6$ and $x = -2, -1, 0, 1$.
29. The set of (x, y) such that $y = x^2$ and $x = 0, 1, 1.5, 2$.
30. The set of (x, y) such that $y = x^2 - 4$ and $x = -2, -1, 0, 1, 2$.
31. The set of (x, y) such that $y = 1/x$ and $x = -2, -1, 1, 2$.
32. The set of (x, y) such that $y = -4x^2 + 8x - 3$ and $x = -1, 0, 1$.
33. The set of (x, y) such that $x \geq 0$.
34. The set of (x, y) such that $x < 0$.
35. The set of (x, y) such that $x > 0$ and $y < 0$.
36. The set of (x, y) such that $x \leq 0$ and $y \leq 0$.
37. The set of (x, y) such that $x \geq 1$ and $y \leq 0$.
38. The set of (x, y) such that $x \leq 1$ and $y \leq 1$.
39. The set of (x, y) such that $x \leq -1$ and $y \geq 1$.
40. The set of (x, y) such that $x \leq -2$ and $y \leq -2$.

section 3 • Graphing Functions

In this section we consider graphing functions in the Cartesian coordinate system. We use the method of *plotting points*; that is, graphing sufficiently many ordered pairs to see what the entire graph looks like. We consider several types of functions.

DEFINITION OF THE GRAPH OF A FUNCTION. The *graph of a function f* is the graph of the set of ordered pairs of the form $(x, f(x))$.

SOME LINEAR FUNCTIONS

First we consider the graphs of some linear functions.

DEFINITION OF A LINEAR FUNCTION. A function f is a *linear function* if there are constants a and b so that

$$f(x) = ax + b.$$

Example 1. Graph the function $f(x) = 2x$.

Solution. The graph of f is the graph of the ordered pairs $(x, f(x))$; that is, the graph of the ordered pairs $(x, 2x)$. Some specific ordered pairs of this form are:

$$(-3, -6); (-2, -4); (-1, -2); (0, 0); (1, 2); (2, 4); (3, 6)$$

When we graph these ordered pairs in Figure 7.7 we see that they lie on a straight line. Thus we would expect that the graph of the function f is the entire straight line, and this is actually true. *Note:* We have graphed only seven ordered pairs for this function. You may wish to graph additional ordered pairs and see that they also lie on the line.

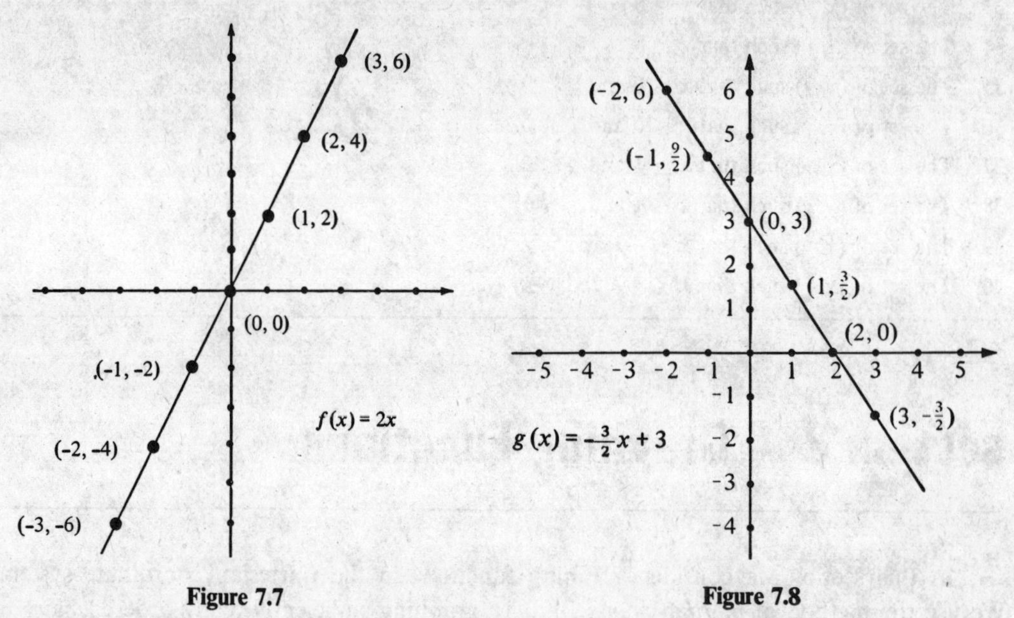

Figure 7.7

Figure 7.8

Example 2. Graph the function $g(x) = -\dfrac{3}{2}x + 3$.

Solution. To graph this function, we graph ordered pairs of the form $(x, g(x))$. By choosing some values of x and finding the corresponding values of $g(x)$, we find the following ordered pairs:

$$(-2, 6); (-1, 9/2); (0, 3); (1, 3/2); (2, 0); (3, -3/2)$$

When we graph these ordered pairs in Figure 7.8, we see that they lie on a straight line. The graph of the function is the entire straight line.

Note: You may wish to graph some additional ordered pairs and see that they lie on the line. In Section 7.5 we will show that the graph of a linear function is always a straight line.

Exercise Set 7.3.

In exercises 1–8, graph each given linear function.

1. $f(x) = x$
2. $f(x) = 3x$
3. $f(x) = -x$
4. $f(x) = -3x$
5. $g(x) = x - 3$
6. $g(x) = 3 - x$
7. $f(x) = -\frac{1}{2}x + 1$
8. $f(x) = \frac{1}{2}x - 1$

SOME QUADRATIC FUNCTIONS

DEFINITION OF A QUADRATIC FUNCTION. A function f is a *quadratic function* if there are constants a, b, and c such that

$$f(x) = ax^2 + bx + c.$$

Example 3. Graph the function $f(x) = x^2 - 4$.

Solution. We graph ordered pairs $(x, f(x))$. Some particular ordered pairs of this form are:

$$(-3, 5); (-2, 0); (-1, -3); (0, -4); (1, -3); (2, 0); (3, 5)$$

We have graphed these ordered pairs in Figure 7.9. We have then drawn a smooth curve through these points; this curve is the graph of the function f.

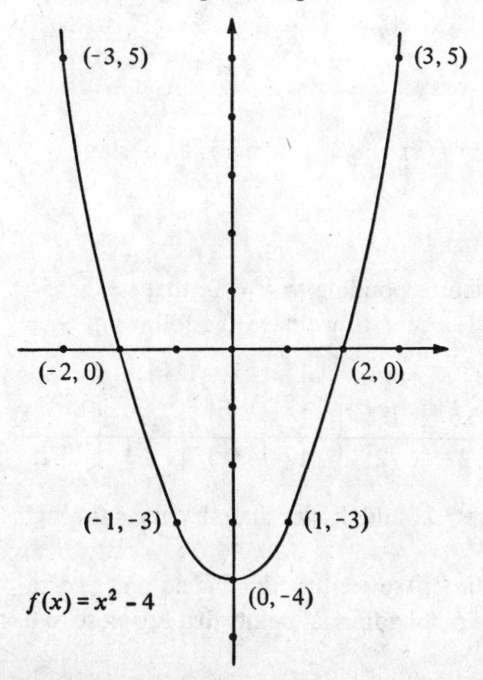

Figure 7.9

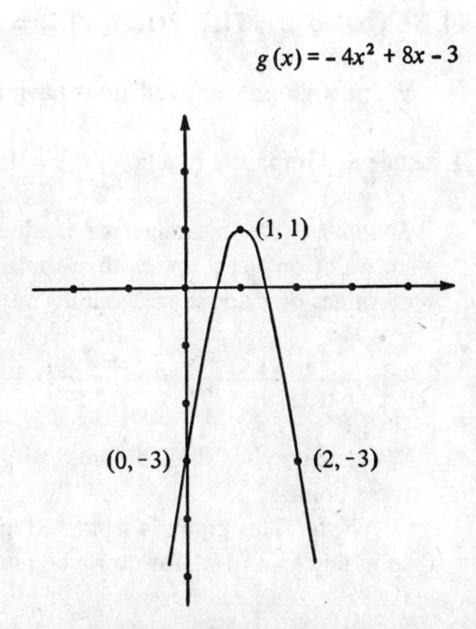

Figure 7.10

Note: The graph in Figure 7.9 is a parabola. If it is not clear from the seven points we have graphed what the shape of the graph is, then you should graph additional ordered pairs. For this function, use some fractional values of x between 3 and -3 and check that the ordered pair lies on the graph we have drawn. If you have a calculator, you can use it to find the value of $f(x)$ corresponding to a value of x.

Example 4. Graph the function $g(x) = -4x^2 + 8x - 3$.

Solution. Some ordered pairs of the form $(x, g(x))$ are

$$(0, -3);\ (0.5, 0);\ (1, 1);\ (1.5, 0);\ (2, -3)$$

We have graphed these ordered pairs in Figure 7.10 and have drawn a smooth curve through these points. This curve is the graph of the function g.

Note: You should graph additional ordered pairs for this function and see that they lie on the graph. Again, the graph in Figure 7.10 is a parabola.

Exercise Set 7.3 (continued)

In exercises 9–16, graph the given quadratic function.

9. $f(x) = x^2$
10. $f(x) = -x^2$
11. $g(x) = x^2 - 1$
12. $g(x) = x^2 + 1$
13. $f(x) = x^2 - 2x$
14. $f(x) = x^2 + 2x$
15. $g(x) = -x^2 - 6x - 4$
16. $g(x) = x^2 + 6x + 9$

FUNCTIONS OF THE FORM F(X) = A/X

We now graph some functions having the form $f(x) = a/x$, where a is a constant.

Example 5. Graph the function $f(x) = 1/x$.

Solution. Note that there is no value of $f(x)$ corresponding to $x = 0$; that is, there is no point on the graph with x-coordinate 0. For $x \neq 0$, we have the following chart of values of x and corresponding values of $f(x)$:

x	-4	-3	-2	-1	-0.5	0.5	1	2	3	4
$f(x)$	-0.25	-0.3	-0.5	-1	-2	2	1	0.5	0.3	0.25

We have graphed these ordered pairs in Figure 7.11 and drawn smooth curves through these points.

Note: This graph is separated into two parts. We see that there is no point corresponding to $x = 0$. However, you may wish to plot additional points that are close to 0.

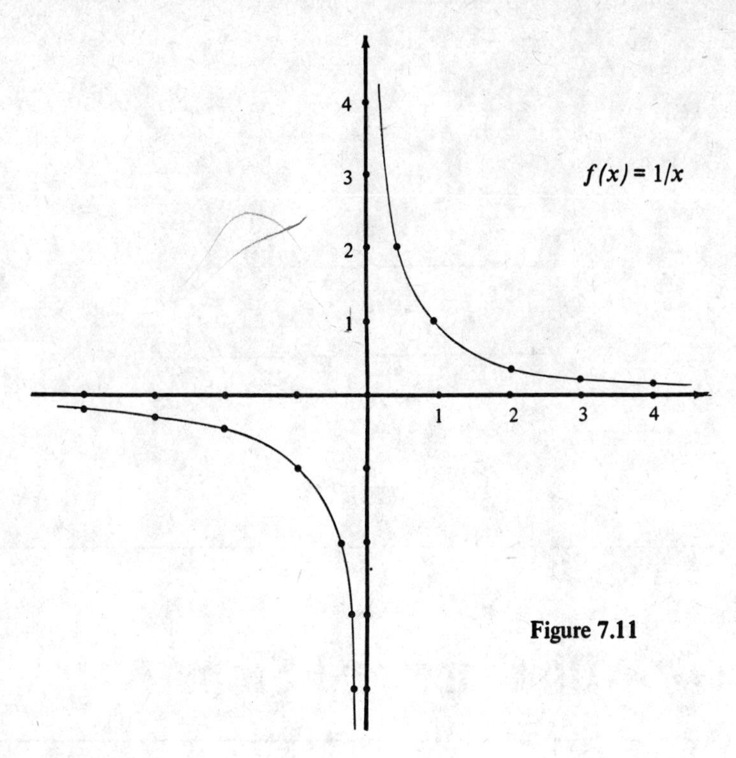

Figure 7.11

Exercise Set 7.3 (continued)

In exercises 17–20, graph each given function.

17. $f(x) = 4/x$
18. $f(x) = 6/x$
19. $f(x) = 3/x$
20. $f(x) = 5/x$

SOME CONSTANT FUNCTIONS

DEFINITION OF A CONSTANT FUNCTION. A function f is a *constant function* if there is a constant c such that

$$f(x) = c.$$

Example 6. Graph the function $f(x) = -2$.

Solution. For every value of x the corresponding value of $f(x)$ is -2. So we graph all ordered pairs of the form $(x, -2)$. The graph is a horizontal straight line (Figure 7.12).

Exercise Set 7.3 (continued)

In exercises 21–24, graph the given constant function.

21. $f(x) = 3$
22. $f(x) = 1$
23. $f(x) = -1$
24. $f(x) = -4$

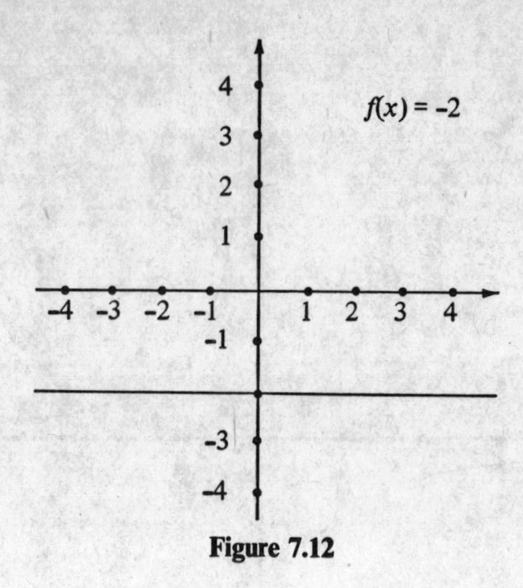

Figure 7.12

section 4 • Distance and Slope

In order to study graphs more effectively in the Cartesian coordinate system, we look at the concepts of distance and slope. Both distance and slope are geometric concepts, and we wish to find algebraic formulas for them. We will use the distance formula to study circles and we will use slope to study straight lines. Other concepts used to study straight lines are the x-intercept and y-intercept.

DISTANCE

Now we derive a formula for the distance between two points in the Cartesian coordinate system.

Suppose P_1 and P_2 are two points in the Cartesian coordinate system and d is the distance between these points. To find a formula for d, let the ordered pair corresponding to P_1 be (x_1, y_1) and the ordered pair corresponding to P_2 be (x_2, y_2) (Figure 7.13). Let Q be the point corresponding to (x_2, y_1). Then triangle P_1QP_2 is a right triangle, and we wish to use the Pythagorean Theorem to find d. First we have to find the length of the two sides. Since P_1Q is horizontal, the length of P_1Q is either $x_2 - x_1$ or $x_1 - x_2$, depending on the position of the points (Figure 7.13). Since QP_2 is vertical, the length of QP_2 is either $y_2 - y_1$ or $y_1 - y_2$, again depending on the position of the points. In both cases, $(x_2 - x_1)^2 = (x_1 - x_2)^2$ and $(y_2 - y_1)^2 = (y_1 - y_2)^2$. Therefore, by the Pythagorean Theorem,

$$d^2 = (x_2 - x_1)^2 + (y_2 - y_1)^2.$$

Therefore, $d = \sqrt{(x_2 - x_1)^2 + (y_2 - y_1)^2}$.
We choose the positive square root because the distance is positive.

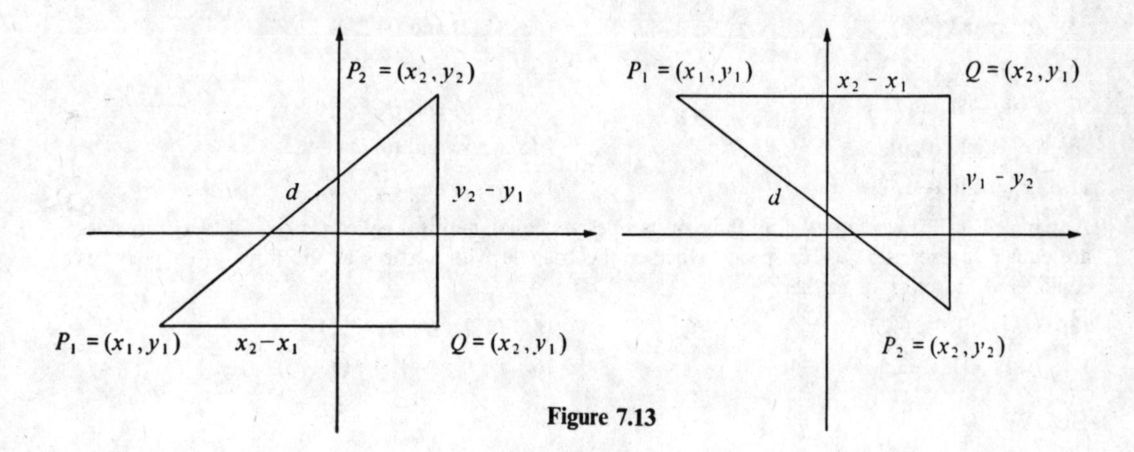

Figure 7.13

THE DISTANCE FORMULA. If $P_1 = (x_1, y_1)$ and $P_2 = (x_2, y_2)$ are two points in the Cartesian coordinate system and d is the distance between P_1 and P_2, then

$$d = \sqrt{(x_2 - x_1)^2 + (y_2 - y_1)^2}.$$

Example 1. Find the distance between $(-3, 2)$ and $(1, -4)$.

Solution. We can label the points as we wish, so suppose $P_1 = (-3, 2)$ and $P_2 = (1, -4)$. Using the distance formula, we find

$$d = \sqrt{(1 - (-3))^2 + ((-4) - 2)^2} = \sqrt{4^2 + (-6)^2} = \sqrt{52} = 2\sqrt{13}.$$

Let us show that we can label the points as we wish. Suppose $P_1 = (1, -4)$ and $P_2 = (-3, 2)$. Then

$$d = \sqrt{((-3) - 1)^2 + (2 - (-4))^2} = \sqrt{(-4)^2 + 6^2} = \sqrt{52} = 2\sqrt{13}.$$

The distance we have here is always a non-negative real number; that is, $d \geq 0$. So we say we have an *undirected* distance.

Example 2. Find the distance between (x, y) and $(-2, 2)$.

Solution. Using the distance formula, we have

$$d = \sqrt{(x - (-2))^2 + (y - 2)^2} = \sqrt{(x + 2)^2 + (y - 2)^2}.$$

Exercise Set 7.4

In exercises 1–12, find the distance between each pair of given points.

1. (1, 2) and (5, 5)
2. (5, 6) and (2, 2)
3. (0, 3) and (2, 5)
4. (1, 2) and (−2, 5)
5. (3, 8) and (−2, −2)
6. (2, 4) and (3, 1)
7. (−1, −5) and (−6, 2)
8. (3, 0) and (−6, 1)
9. (x, y) and (0, 0)
10. (x, y) and (0, 1)
11. (x, y) and (−1, 2)
12. (x, y) and (3, −2)

A triangle is called *equilateral* if all three of its sides are equal, and it is called *isosceles* if two of its sides are equal. In exercises 13–16, specify whether the triangle with vertices at the three given points is equilaterial, isosceles, or neither.

13. (1, 1); (3, 4); (4, 3)
14. (2, 1); (4, 2); (3, 3)
15. (−1, −1); (−3, −4); (−4, −3)
16. (1, 0); (−1, 0); $(0, \sqrt{3})$

SLOPE

Now we define the slope of a straight line in the Cartesian coordinate system. If we choose two points P_1 and P_2 on the line, these points determine a *rise* and a *run*; that is, the points measure how much the line rises with respect to the run. If we choose two other points P'_1 and P'_2 on the same line, we obtain a different rise and a different run (see Figure 7.14). However, since the triangle P_1QP_2 is similar to the triangle $P'_1Q'P'_2$, the *ratio* of the rise to the run is the same in each case. That is, the *ratio* of the rise to the run is independent of which two points we choose on the line.

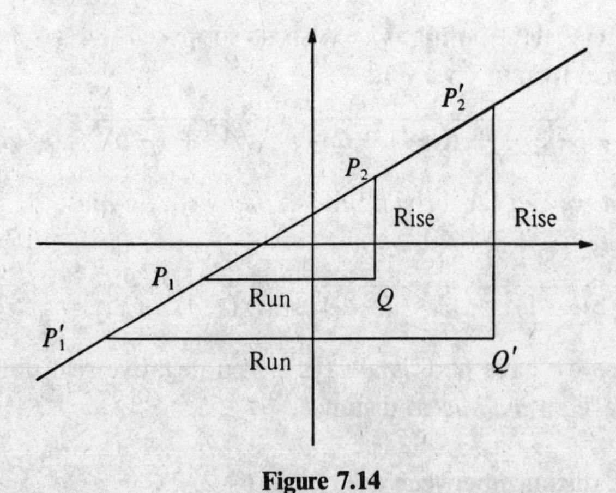

Figure 7.14

DEFINITION OF SLOPE. The slope m of a straight line is

$$m = \text{rise/run}.$$

To calculate the slope of a given straight line, refer to Figure 7.15. Choose any two different points $P_1 = (x_1, y_1)$ and $P_2 = (x_2, y_2)$ on the line whose slope we wish to determine. Look at the point $Q = (x_2, y_1)$, so triangle P_1QP_2 is a right triangle. We wish to

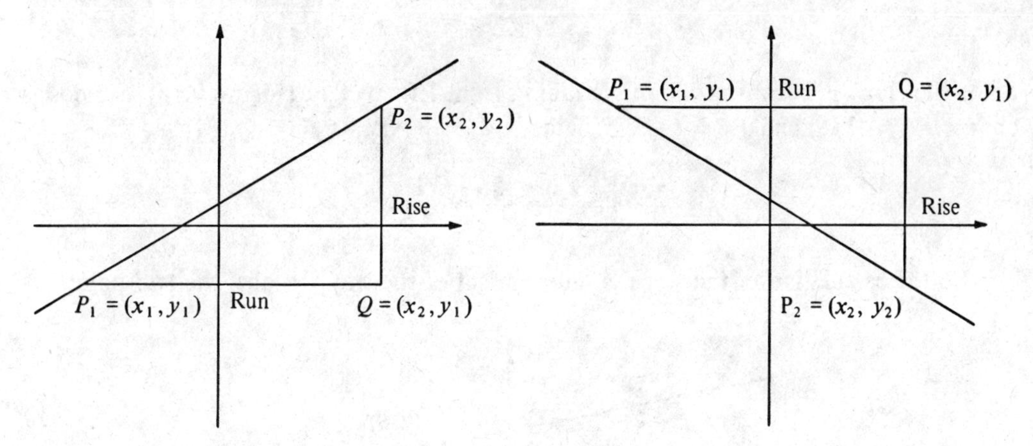

Figure 7.15

calculate $\frac{rise}{run}$. Then rise $= QP_2 = y_2 - y_1$ and run $= P_1Q = x_2 - x_1$, where the rise and run are directed quantities that might be either positive or negative (Figure 7.15). Thus,

$$m = \frac{rise}{run} = \frac{y_2 - y_1}{x_2 - x_1}.$$

SLOPE OF A STRAIGHT LINE. If $P_1 = (x_1, y_1)$ and $P_2 = (x_2, y_2)$ are two points on a straight line, then the *slope m* of the line is

$$m = \frac{y_2 - y_1}{x_2 - x_1}.$$

Note: If the line is rising from left to right, then $y_2 - y_1$ and $x_2 - x_1$ have the same sign and $m > 0$ (Figure 7.15). If the line is falling from left to right, then $y_2 - y_1$ and $x_2 - x_1$ have opposite signs and $m < 0$. If the line is horizontal, then $y_2 = y_1$ and $m = 0$. If the line is vertical, then $x_1 = x_2$; in this case, because we cannot divide by 0, we say that the *slope of a vertical line is undefined.*

Note: The ratio $(y_1 - y_2)/(x_1 - x_2)$ can also be used to calculate the slope because

$$\frac{y_1 - y_2}{x_1 - x_2} = \frac{-(y_2 - y_1)}{-(x_2 - x_1)} = \frac{y_2 - y_1}{x_2 - x_1} = m.$$

It doesn't matter in what order we take the points P_1 and P_2, so long as we subtract the coordinates in the same *order* in both numerator and denominator.

Example 3. Find the slope of the line passing through the two points $(-3, 6)$ and $(2, 5)$.

Solution. We choose $P_1 = (-3, 6)$ and $P_2 = (2, 5)$. Then

$$m = \frac{5-6}{2-(-3)} = \frac{-1}{5}$$

Since the slope is negative, the line is falling from left to right (Figure 7.16). Suppose we choose $P_1 = (2, 5)$ and $P_2 = (-3, 6)$. Then

$$m = \frac{6-5}{-3-2} = \frac{1}{-5}.$$

So the calculation of the slope is independent of the way we label the two points.

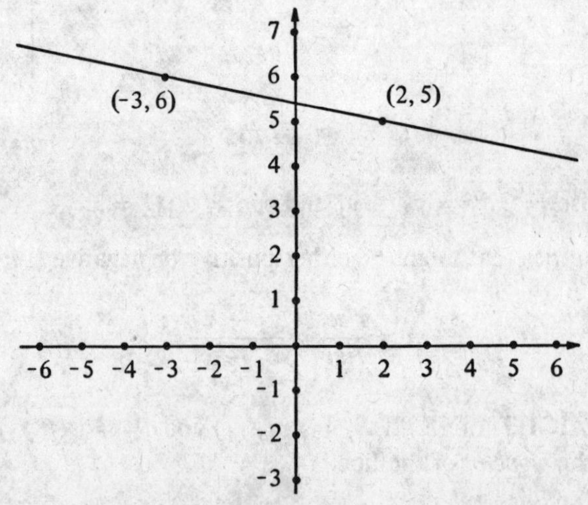

Figure 7.16

Example 4. Find the slope of the line passing through (x, y) and $(-2, 2)$.

Solution. Using the formula for slope,

$$m = \frac{y-2}{x-(-2)} = \frac{y-2}{x+2}.$$

Exercise Set 7.4 (continued)

17. Graph the three points $(-2, -2)$, $(2, 0)$, and $(4, 1)$ to see that they lie on the same straight line Calculate the slope of this line by:
    a. Using the points $(2, 0)$ and $(4, 1)$.
    b. Using the points $(-2, -2)$ and $(4, 1)$.
    c. Using the points $(-2, -2)$ and $(2, 0)$.

18. Graph the three points $(-1, 5)$, $(1, 1)$, and $(3, -3)$ to see that they lie on the same straight line. Calculate the slope of this line by:
    a. Using the points $(-1, 5)$ and $(1, 1)$.
    b. Using the points $(-1, 5)$ and $(3, -3)$.
    c. Using the points $(1, 1)$ and $(3, -3)$.

In exercises 19–30, find the slope of the line through the two given points.

19. $(1, 3)$ and $(3, 5)$
20. $(1, -2)$ and $(4, 1)$
21. $(-1, 1)$ and $(2, -5)$
22. $(-2, 8)$ and $(2, -4)$
23. $(0, -2)$ and $(4, 1)$
24. $(0, -5)$ and $(3, -2)$
25. $(3, 2)$ and $(-3, 10)$
26. $(4, -7)$ and $(-2, -4)$
27. (x, y) and $(1, 1)$
28. (x, y) and $(0, 2)$
29. (x, y) and $(-2, -1)$
30. (x, y) and $(3, -2)$

INTERCEPTS

In Section 7.5 we will see that to graph an equation, it is frequently useful to know where the graph crosses the x-axis and the y-axis. These points are called intercepts of the graph.

DEFINITION OF INTERCEPTS. The *x-intercept* of a graph is the point (or points) where the graph crosses the x-axis. The *y-intercept* of a graph is the point (or points) where the graph crosses the y-axis.

Example 5. Find the intercepts of the graph of

$$3x + 4y = 12.$$

Solution. The graph crosses the x-axis when $y = 0$ and crosses the y-axis when $x = 0$. Thus,

x-intercept	y-intercept
Set $y = 0$	Set $x = 0$
$3x + 4(0) = 12$	$3(0) + 4y = 12$
$3x = 12$	$4y = 12$
$x = 4$	$y = 3$

The x-intercept is $(4, 0)$ and the y-intercept is $(0, 3)$.

Note: A graph need not have an intercept. For example, a straight line parallel to the x-axis (but not the x-axis itself) has a y-intercept but no x-intercept.

Exercise Set 7.4 (continued)

In exercises 31–40, find any x-intercepts and y-intercepts of the graph of each given equation.

31. $2x + 3y - 6 = 0$
32. $y = 2x + 5$
33. $x + y = 4$
34. $x - y = 3$
35. $3x + 2y = 6$
36. $4x + 3y = 12$
37. $4x - 3y = 12$
38. $5x + 2y = 10$
39. $3x - y - 4 = 0$
40. $4x - y - 1 = 0$

section 5 • Straight Lines

In this section we study straight lines in the Cartesian coordinate system. First we define a linear equation in two variables. Then we show that:

1. The equation of a straight line is a linear equation; and
2. The graph of a linear equation is a straight line.

Finally, we show how to graph linear equations by finding two points on the line.

DEFINITION OF LINEAR EQUATIONS

We have found a way to graph the solution set of an equation in two variables: we choose some ordered pairs in the solution set and graph enough of them to detect what the entire graph looks like. This method depends on graphing a large number of ordered pairs, however, and we naturally ask if there is some way to graph equations more easily. What we want to do is to be able to recognize the general *shape* of the graph by recognizing the *form* of the equation. Some classes of equations always lead to the same type of graph. In this section we look at the class of equations whose graphs are straight lines in the Cartesian coordinate system. Because the graphs are straight lines, these equations are called *linear equations.*

DEFINITION OF A LINEAR EQUATION IN TWO VARIABLES. A *linear equation* in two variables x and y is any equation equivalent to one of the form

$$Ax + By + C = 0$$

where A, B, C are constants and not both A and B are 0.

The basic thing about a linear equation is that there are first-degree terms but no terms of degree higher than one—there are no terms $x^2, x^3, \ldots$ or $y^2, y^3, \ldots$ or even *mixed terms*

xy. So the terms are all first-degree. It's possible that one, but not both, of the variables may be missing from the equation; if $A = 0$, then x is missing; and if $B = 0$, then y is missing.

Example 1. Linear equations in two variables.

a. $3x + 2y - 6 = 0$ is a linear equation in x and y; here $A = 3$, $B = 2$, and $C = -6$.

b. $y = 2x + 3$ is a linear equation in x and y because it is equivalent to $-2x + y - 3 = 0$; so $A = -2$, $B = 1$, $C = -3$.

c. $y - 4 = 0$ is a linear equation in x and y, even though the x is missing; $A = 0$, $B = 1$, $C = -4$.

Exercise Set 7.5

In exercises 1–8, each given equation is linear. Specify the constants A, B, and C.

1. $2x - 2y + 3 = 0$
2. $4x + 3y - 5 = 0$
3. $3x + 4y = 12$
4. $2x + 3y = 6$
5. $y = 2x - 5$
6. $y = -x + 4$
7. $y - 4 = -\frac{3}{5}(x - 4)$
8. $y - 7 = -\frac{3}{5}(x + 1)$

POINT–SLOPE EQUATION

Now we derive an equation of a given straight line in the Cartesian coordinate system. Suppose for the moment that the line is not vertical (so its slope is defined).

To derive the equation of the line, we must have some information about the line. So suppose we know a fixed point $P_1 = (x_1, y_1)$ on the line, and we also know the slope m of the line. We will find an equation of the line in terms of the three *constants* x_1, y_1, and m.

If we have a point $P = (x, y)$ in the Cartesian coordinate system, what condition must P satisfy to be on the given straight line? One way to state the condition is to say that the slope of the line through P and P_1 must be equal to m, the slope of the given line (Figure 7.17).

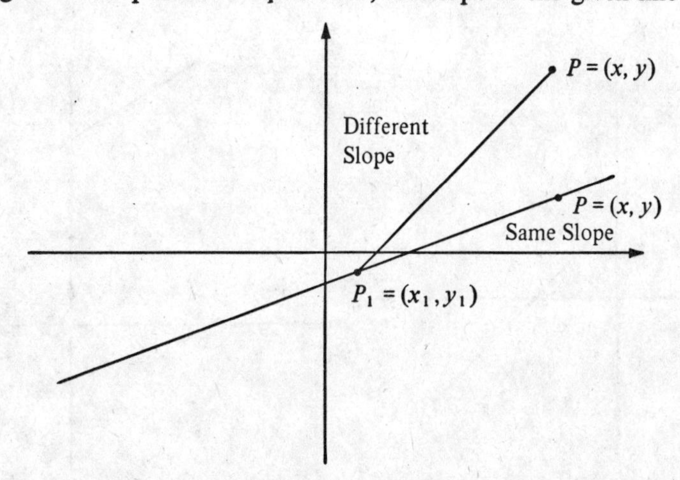

Figure 7.17

If the point P is not on the line, there will be a different slope. From Section 7.4, assuming $P \neq P_1$, we see that the slope of the line through P and P_1 is

$$m = \frac{y - y_1}{x - x_1}.$$

Multiplying this equation by $x - x_1$, we obtain the equation

$$y - y_1 = m(x - x_1).$$

This second equation is satisfied by (x_1, y_1) as well as by all other points on the line, so we have shown that a point (x, y) is on the line if and only if it satisfies this equation. The equation is known as the *point–slope form* of the equation of a straight line.

POINT–SLOPE FORM OF THE EQUATION OF A STRAIGHT LINE. An equation of the straight line passing through the point (x_1, y_1) and having slope m is

$$y - y_1 = m(x - x_1).$$

Example 2. Find an equation of the line passing through the point $(-3, 6)$ and having slope -1 (Figure 7.18).

Solution. We have $x_1 = -3$, $y_1 = 6$, and $m = -1$; so the point–slope form of the equation is

$$y - 6 = -1(x + 3).$$

This equation is equivalent to the linear equation

$$x + y - 3 = 0.$$

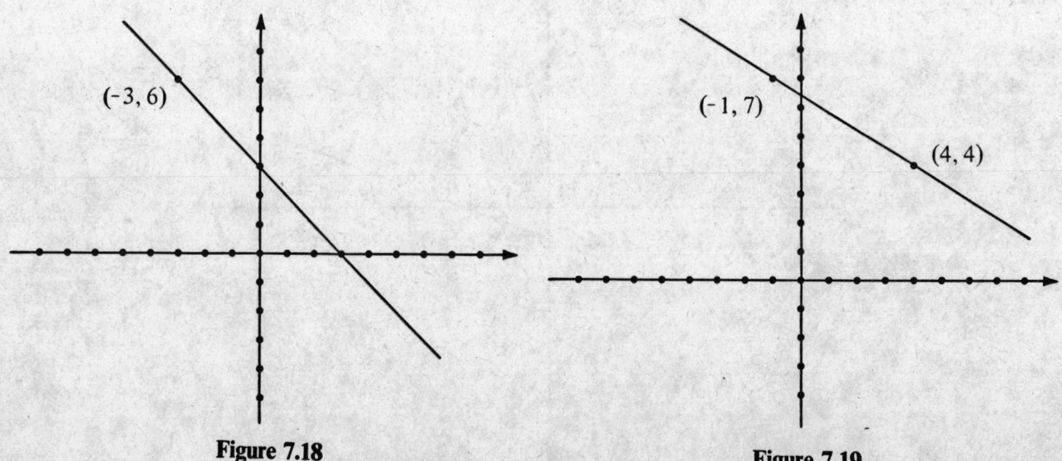

Figure 7.18 **Figure 7.19**

Example 3. Find an equation of the line passing through the two points (4, 4) and (−1, 7) (Figure 7.19).

Solution. First we calculate the slope of this line:

$$m = \frac{7 - 4}{-1 - 4} = \frac{3}{-5} = -\frac{3}{5}.$$

So a point-slope form equation of the line is

$$y - 4 = -\frac{3}{5}(x - 4).$$

This equation uses the point (4, 4). We could also use the point (−1, 7) and then obtain the point-slope equation

$$y - 7 = -\frac{3}{5}(x + 1).$$

So there are different point–slope forms for the same line. These equations are equivalent, however, and they are both equivalent to the equation

$$3x + 5y - 32 = 0.$$

Exercise Set 7.5 (continued)

In exercises 9–18, find a point–slope form equation of the line with the given slope and passing through the given point. Then find an equivalent equation in the form $Ax + By + C = 0$.

9. $m = 2$; (1, 1)
10. $m = 3$; (2, 3)
11. $m = -1$; (−3, 4)
12. $m = -2$; (−2, −5)
13. $m = 4/3$; (2, −5)
14. $m = -3/2$; (−1, 1)
15. $m = -3/5$; (0, 0)
16. $m = 2/3$; (0, 0)
17. $m = 0$; (2, 3)
18. $m = 0$; (−1, 3)

In exercises 19–28, find a point–slope equation of the line passing through the two given points, and then write the equation in the form $Ax + By + C = 0$.

19. (3, 8) and (−2, −2)
20. (2, 4) and (3, 1)
21. (−1, 1) and (1, −1)
22. (−6, 4) and (−6, 6)
23. (1, 3) and (0, 0)
24. (−2, 4) and (0, 0)
25. (−2, 2) and (4, −4)
26. (1, −5) and (−1, 5)
27. (−2, 1) and (1, −2)
28. (3, −4) and (−4, 3)

SLOPE–INTERCEPT EQUATION

If we are given a point (x_1, y_1) on a straight line and the slope m of the line, a point–slope equation of the line is

$$y - y_1 = m(x - x_1).$$

Recall from Section 7.4 that the y-intercept of the line is the point at which the line crosses the y-axis. If now we are given the y-intercept $(0, b)$ and the slope m of the line, the point–slope equation becomes

$$y - b = m(x - 0)$$
$$y = mx + b$$

This form of the equation is called the *slope–intercept form.*

SLOPE–INTERCEPT FORM OF THE EQUATION OF A STRAIGHT LINE. An equation of the straight line having y-intercept $(0, b)$ and slope m is

$$y = mx + b.$$

Note that the slope–intercept form expresses y as a linear function of x.

Example 4. Find an equation of the line with y-intercept $(0, -2)$ and slope $3/2$.

Solution. Writing a slope–intercept equation, we have

$$y = \frac{3}{2}x - 2.$$

This equation is equivalent to the linear equation

$$3x - 2y - 4 = 0.$$

Note: So far we have assumed that the straight line is not vertical in the Cartesian coordinate system. A vertical line does not possess a y-intercept, and the slope of a vertical line is not defined. However, the equation of a vertical line with x-intercept $(c, 0)$ is $x = c$. This equation is a linear equation in which the y-term is missing.

To summarize what we have so far:

A STRAIGHT LINE HAS A LINEAR EQUATION. Any equation of a straight line in the Cartesian coordinate system is a linear equation in two variables—that is, if the variables are x and y, the equation is equivalent to an equation of the form

$$Ax + By + C = 0,$$

where A and B are not both 0.

Now, conversely, suppose we have a linear equation

$$Ax + By + C = 0$$

Is its graph a straight line? First, if the constant $B = 0$, then $A \neq 0$ and

$$Ax + C = 0 \text{ is equivalent to } x = -C/A.$$

The graph is then a vertical straight line. If $B \neq 0$ we can solve for y as a function of x:

$$By = -Ax - C$$

$$y = -\frac{A}{B}x - \frac{C}{B}.$$

This equation is the slope–intercept form of a line with slope $-A/B$ and y-intercept $(0, -C/B)$. Therefore, the graph is a straight line.

THE GRAPH OF A LINEAR EQUATION IS A STRAIGHT LINE. The graph in the Cartesian coordinate system of the linear equation

$$Ax + By + C = 0$$

is a straight line. If $B = 0$, the line is vertical; if $B \neq 0$, the line has slope $-A/B$ and y-intercept $-C/B$.

Example 5. Find the slope and y-intercept of the graph of

$$2x + 3y - 6 = 0.$$

Solution. We solve for y as a function of x:

$$3y = -2x + 6$$

$$y = -\frac{2}{3}x + 2$$

The slope of the line is $-2/3$ and the y-intercept is $(0, 2)$.

Exercise Set 7.5 (continued)

In exercises 29–34, find the slope-intercept equation of the line with given slope and y-intercept.

29. $m = 1$; $(0, 3)$
30. $m = 2$; $(0, -1)$
31. $m = -2$; $(0, -3)$
32. $m = -3$; $(0, 2)$
33. $m = 1/2$; $(0, -4)$
34. $m = -3/2$; $(0, 0)$

In exercises 35–40, find the slope and y-intercept of the graph of each given equation.

35. $y = 4x + 6$
36. $y = x/4$
37. $2x + 2y = 12$
38. $2x + 2y = 24$
39. $3x + 2y = 6$
40. $4x - 3y = 12$

GRAPHING LINEAR EQUATIONS

The graph of a linear equation is a straight line. Therefore, to graph a linear equation, we need only find two points on the graph and draw the line determined by these points. Usually it is easy to find the x-intercept and y-intercept.

Example 6. Graph the equation $2x + 3y - 6 = 0$.

Solution. Since this equation is linear, we find the x-intercept and y-intercept:

x-intercept	y-intercept
Set $y = 0$	Set $x = 0$
$2x + 3(0) - 6 = 0$	$2(0) + 3y - 6 = 0$
$2x = 6$	$3y = 6$
$x = 3$	$y = 2$

The x-intercept is (3, 0) and the y-intercept is (0, 2). The graph is Figure 7.20.

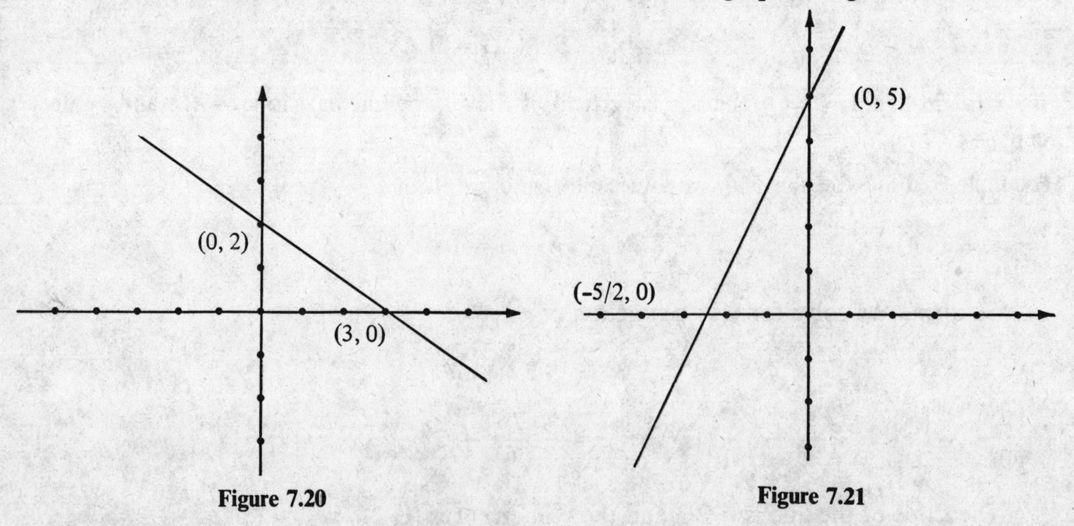

Figure 7.20

Figure 7.21

Example 7. Graph the equation $y = 2x + 5$.

Solution. Since this equation is linear, we find the x-intercept and y-intercept:

x-intercept	y-intercept
Set $y = 0$	Set $x = 0$
$0 = 2x + 5$	$y = 2(0) + 5$
$x = -5/2$	$y = 5$

The x-intercept is $(-5/2, 0)$ and the y-intercept is $(0, 5)$. The graph is Figure 7.21.

Exercise Set 7.5 (continued)

In exercises 41–48, graph each given equation.

41. $x + y = 4$
42. $x - y = 3$
43. $3x + 2y = 6$
44. $4x + 3y = 12$
45. $4x - 3y = 12$
46. $5x + 2y = 10$
47. $3x - y - 4 = 0$
48. $4x - y - 1 = 0$

section 6 • Linear Inequalities

In the previous section we saw that the graph of a linear equation in two variables is a straight line in the Cartesian coordinate system. In this section we examine the graph of a linear inequality in two variables.

DEFINITION OF A LINEAR INEQUALITY. A *linear inequality* in the two variables x and y is an inequality equivalent to one of the form

$$Ax + By + C > 0 \quad \text{or} \quad Ax + By + C < 0$$

or

$$Ax + By + C \geq 0 \quad \text{or} \quad Ax + By + C \leq 0,$$

where A, B, and C are constants and not both A and B are 0.

GRAPHING LINEAR INEQUALITIES

The technique we discuss for graphing a linear inequality reduces to graphing a linear equation.

Example 1. Graph the inequality $3x + 2y \geq 6$.

Solution. The Cartesian coordinate system is divided into two regions—one region satisfies the given inequality $3x + 2y \geq 6$ and the other region satisfies the opposite inequality $3x + 2y < 6$. The border separating these two regions comes from the equation $3x + 2y = 6$.

So first we graph the linear equation $3x + 2y = 6$. We see that the x-intercept is $(2, 0)$ and the y-intercept is $(0, 3)$, so the graph is the straight line passing through these points. Since this straight line is the border between those points that satisfy the inequality and those that do not, the region satisfying the inequality is either above or below the line (in either case it includes the line). So we simply try a test point. The origin lies below the line and does *not* satisfy the given inequality because

$$3(0) + 2(0) = 0 < 6.$$

Therefore, the region satisfying the inequality lies above the line (and including the line). We have graphed this inequality in Figure 7.22.

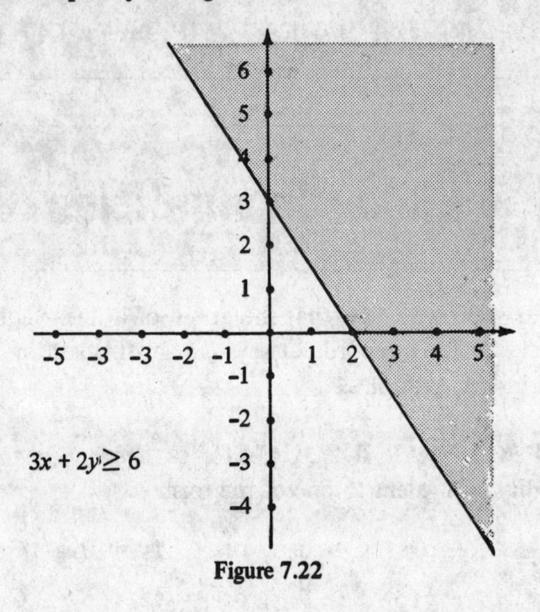

Figure 7.22

Note: We can try any test point from either region. We may wish to try a test point from the opposite region as a check. For example, (5, 5) lies above the line and satisfies the given inequality because

$$3(5) + 2(5) = 25 \geq 6.$$

Example 2. Graph the inequality $y < 2x + 5$.

Solution. The Cartesian coordinate system is divided into two regions—one region satisfies the given inequality $y < 2x + 5$ and the other region satisfies the opposite inequality $y \geq 2x + 5$. The border separating these two regions comes from the linear equation $y = 2x + 5$.

So first we graph the equation $y = 2x + 5$. We see that the x-intercept is $(-5/2, 0)$ and the y-intercept is $(0, 5)$, so the graph is the straight line passing through these points. (We graphed this equation in Example 7 of Section 7.5; the graph is Figure 7.21.) Since this straight line is the border between those points that satisfy the inequality and those that do not, the region satisfying the inequality is either to the right of the line or to the left of the line (not including the line). We try a test point. The origin lies to the right and satisfies the given inequality because

$$0 < 2(0) + 5.$$

Therefore, the region satisfying the inequality lies to the right of the line (not including the line). Figure 7.23 shows the graph of this inequality.

Note: To indicate that the line is not included in the graph, we have drawn it as a dashed line.

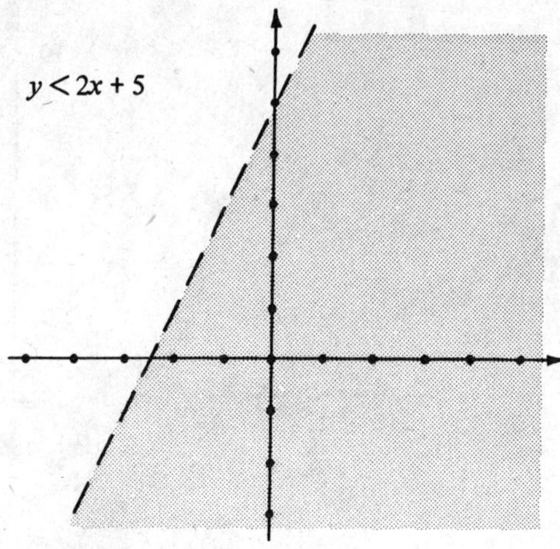

Figure 7.23

Exercise Set 7.6

In exercises 1–8, graph the given linear inequality.

1. $x + y \leq 4$
2. $x - y \leq 3$
3. $3x + 2y > 6$
4. $4x - 3y < 12$
5. $4x - 3y \leq 12$
6. $5x + 2y \geq 10$
7. $3x + y > 3$
8. $2x - 5y < 10$

FIRST QUADRANT INEQUALITIES

Frequently in applications we have a linear inequality in two variables x and y, and neither x nor y can be negative. That is, in addition to the given linear inequality we have the two inequalities

$$x \geq 0 \quad \text{and} \quad y \geq 0.$$

The effect of these two additional inequalities is to confine the graph to the first quadrant of the Cartesian coordinate system.

Example 3. Graph the inequalities $4x + 5y \leq 50$, $x \geq 0$, and $y \geq 0$.

Solution. The graph is in the first quadrant. Consider the linear equation $4x + 5y = 50$. Its x-intercept is $(25/2, 0)$ and its y-intercept is $(0, 10)$. We draw the line in the first quadrant between these two points. To see if the inequality is satisfied above the line or below the line, we try a test point. The point $(2, 2)$ is below the line and satisfies the inequality because

$$4(2) + 5(2) = 18 \leq 50.$$

So the graph is below the line (including the line). We have graphed these inequalities in Figure 7.24.

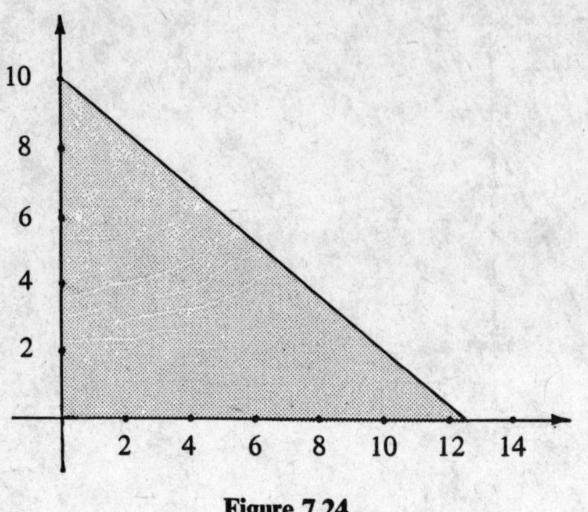

Figure 7.24

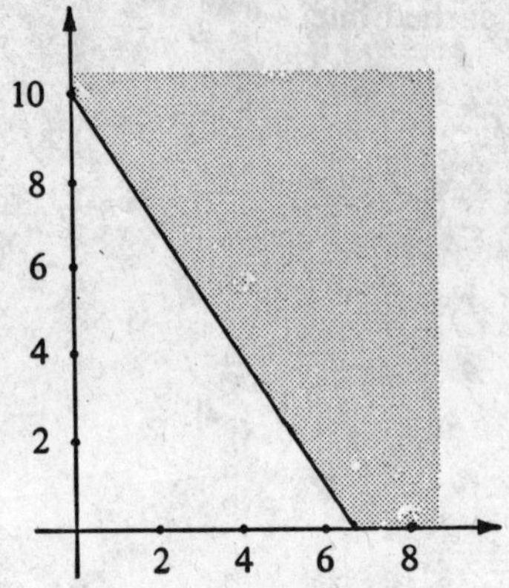

Figure 7.25

Example 4. Graph the inequalities $3x + 2y \geq 20$, $x \geq 0$, and $y \geq 0$.

Solution. The graph is in the first quadrant. Consider the linear equation $3x + 2y = 20$. Its x-intercept is $(20/3, 0)$ and its y-intercept is $(0, 10)$. We draw the line in the first quadrant between these two points. To see whether the inequality is satisfied above the line or below the line, we try a test point. The point $(2, 2)$ is below the line and does not satisfy the inequality because

$$3(2) + 2(2) = 10 < 20.$$

So the graph is above the line (including the line). We have graphed these inequalities in Figure 7.25.

Exercise Set 7.6 (continued)

In exercises 9–16, graph the given linear inequality with the inequalities $x \geq 0$ and $y \geq 0$.

9. $4x + 5y \leq 60$
10. $4x + 5y \leq 40$
11. $200x + 75y \leq 3000$
12. $8x + 25y \leq 20{,}000$
13. $2x + 2y \geq 24$
14. $2x + 3y \geq 15$
15. $2x + 3y \geq 30$
16. $x + 3y \geq 21$

section 7 • Circles

In this section we consider circles in the Cartesian coordinate system. Basically we do two things:

1. Given a circle in the Cartesian coordinate system, we find an equation of the circle.
2. Given an equation in x and y, we determine whether the graph is a circle and, if so, we graph the equation.

We will use the distance formula from Section 7.4 and the technique of completing the square from Section 5.2.

FINDING THE EQUATION OF A CIRCLE

In geometry, a circle is defined to be the set of points that are a fixed distance r (called the *radius*) from a fixed point C (called the *center*). In the Cartesian coordinate system, suppose the point C is denoted by (h, k). Then a point (x, y) is on the circle if and only if the distance between (x, y) and (h, k) is r. Using the distance formula (Section 7.4), we find

$$\sqrt{(x - h)^2 + (y - k)^2} = r.$$

Since both sides of this equation represent positive quantities, we can square both sides of the equation and write

$$(x - h)^2 + (y - k)^2 = r^2.$$

EQUATION OF A CIRCLE. An equation of the circle in the Cartesian coordinate system with center at (h, k) and radius r is

$$(x - h)^2 + (y - k)^2 = r^2.$$

Example 1. Find an equation of the circle with center at $(-3, 4)$ and radius 5 (Figure 7.26).

Solution. Here $h = -3$, $k = 4$, and $r = 5$. So an equation is

$$(x + 3)^2 + (y - 4)^2 = 25.$$

This equation is equivalent to

$$x^2 + 6x + 9 + y^2 - 8y + 16 = 25$$
$$x^2 + y^2 + 6x - 8y = 0$$

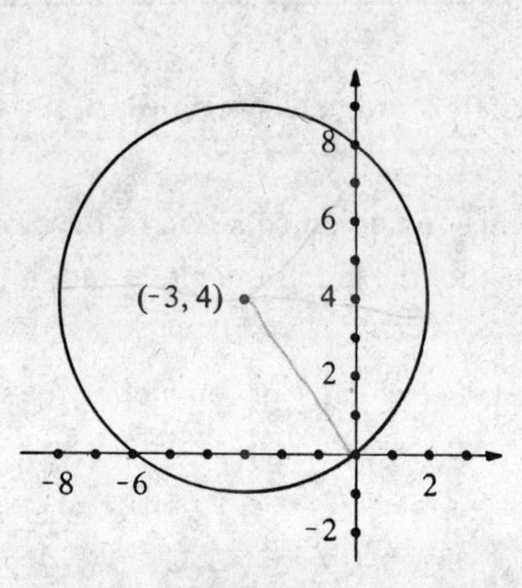

Figure 7.26

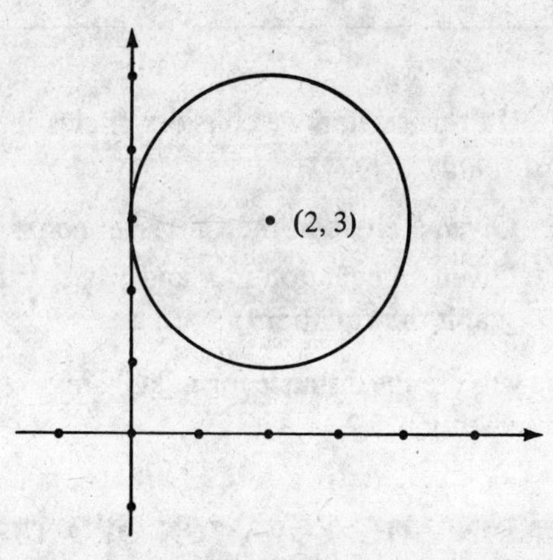

Figure 7.27

Example 2. Find an equation of the circle with center at (2, 3) and tangent to the y-axis.

Solution. We have $h = 2$, $k = 3$, and we must find the radius r. It is useful to draw a picture as in Figure 7.27. We see that $r = 2$, so the equation is

$$(x - 2)^2 + (y - 3)^2 = 2^2$$

This equation is equivalent to

$$x^2 - 4x + 4 + y^2 - 6y + 9 = 4$$
$$x^2 + y^2 - 4x - 6y + 9 = 0$$

Exercise Set 7.7

In exercises 1–10, find an equation of the circle with the given center and radius.

1. Center at (5, 5); radius 1
2. Center at (1, −3); radius 6
3. Center at (−1, −2); radius 5
4. Center at (0, 4); radius 4
5. Center at (0, 0); radius 2
6. Center at (0, 0); radius 3
7. Center at (−2, 1); radius $\sqrt{2}$
8. Center at (3, −3); radius $\sqrt{3}$
9. Center at (2, 2); radius $2\sqrt{2}$
10. Center at (−3, −3); radius $3\sqrt{3}$

In exercises 11–20, find an equation of the circle satisfying the given conditions.

11. Center at (4, 3) and tangent to the y-axis
12. Center at $(-2, -3)$ and tangent to the y-axis
13. Center at $(3, -3)$ and tangent to the x-axis
14. Center at $(-1, 1)$ and tangent to the x-axis
15. Center at the origin and passing through (3, 4)
16. Center at the origin and passing through $(-1, -1)$
17. Center at (1, 1) and passing through the origin
18. Center at $(2, -2)$ and passing through the origin
19. Center at $(-1, 3)$ and passing through the origin
20. Center at $(4, -1)$ and passing through the origin

WHEN THE GRAPH IS A CIRCLE

We can graph a circle if we know its center and radius. Therefore, we now show how to recognize when the graph of an equation in x and y is a circle and how to find the center and radius of the graph.

We have shown that the equation of a circle is always equivalent to an equation of the form

$$(x - h)^2 + (y - k)^2 = r^2.$$

We can also write this equation as

$$x^2 - 2hx + h^2 + y^2 - 2ky + k^2 = r^2$$

$$x^2 + y^2 - 2hx - 2ky + h^2 + k^2 - r^2 = 0$$

So the equation of a circle possesses the form

$$x^2 + y^2 + Ax + By + C = 0,$$

where A, B, and C are constants.

Suppose now we have an equation of the above form. Is the graph always a circle, and if so, what is the center and radius?

Example 3. Find the center and radius of the graph of

$$x^2 + y^2 = 16.$$

Solution. We can write the equation in the form

$$x^2 + y^2 = 4^2.$$

The graph of the equation is a circle with center at the origin and radius 4.

Example 4. Find the center and radius of the circle with equation

$$x^2 + y^2 - 6x + 4y + 1 = 0.$$

Solution. We wish to write the equation in the form $(x - h)^2 + (y - k)^2 = r^2$ for some h, k, and r. We write the equation in the form

$$x^2 - 6x + y^2 + 4y = -1.$$

Now we complete the square on both x and y (Section 5.2) to obtain

$$x^2 - 6x + 9 + y^2 + 4y + 4 = -1 + 9 + 4$$
$$(x - 3)^2 + (y + 2)^2 = 12$$

We see that the center is at $(3, -2)$ and the radius is $\sqrt{12} = 2\sqrt{3}$.

Example 5. Find the center and radius of the circle with equation

$$x^2 + y^2 = 4x - 2y - 6.$$

Solution. We wish to write the equation in the form $(x - h)^2 + (y - k)^2 = r^2$. To do so, we write

$$x^2 - 4x + y^2 + 2y = -6.$$

Completing the square on both x and y; we obtain

$$x^2 - 4x + 4 + y^2 + 2y + 1 = 4 + 1 - 6$$
$$(x - 2)^2 + (y + 1)^2 = -1$$

The graph of this equation is not a circle because we would have to have $r^2 = -1$, and this is impossible for real numbers r.

Note: The left-hand side of this equation is a sum of two squares and so cannot be equal to the *negative* number -1. Therefore, there are *no* ordered pairs (x, y) of real numbers that satisfy this equation; that is, this equation does not have a graph in the Cartesian coordinate system. This example shows that an equation may possess the form of the equation of a circle and still have no graph.

Exercise Set 7.7 (continued)

In exercises 21–40, specify whether the given equation has a circle for its graph and, if it does, find the center and radius of the circle.

21. $x^2 + y^2 = 25$
22. $x^2 + y^2 = 100$
23. $x^2 + y^2 = 25/4$
24. $x^2 + y^2 = 8$

25. $x^2 + y^2 + 4 = 0$

26. $x^2 + y^2 + 16 = 0$

27. $x^2 + y^2 - 6y - 7 = 0$

28. $x^2 + y^2 + 8x + 15 = 0$

29. $x^2 + y^2 - 4x - 4y - 1 = 0$

30. $x^2 + y^2 + 8x - 2y + 15 = 0$

31. $x^2 + y^2 + 6x + 10y - 2 = 0$

32. $x^2 + y^2 + 2x + 2y + 2 = 0$

33. $x^2 + y^2 = 2x + 2y$

34. $x^2 + y^2 = 10x$

35. $x^2 + y^2 = 2x + 2y - 4$

36. $x^2 + y^2 = 10x - 36$

37. $x^2 + y^2 = 8x + 6y$

38. $x^2 + y^2 = 6x - 6y + 207$

39. $x^2 + y^2 = 8x + 6y - 50$

40. $x^2 + y^2 = 6x - 6y - 207$

section 8 • Applications

In this section we give applications of graphing and functions to geometry, business, diet problems, budget problems, grade and pitch considerations, and linear correlation.

GEOMETRY

Example 1. A rectangle has sides of lengths x and y. If the perimeter of this rectangle is 20 meters, express y as a function of x. Now if this function is called f, so $y = f(x)$, find $f(1)$, $f(2)$, $f(5)$, and $f(7.5)$.

Solution. Since the perimeter is 20, we have

$$\begin{aligned} 2x + 2y &= 20 \\ 2y &= 20 - 2x \\ y &= 10 - x \end{aligned}$$

Therefore, y as a function of x is expressed by

$$y = f(x) = 10 - x.$$

$$\begin{aligned} f(1) &= 10 - 1 = 9 \\ f(2) &= 10 - 2 = 8 \\ f(5) &= 10 - 5 = 5 \qquad \text{(a square)} \\ f(7.5) &= 10 - 7.5 = 2.5 \end{aligned}$$

Exercise Set 7.8

1. A rectangle has sides of lengths x and y. If the perimeter of the rectangle is 12 meters, express y as a function of x. Now if this function is called f, so $y = f(x)$, find $f(1)$, $f(2)$, $f(3)$, and $f(5.5)$.
2. A rectangle has sides of lengths x and y. If the perimeter of the rectangle is 24 meters, express y as a function of x. Now if this function is called f so $y = f(x)$, find $f(1)$, $f(5)$, $f(6)$, and $f(10)$.
3. A rectangle has sides of lengths x and y and perimeter 8 meters.
   a. Express y as a function of x.
   b. Express the area A of the rectangle as a function of x.
4. A rectangle has sides of length x and y and perimeter 30 meters.
   a. Express y as a function of x.
   b. Express the area A of the rectangle as a function of x.
5. If a rectangle has sides of lengths x and y and area 4 square meters, express y as a function of x. If this function is called g, so $y = g(x)$, find $g(1)$, $g(2)$, $g(3)$, and $g(0.5)$.
6. If a rectangle has sides of lengths x and y and area 24 square meters, express y as a function of x. If this function is called g, so $y = g(x)$, find $g(1)$, $g(2)$, $g(3)$, and $g(0.5)$.
7. One side of a rectangle is 3 meters longer than the adjacent side.
   a. Express the perimeter P as a function of the shorter side x.
   b. Express the area A as a function of the shorter side x.
8. One side of a rectangle is 4 meters longer than the adjacent side x.
   a. Express the perimeter P as a function of the shorter side x.
   b. Express the area A as a function of the shorter side x.
9. a. Express the perimeter P of a square as a function of its side x.
   b. Express the area A of a square as a function of its side x.
10. a. Express the circumference C of a circle as a function of its radius r.
    b. Express the area A of a circle as a function of its radius r.
11. a. Express the side x of a square as a function of its perimeter P.
    b. Express the side x of a square as a function of its area A.
12. a. Express the radius r of a circle as a function of its circumference C.
    b. Express the radius r of a circle as a function of its area A.

BUSINESS: PROFIT FUNCTIONS

A business makes a product which it then sells. It is important to analyze the profit P obtained from selling n units of the product; the profit P is a function of n. The total income obtained from selling n units of the product is pn, where p is the selling price of each unit. However, the total cost of making n units of the product is the sum of two types of costs—*fixed costs* and *variable costs*. Fixed costs may include office rent, secretaries' salaries, telephone bills, and so on. These costs do not depend on the number of units manufactured. Variable costs are those directly related to the manufacture of the product, such as material and labor used in its manufacture. If we let F denote the fixed costs and c the cost of manufacturing each unit, then the total cost of manufacturing n units is $F + cn$. Therefore, the profit P obtained on n units is

$$P = pn - (F + cn) = (p - c)n - F.$$

Example 2. A publisher of paperback novels publishes a book called *Watership Down*. The publisher finds that the fixed costs (typesetting, editor's salary, etc.) are \$32,000. The cost of each book (paper, printing, author's royalties, etc.) is \$1.95. The publisher prices the book at \$2.45. The profit P obtained from selling n books is

$$P = 2.45n - (1.95n + 32{,}000) = 0.50n - 32{,}000.$$

If we call this function f, so that $P = f(n)$, then $f(50{,}000)$ is the profit obtained from selling 50,000 books and $f(100{,}000)$ is the profit obtained from selling 100,000 books. Thus,

$$f(50{,}000) = 0.50(50{,}000) - 32{,}000 = -7{,}000$$

$$f(100{,}000) = 0.50(100{,}000) - 32{,}000 = 18{,}000$$

The publisher loses \$7,000 on 50,000 books but realizes a profit of \$18,000 on 100,000 books.

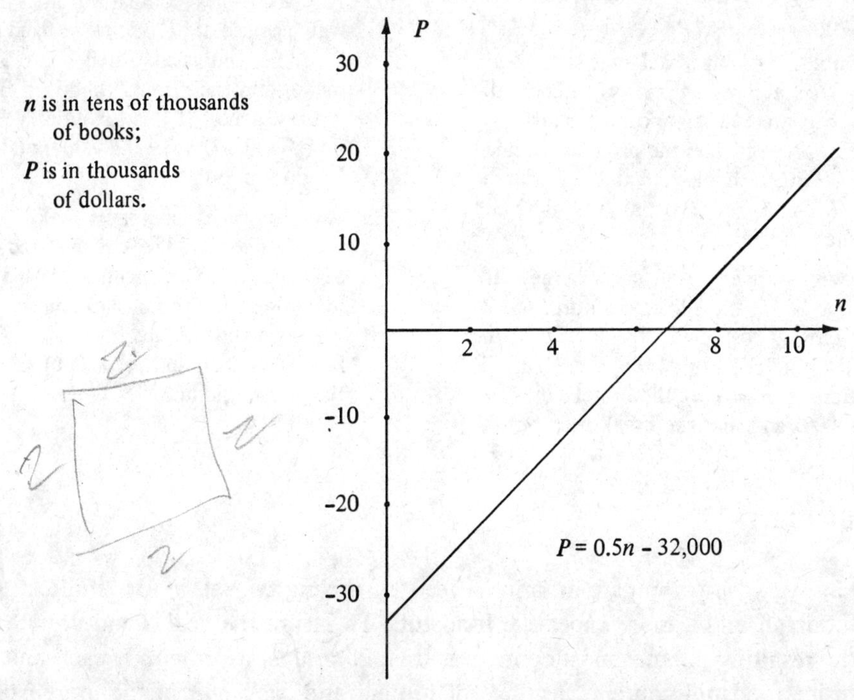

Figure 7.28

We have graphed P as a function of n in Figure 7.28; n is along the horizontal axis and P is along the vertical axis. Since the publisher cannot sell a negative number of books, it makes sense only to graph non-negative values of n. We see that the graph is a straight line for which the n-intercept represents the *break-even point*. We find the n-intercept by setting $P = 0$:

$$0 = f(n) = 0.50n - 32{,}000$$

$$0.50n = 32{,}000$$

$$n = 64{,}000$$

That is, the publisher loses money when selling less than 64,000 books, breaks even at 64,000 books, and makes money when selling more than 64,000 books.

Exercise Set 7.8 (continued)

13. Suppose the publisher of paperback novels publishes *For Whom the Bell Tolls* at a selling price of \$2.95. If the fixed costs are \$28,000 and each book costs \$2.25 to produce, find the profit P obtained from selling n books. If this function is $P = f(n)$, find $f(50{,}000)$ and $f(100{,}000)$ and find the break-even point.
14. The publisher of paperback novels publishes *Moby Dick* at a selling price of \$2.25. If the fixed costs are \$22,000 and each book costs \$1.80 to produce, find the profit P obtained from selling n books. If this function is $P = f(n)$, find $f(50{,}000)$ and $f(100{,}000)$ and find the break-even point.
15. A record company brings out a new LP priced at \$6.75. If the fixed costs are \$46,000 and each LP costs \$4.35 to make, find the profit P obtained from selling n LPs. If this function is $P = f(n)$, find $f(50{,}000)$ and $f(100{,}000)$ and find the break-even point.
16. A record company brings out a new LP priced at \$5.95. If the fixed costs are \$39,000 and each LP costs \$4.15 to make, find the profit P obtained from selling n LPs. If this function is $P = f(n)$, find $f(50{,}000)$ and $f(100{,}000)$ and find the break-even point.
17. An electronics company makes a digital watch priced at \$19.95. If the fixed costs are \$143,000 and each watch costs \$18.85 to make, find the profit P obtained from selling n watches. If this function is $P = f(n)$, find $f(100{,}000)$ and $f(200{,}000)$ and find the break-even point.
18. An electronics company makes a digital watch priced at \$32.50. If the fixed costs are \$186,000 and each watch costs \$30.75 to make, find the profit P obtained from selling n watches. If this function is $P = f(n)$, find $f(100{,}000)$ and $f(200{,}000)$ and find the break-even point.

DIET PROBLEMS

Example 3. We wish to prepare an animal feed by mixing two basic foodstuffs, a less expensive foodstuff and a more expensive foodstuff. To insure the health and growth of the animal, the resulting mixture must contain at least 20 grams of protein. If each unit of the less expensive foodstuff contains 3 grams of protein and each unit of the more expensive foodstuff contains 2 grams of protein, graph the region of the plane that satisfies this requirement.

Solution. If we mix x units of the less expensive foodstuff with y units of the more expensive foodstuff, then the mixture contains $3x + 2y$ grams of protein. So the protein requirement is expressed by the inequalities

$$3x + 2y \geq 20,$$

$$x \geq 0, \qquad y \geq 0.$$

The inequalities $x \geq 0$ and $y \geq 0$ say that we cannot mix a negative amount of foodstuffs. The graph is Figure 7.29.

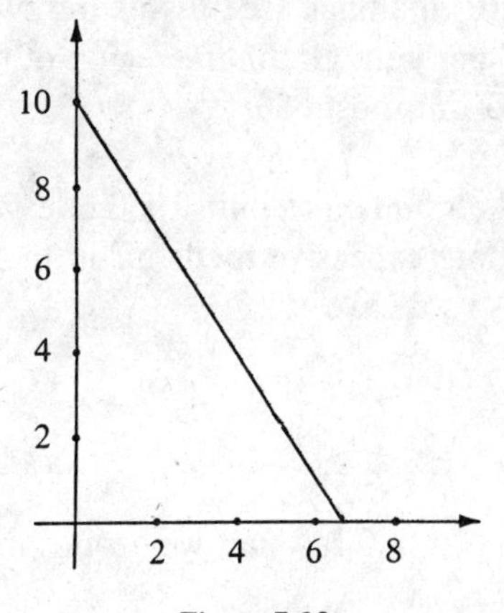

Figure 7.29

Exercise Set 7.8 (continued)

19. We prepare an animal feed by mixing a less expensive foodstuff with a more expensive foodstuff. The mixture must contain at least 24 grams of protein. Graph the region of the plane satisfying this requirement if each unit of the less expensive foodstuff contains 2 grams of protein and each unit of the more expensive foodstuff contains 2 grams of protein. (Let x denote the number of units of the less expensive foodstuff and y denote the number of units of the more expensive foodstuff.)

20. We prepare an animal feed by mixing a less expensive foodstuff with a more expensive foodstuff. The mixture must contain at least 15 grams of fats. Graph the region of the plane satisfying these requirements if each unit of the less expensive foodstuff contains 2 grams of fats and each unit of the more expensive foodstuff contains 3 grams of fats. (Let x denote the number of units of the less expensive foodstuff and y denote the number of units of the more expensive foodstuff.)

21. A food company plans to come out with a drink consisting of a mixture of orange juice and pineapple juice. To advertise the nutritional value of the drink, the management requires each can to contain at least 30 milligrams of Vitamin C. Graph the region of the plane satisfying this requirement if each fluid ounce of orange juice contains 2 mg of Vitamin C and each fluid ounce of pineapple juice contains 3 mg of Vitamin C. (Let x denote the number of fluid ounces of orange juice and y the number of fluid ounces of pineapple juice.)

22 A food company plans to come out with a drink consisting of a mixture of orange juice and pineapple juice. To advertise the nutritional value of the drink, the management requires each can to contain at least 21 milligrams of Vitamin D. Graph the region of the plane satisfying this requirement if each fluid ounce of orange juice contains 1 mg of Vitamin D and each fluid ounce of pineapple juice contains 3 mg of Vitamin D. (Let x denote the number of fluid ounces of orange juice and y the number of fluid ounces of pineapple juice.)

BUDGET PROBLEMS

Example 4. We wish to prepare an animal feed by mixing two basic foodstuffs, one costing \$4 per unit and one costing \$5 per unit. Graph the region of the plane satisfying the requirement that the mixture must cost at most \$50.

Solution. Let x denote the number of units of the less expensive foodstuff and y the number of units of the more expensive foodstuff in the mixture. Then the cost of the mixture is $4x + 5y$, and so we have

$$4x + 5y \leq 50,$$

$$x \geq 0, \qquad y \geq 0.$$

The inequalities $x \geq 0$ and $y \geq 0$ say that we cannot mix a negative amount of foodstuffs. The graph is Figure 7.30.

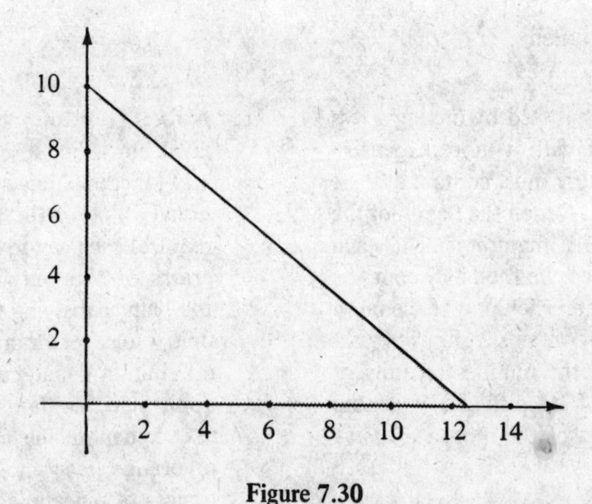

Figure 7.30

Exercise Set 7.8 (continued)

23. We prepare an animal feed by mixing two basic foodstuffs, one costing \$4 per unit and the other costing \$5 per unit. Graph the region of the plane satisfying the requirement that the mixture must cost at most \$60. (Let x denote the number of units of the less expensive foodstuff and y denote the number of units of the more expensive foodstuff.)

24. We prepare an animal feed by mixing two basic foodstuffs, one costing \$4 per unit and the other costing \$5 per unit. Graph the region of the plane satisfying the requirement that the mixture must cost at most \$40. (Let x denote the number of units of the less expensive foodstuff and y denote the number of units of the more expensive foodstuff.)

25. A food company plans to come out with a drink consisting of a mixture of orange juice and pineapple juice. Each fluid ounce of orange juice costs \$0.04 and each fluid ounce of pineapple juice costs \$0.05. Graph the region of the plane satisfying the requiremet that the cost of the mixture is at most \$0.60. (Let x denote the number of fluid ounces of orange juice and y denote the number of fluid ounces of pineapple juice.)
26. A food company plans to come out with a drink consisting of a mixture of orange juice and pineapple juice. Each fluid ounce of orange juice costs \$0.03 and each fluid ounce of pineapple juice costs \$0.03. Graph the region of the plane satisfying the requirement that the cost of the mixture is at most \$0.54. (Let x denote the number of fluid ounces of orange juice and y denote the number of fluid ounces of pineapple juice.)
27. A city has at most \$3000 to spend on street lights and parking meters. If each street light costs \$200 and each parking meter costs \$75, graph the region of the plane satisfying this requirement. (Let x denote the number of street lights and y denote the number of parking meters.)
28. A school board has at most \$20,000 to spend on textbooks and audiovisual material. If each textbook costs \$8 and each unit of audiovisual material costs \$25, graph the region of the plane satisfying this requirement. (Let x denote the number of textbooks and y denote the number of units of audiovisual material.)

GRADE AND PITCH

Example 5. The notion of the *grade* of a street or highway is analogous to the slope of a line. The grade of a street is the distance the street rises per unit length along the *surface* of the street. The steepest cable car street in San Francisco is on the Hyde Street Line; it has a grade of 0.213. How far does the street rise for 120 meters along the surface?

Solution. If r denotes the rise, then

$$r/120 = 0.213$$

$$r = (120)(0.213) = 25.56 \text{ meters}$$

Exercise Set 7.8 (continued)

29. The steepest cable car street on the California Street Line in San Francisco has grade 0.182. How many meters does the street rise for 120 meters along the surface?
30. The steepest cable car street on the Powell Street Line in San Francisco has grade 0.175. How many meters does the street rise for 120 meters along the surface?
31. The steepest street in San Francisco is Filbert Street, which has a grade of 0.522 (about 31.5 degrees). How many meters does the street rise for 120 meters along the surface?
32. Railroad lines are built as level as possible because the tractive effort required to pull a load up a grade of 0.01 is about 5 times that required on level track. If a railroad line has a grade of 0.01 for 3 miles along the surface, how many feet does the line rise? *Note:* One mile is 5280 feet.
33. The notion of the *pitch* of a roof is a special case of the slope of a line. The pitch of a roof is the distance the roof rises per unit of horizontal distance. Find the pitch of a roof if the horizontal distance is 24 feet and the roof rises 10 feet.
34. Find the pitch of a roof (see exercise 33) if the rise is 4 meters and the run is 24 meters.

LINEAR CORRELATION

Frequently an experimenter graphs data in the Cartesian coordinate system and then approximates the graph using a straight line. Such a straight-line approximation is called a *linear correlation.*

Example 6. In Figure 7.31 we have graphed, for various countries, the energy consumption of the country and the GNP (gross national product) per capita for the country. The energy consumption (vertical axis) is in kilograms per person per year (coal equivalent) and the GNP (horizontal axis) is the gross national product in U.S. dollars per person per year. Find an equation of the linear correlation line.

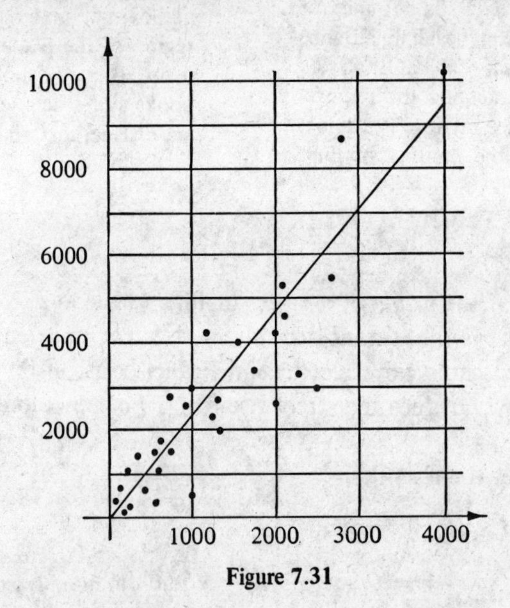

Figure 7.31

Solution. The linear correlation line passes through the two points (0, 0) and (3,000, 7000). Therefore, its slope is

$$m = \frac{7000 - 0}{3000 - 0} = \frac{7000}{3000} = \frac{7}{3}$$

Therefore, a point–slope equation is

$$y - 0 = \frac{7}{3}(x - 0)$$

$$3y = 7x$$

Exercise Set 7.8 (continued)

35. Figure 7.32 is a graph, for a sample of boys, of the average weight of the boys in the sample as a function of age. The average weight (vertical axis) is in kilograms and the age (horizontal axis) is in years. Find an equation of the linear correlation line.

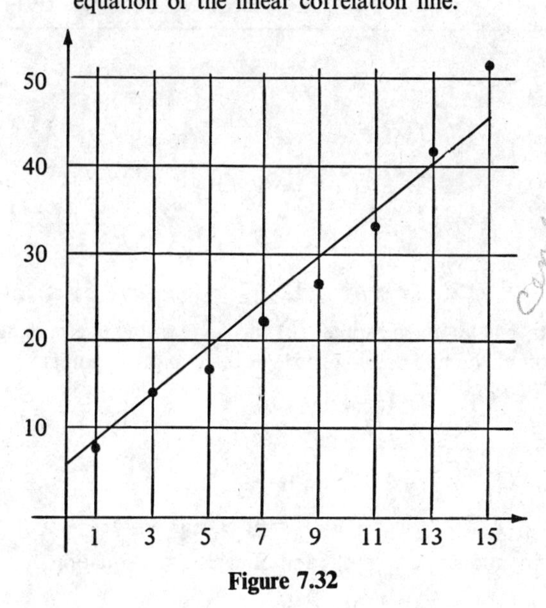

Figure 7.32

36. Figure 7.33 is a graph, for an experiment on soybean plants, of the average height of the plants as a function of their age. The average height (vertical axis) is in centimeters and the age (horizontal axis) is in weeks. Find an equation of the linear correlation line.

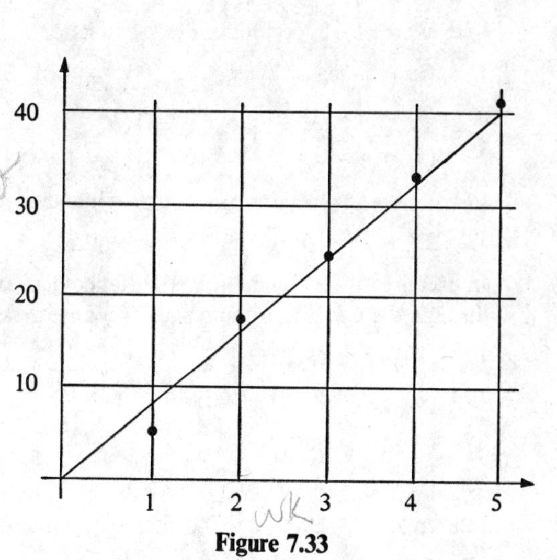

Figure 7.33

37. Figure 7.34 is a graph of the number of red blood corpuscles of blood in a person as a function of elevation above sea level. The number of red blood corpuscles (vertical axis) is in millions per cubic millimeter of blood and the elevation (horizontal axis) is in kilometers. Find an equation of the linear correlation line.

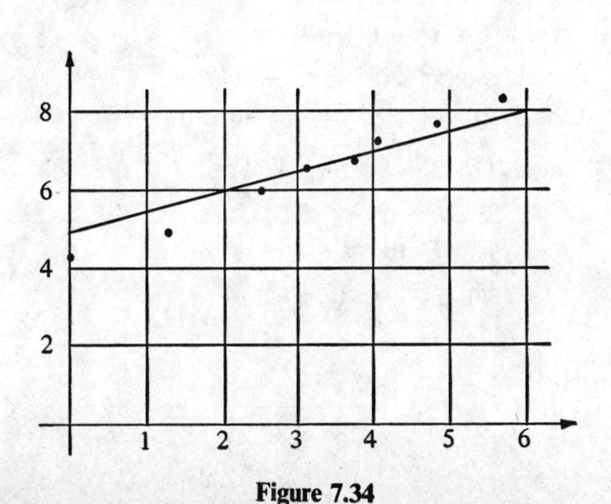

Figure 7.34

38. Figure 7.35 is a graph of the weight of potassium bromide which will dissolve in 100 grams of water as a function of temperature. The weight (vertical axis) is in grams and the temperature (horizontal axis) is in degrees Centigrade. Find an equation of the linear correlation line.

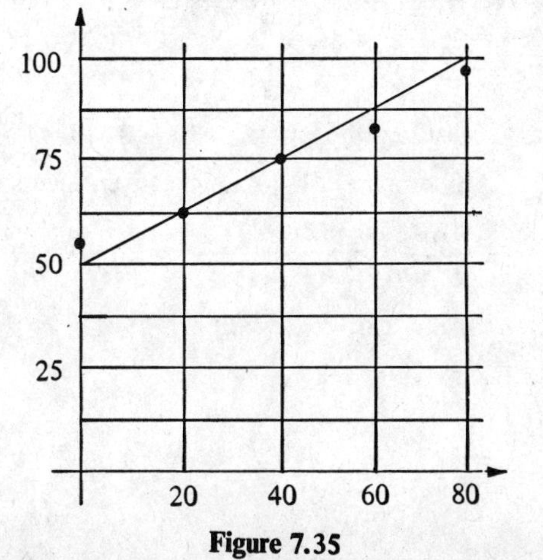

Figure 7.35

Review Exercises

In exercises 1–12, graph each given function.

1. $f(x) = -2x$
2. $f(x) = 4x$
3. $f(x) = 3$
4. $f(x) = -2$
5. $g(x) = x + 1$
6. $g(x) = x - 1$
7. $g(x) = 2/x$
8. $g(x) = 5/x$
9. $f(x) = 4 - x^2$
10. $f(x) = 1 + x^2$
11. $f(x) = 2x - 2$
12. $f(x) = 3x + 1$

In exercises 13–16, graph each linear inequality.

13. $2x + 3y \leq 6$
14. $x + y \geq 2$
15. $3x + 4y \geq 12$
16. $4x - 3y \leq 12$

In exercises 17–24, find: (a) the distance between the two given points; (b) the slope of the line passing through the two given points; and (c) an equation of the line passing through the two given points.

17. (1, 1) and (3, 3)
18. (−1, −1) and (2, 2)
19. (4, 3) and (0, 0)
20. (−1, 3) and (0, 0)
21. (−3, −2) and (4, −1)
22. (−2, 1) and (−3, −4)
23. (1, −3) and (−3, 1)
24. (2, −4) and (−4, 2)

In exercises 25–28, find the x-intercept, y-intercept, and slope of the graph of each given equation.

25. $3x - 5y = 15$
26. $2x + 5y = 10$
27. $6x + 4y = 9$
28. $3x - 5y = 12$

In exercises 29–34, find an equation of the circle satisfying the given conditions.

29. Center at (−1, 2); radius 3
30. Center at (1, −1); radius 4
31. Center at origin; radius $\sqrt{3}$
32. Center at origin; radius $\sqrt{5}$
33. Center at (−3, 5) and tangent to the y-axis
34. Center at (4, −1) and tangent to the x-axis

In exercises 35–40, specify whether each given equation has a circle for its graph and, if it does, find the center and radius of the circle.

35. $x^2 + y^2 = 25$
36. $x^2 + y^2 = 20$
37. $x^2 + y^2 = 2x + 2y + 2$
38. $x^2 + y^2 = 4x - 2y$
39. $x^2 + y^2 + 2x - 2y + 3 = 0$
40. $x^2 + y^2 - 6x + 10 = 0$

In exercises 41–46, for the given function f, find $f(2)$, $f(1)$, $f(0)$, $f(-1)$, and $f(-2)$.

41. $f(x) = 2x - 1$
42. $f(x) = x + 4$
43. $f(x) = x + x^2$
44. $f(x) = x - x^2$
45. $f(x) = x(2 - x)$
46. $f(x) = x(5 + x)$

47. A rectangle has sides of lengths x and y and perimeter 2.
    a. Express y as a function of x.
    b. Express the area A of the rectangle as a function of x.

48. A rectangle has sides of length x and y and area 1 square meter. Express y as a function of x.

49. A publisher of paperback books publishes a novel called *Catch-22*. The publisher finds that the fixed costs are \$24,000 and the cost of each book is \$1.05. If the selling price is \$1.45, find the profit P obtained from selling n books. If this function is $P = f(n)$, find $f(50{,}000)$ and $f(100{,}000)$ and find the break-even point.

50. We prepare an animal feed by mixing a less expensive foodstuff with a more expensive foodstuff. The resulting mixture must contain at least 18 grams of protein. Graph the region of the plane satisfying this requirement if each unit of the less espensive foodstuff contains 2 grams of protein and each unit of the more expensive foodstuff contains 3 grams of protein.

chapter 8

Exponential and Logarithmic Functions

1. Exponential Functions

2. Logarithmic Functions

3. Properties of Logarithms

4. Use of Tables

5. Computations with Logarithms

6. Applications

The starting points for the concepts of this chapter are the definition and properties of exponents. We use these concepts to define exponential functions and logarithmic functions.

In Section 8.1 we define exponential functions, examine their properties, and graph exponential functions. In Section 8.2 we define and graph logarithmic functions. We examine properties of logarithmic functions in Section 8.3. We show how to use logarithm tables in Section 8.4, and we perform computations with logarithms in Section 8.5. In Section 8.6 we apply exponential and logarithmic functions to exponential growth, exponential decay, and compound interest.

section 1 • Exponential Functions

Now we define the class of exponential functions and examine their properties and graphs. You may wish to review the definition and properties of exponents studied in Chapter 3 for this section.

DEFINITION OF EXPONENTIAL FUNCTIONS

We will use functional notation introduced in Section 7.1.

DEFINITION OF EXPONENTIAL FUNCTIONS. If b is a positive real number, the function

$$f(x) = b^x$$

is called the *exponential function with base b.*

Note: The thing that distinguishes an exponential function is that the variable occurs as an exponent. The base b is a constant but the exponent x is a variable. Don't confuse this function with functions in which the variable is the base; that is, $b^x \neq x^b$.

Note: We have defined an exponential function only for rational number values of the variable x because we have only defined rational number exponents. It is possible to define an exponential function for irrational values of the variable; this definition is usually covered in calculus.

The thing that distinguishes one exponential function from another exponential function is the base. We use different bases for different applications. For example, when we consider common logarithms, we will use an exponential function with base 10; that is, $f(x) = 10^x$. Applications to half-life use an exponential function with base 1/2; that is, $f(x) = (1/2)^x$. Interest rates use an exponential function with base $1 + r$; that is, $f(x) = (1 + r)^x$.

Example 1. If $f(x) = 2^x$, find $f(0), f(1), f(-1), f(2), f(-2), f(3)$, and $f(-3)$.

Solution.

$f(0) = 2^0 = 1$

$f(1) = 2^1 = 2$ $\qquad$ $f(-1) = 2^{-1} = 1/2 = 0.5$

$f(2) = 2^2 = 4$ $\qquad$ $f(-2) = 2^{-2} = 1/4 = 0.25$

$f(3) = 2^3 = 8$ $\qquad$ $f(-3) = 2^{-3} = 1/8 = 0.125$

Example 2. If $g(x) = 2.5(1/2)^x$, find $g(0)$, $g(1)$, $g(-1)$, $g(2)$, and $g(-2)$.

Solution.
$$g(0) = 2.5(1/2)^0 = 2.5(1) = 2.5$$
$$g(1) = 2.5(1/2)^1 = 2.5(1/2) = 1.25$$
$$g(-1) = 2.5(1/2)^{-1} = 2.5(2) = 5$$
$$g(2) = 2.5(1/2)^2 = 2.5(1/4) = 0.625$$
$$g(-2) = 2.5(1/2)^{-2} = 2.5(4) = 10$$

Exercise Set 8.1

In exercises 1–6, find $f(0)$, $f(1)$, $f(-1)$, $f(2)$, $f(-2)$, $f(3)$, and $f(-3)$ for each given exponential function f.

1. $f(x) = 3^x$
2. $f(x) = 5^x$
3. $f(x) = 10^x$
4. $f(x) = 100^x$
5. $f(x) = (1/3)^x$
6. $f(x) = (0.1)^x$
7. If $f(x) = 10^x$, find $f(6)$, $f(-6)$, $f(9)$, and $f(-9)$.
8. If $f(x) = (0.1)^x$, find $f(6)$, $f(-6)$, $f(9)$, and $f(-9)$.
9. If $g(x) = 10(1/2)^x$, find $g(0)$, $g(1)$, $g(-1)$, $g(2)$, and $g(-2)$.
10. If $g(x) = 6.5(1/2)^x$, find $g(0)$, $g(1)$, $g(-1)$, $g(2)$, and $g(-2)$.
11. If $g(x) = 1000(1.05)^x$, find $g(0)$, $g(1)$, $g(2)$, and $g(3)$.
12. If $g(x) = 1000(1.08)^x$, find $g(0)$, $g(1)$, $g(2)$, and $g(3)$.

CALCULATOR EXERCISES: In exercises 13–16, use an electronic calculator to find $f(x)$.

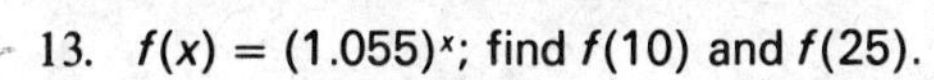

13. $f(x) = (1.055)^x$; find $f(10)$ and $f(25)$.
14. $f(x) = (1.014)^x$; find $f(10)$ and $f(40)$.
15. $f(x) = (1.01625)^x$; find $f(8)$ and $f(20)$.
16. $f(x) = 211.9(1.008)^x$; find $f(11)$ and $f(26)$.

PROPERTIES OF EXPONENTIAL FUNCTIONS

Exponential functions have three properties that we will use in this chapter. These properties are statements of laws for exponents in functional notation.

PROPERTIES OF EXPONENTIAL FUNCTIONS. If $f(x) = b^x$ is an exponential func- with base b and $b > 0$, then:

	Exponential Notation	Functional Notation
1.	$b^{u+v} = b^u b^v$	$f(u + v) = f(u)f(v)$
2.	$b^{u-v} = b^u/b^v$	$f(u - v) = f(u)/f(v)$

3. $b^{kx} = (b^x)^k$ $\qquad f(kx) = f(x)^k$

Example 3. If $f(x) = 2^x$ and $f(0.3) = 1.23$, find $f(2.3)$ and $f(-0.7)$.

Solution. We use the properties of exponential functions:

$$f(2.3) = f(2 + 0.3) = f(2)f(0.3) = 2^2(1.23) = 4.92$$
$$f(-0.7) = f(0.3 - 1) = f(0.3)/f(1) = 1.23/2 = 0.615$$

Example 4. Write the function $f(x) = (1/2)^x$ using a base of 2.

Solution. Using the properties of exponents, we have

$$1/2 = 2^{-1}$$
$$f(x) = (1/2)^x = (2^{-1})^x = 2^{-x}$$

Exercise Set 8.1 (continued)

17. If $f(x) = 2^x$ and $f(0.5) = 1.4$, find $f(1.5)$, $f(2.5)$, $f(-0.5)$, $f(-1.5)$.
18. If $f(x) = 2^x$ and $f(0.7) = 1.6$, find $f(1.7)$, $f(2.7)$, $f(-0.3)$, $f(-1.3)$.
19. If $f(x) = 5^x$ and $f(0.2) = 1.4$, find $f(0.4)$, $f(0.6)$, $f(1.2)$, $f(-0.2)$.
20. If $f(x) = 5^x$ and $f(0.3) = 1.6$, find $f(0.6)$, $f(0.9)$, $f(1.2)$, $f(-0.3)$.
21. If $f(x) = 10^x$ and $f(0.1) = 1.3$, find $f(0.3)$, $f(0.5)$, $f(-0.1)$, $f(-0.3)$.
22. If $f(x) = (1.05)^x$ and $f(0.5) = 1.02$, find $f(1.5)$, $f(2.5)$, $f(-0.5)$.
23. Write the function $f(x) = (1/5)^x$ using a base of 5.
24. Write the function $f(x) = (0.1)^x$ using a base of 10.
25. Write the function $f(x) = 4^x$ using a base of 2.
26. Write the function $f(x) = (1/4)^x$ using a base of 2.
27. Write the function $f(x) = 100^x$ using a base of 10.
28. Write the function $f(x) = 8^x$ using a base of 2.

GRAPH OF AN EXPONENTIAL FUNCTION

To graph an exponential function with base b, we plot ordered pairs of the form (x, b^x).

Example 5. Graph the function $f(x) = 2^x$.

Solution. We make a chart of ordered pairs $(x, 2^x)$:

x	−3.0	−2.5	−2.0	−1.5	−1.0	−0.5	0	0.5	1.0	1.5	2.0	2.5	3.0
2^x	0.13	0.18	0.25	0.36	0.5	0.7	1.0	1.4	2.0	2.8	4.0	5.6	8.0

When we plot these ordered pairs we obtain the graph shown in Figure 8.1.

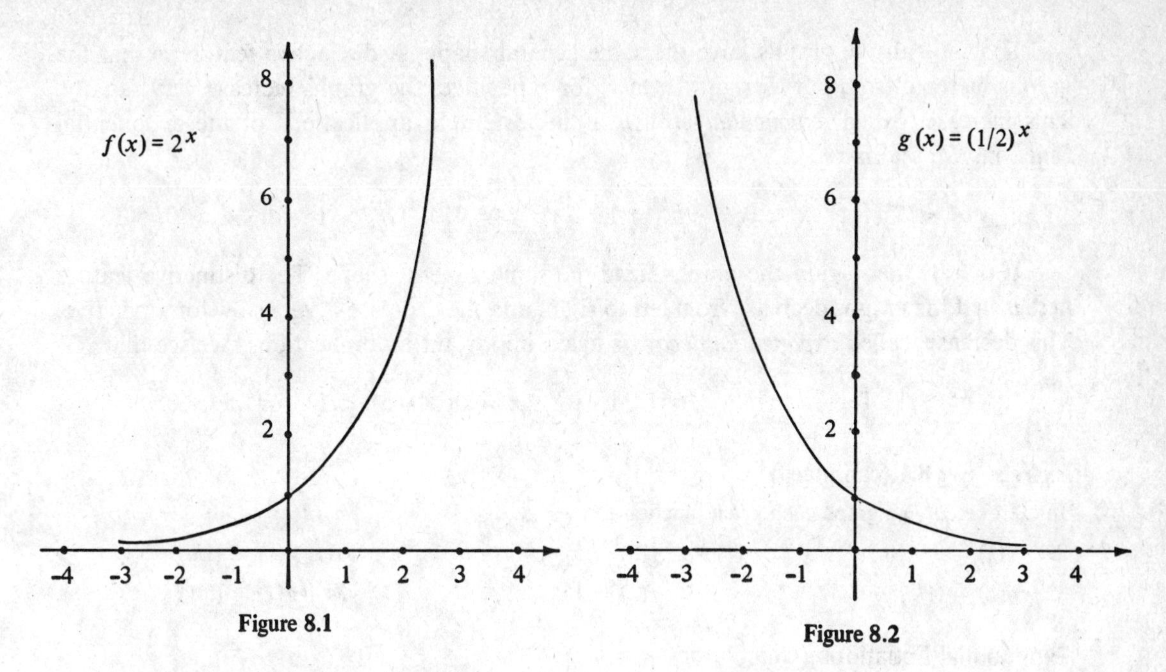

Figure 8.1

Figure 8.2

Example 6. Graph the function $g(x) = (1/2)^x$.

Solution. Using the fact that $g(x) = 2^{-x}$, we make a chart of ordered pairs $(x, 2^{-x})$.

x	−3.0	−2.5	−2.0	−1.5	−1.0	−0.5	0	0.5	1.0	1.5	2.0	2.5	3.0
2^{-x}	8.0	5.6	4.0	2.8	2.0	1.4	1.0	0.7	0.5	0.36	0.25	0.18	0.13

When we plot these ordered pairs we obtain the graphs shown in Figure 8.2.

The basic form of the graph of an exponential function with base b depends on whether $b > 1$ or $b < 1$. Figure 8.3 shows the graphs of various exponential functions for $b > 1$ and also for $b < 1$. All the graphs pass through the point (0, 1), since $f(0) = b^0 = 1$.

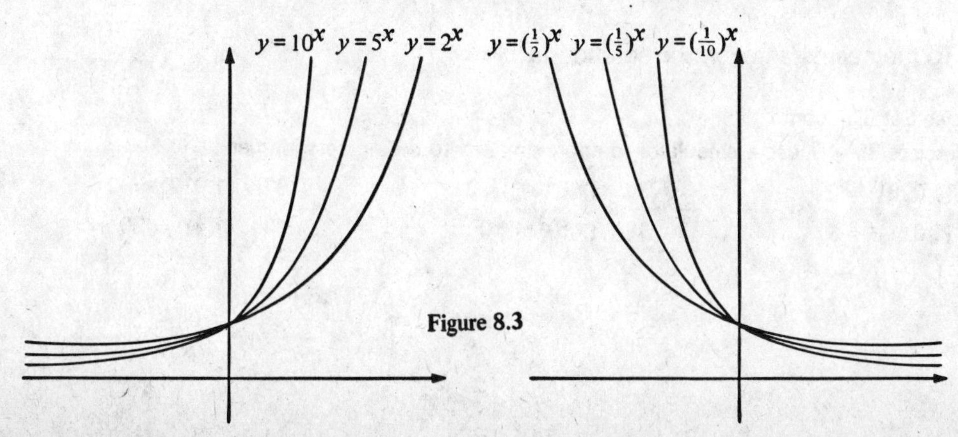

Figure 8.3

If $b > 1$, all the graphs have the same general shape. A distinctive feature is that the graphs increase from left to right. In fact, for x positive, the graphs increase very rapidly. This increase, called *exponential growth*, is important in applications of the exponential function. We see that:

$$b^x < 1 \quad \text{for} \quad x < 0; \qquad b^x = 1 \quad \text{for} \quad x = 0; \qquad b^x > 1 \quad \text{for} \quad x > 0.$$

If $b < 1$, then again the graphs have the same general shape. The distinctive feature here is that the graphs decrease from left to right, and the decrease is very slow for x positive. This decrease, called *exponential decay*, is again important in applications. We see that:

$$b^x > 1 \quad \text{for} \quad x < 0; \qquad b^x = 1 \quad \text{for} \quad x = 0; \qquad b^x < 1 \quad \text{for} \quad x > 0.$$

Exercise Set 8.1 (continued)

In exercises 29–34, graph each given function for $-3 \le x \le 3$.

29. $f(x) = 3^x$
30. $f(x) = 5^x$
31. $g(x) = (1/3)^x$
32. $g(x) = (1/5)^x$
33. $f(x) = 10^x$
34. $g(x) = (0.1)^x$

Exponential Equations (Calculator)

Frequently in applications of exponential functions we wish to solve an equation of the form $b^x = c$. Using an electronic calculator, we can approximate x to the nearest integer.

Example 7. Solve $(1.05)^x = 2$.

Solution. We find using a calculator that (approximately)

$$(1.05)^{12} = 1.79586$$
$$(1.05)^{13} = 1.88565$$
$$(1.05)^{14} = 1.97993$$
$$(1.05)^{15} = 2.07893$$

To the nearest integer, the solution is $x = 14$.

Exercise Set 8.1 (contd.)

In exercises 35–40, use a calculator to approximate x to the nearest integer.

35. $(1.028)^x = 2$
36. $(1.013)^x = 2$
37. $(1.017)^x = 2$
38. $(1.021)^x = 2$
39. $(1.5)^x = 10$
40. $(1.4)^x = 20$

section 2 • Logarithmic Functions

In this section we define logarithmic functions in terms of exponential functions, and we graph logarithmic functions.

DEFINITION OF LOGARITHMIC FUNCTIONS

The logarithmic function is defined in terms of the exponential function. Just as with the exponential function, we speak of logarithms with base b. We restrict b so that $b > 1$.

DEFINITION OF LOGARITHMIC FUNCTIONS. If b is a real number so that $b > 1$, for all $x > 0$,

$$y = \log_b x \quad \text{means} \quad x = b^y.$$

The function $f(x) = \log_b x$ is called the *logarithmic function with base b.*

Note: It is possible to define logarithmic functions with base b so that $0 < b < 1$, but we will not do so in this book. Two logarithmic functions are used frequently in applications: the common logarithm function and the natural logarithm function. The common logarithm function has base 10 and is used for computations. The natural logarithm function has base e and occurs in calculus. The number e is an irrational number; 2.718 is an approximation to e correct to three decimal places. You will see the following notation used:

$$\log x = \log_{10} x \quad \text{common logarithm function}$$

$$\ln x = \log_e x \quad \text{natural logarithm function}$$

Example 1. Logarithmic function with base 2.

a. $4 = \log_2 16$ since $16 = 2^4$.

b. $3 = \log_2 8$ since $8 = 2^3$.

c. $2 = \log_2 4$ since $4 = 2^2$.

d. $1 = \log_2 2$ since $2 = 2^1$.

e. $0 = \log_2 1$ since $1 = 2^0$.

f. $-1 = \log_2 \dfrac{1}{2}$ since $\dfrac{1}{2} = 2^{-1}$.

g. $-2 = \log_2 \dfrac{1}{4}$ since $\dfrac{1}{4} = 2^{-2}$.

h. $-3 = \log_2 \dfrac{1}{8}$ since $\dfrac{1}{8} = 2^{-3}$.

i. $-4 = \log_2 \dfrac{1}{16}$ since $\dfrac{1}{16} = 2^{-4}$.

Notice that a logarithmic function just reverses what an exponential function does. To make this statement more precise, consider the exponential function with base 2; that is, $f(x) = 2^x$. With this exponential function, given a number x we find the number 2^x. In other words, in an exponential function, given the exponent, we find the power. Now let $g(x) = \log_2 x$ be the logarithmic function with base 2. For the logarithmic function we are given the power and we must find the exponent. That is, given x, we must find y so that $2^y = x$. The number y that we find is the logarithm of x. To summarize:

exponential function	given exponent x find power $y = 2^x$
logarithmic function	given power x find exponent y so that $2^y = x$

Using the functional notation $g(x) = \log_2 x$, we can write the results of Example 1 as follows:

$g(16) = 4$	$g(8) = 3$	$g(4) = 2$
$g(2) = 1$	$g(1) = 0$	$g(1/2) = -1$
$g(1/4) = -2$	$g(1/8) = -3$	$g(1/16) = -4$

It is easier to find the values of exponential functions than to find the values of logarithmic functions. For example, to find $g(3) = \log_2 3$, we must find y so that $2^y = 3$. It is not clear how we would find such a y (although y must be between 1 and 2, because $2^1 = 2$ and $2^2 = 4$).

To find the values of logarithmic functions we rely on tables of logarithms. Two kinds of logarithm tables are frequently used—tables of common logarithms (base 10) and tables of natural logarithms (base e). We discuss the use of tables of common logarithms in Section 8.4.

Example 1 (continued). If $g(x) = \log_2 x$, find $g(\sqrt{2})$, $g(\sqrt{8})$, and $g(1/\sqrt{2})$.

Solution. We wish to find y so that $2^y = \sqrt{2}$. We recognize that $2^{1/2} = \sqrt{2}$, and so $g(\sqrt{2}) = 1/2$. Similarly,

$$g(\sqrt{8}) = 3/2 \quad \text{since} \quad \sqrt{8} = (2^3)^{1/2} = 2^{3/2}$$

$$g(1/\sqrt{2}) = -1/2 \quad \text{since} \quad 1/\sqrt{2} = (2^{1/2})^{-1} = 2^{-1/2}$$

Note: Suppose $b > 1$ and $g(x) = \log_b x$. Then we have $x = b^y$ for some number y. But b^y cannot be a negative number or 0; therefore, x must be positive. That is, a logarithmic function $\log_b x$ is only defined for positive numbers x.

Exercise Set 8.2

In exercises 1–6, find the given logarithms.

1. $\log_5 25$, $\log_5 125$, $\log_5 5$, $\log_5 1$, $\log_5 (1/5)$.
2. $\log_3 9$, $\log_3 27$, $\log_3 81$, $\log_3 (1/9)$, $\log_3 (1/27)$.
3. $\log 10$, $\log 100$, $\log 1000$, $\log 0.1$, $\log 0.01$, $\log 0.001$.
4. $\log_7 7$, $\log_7 49$, $\log_7 1$, $\log_7 (1/7)$, $\log_7 (1/49)$.
5. $\log_9 9$, $\log_9 81$, $\log_9 3$, $\log_9 1$, $\log_9 (1/3)$.
6. $\log_{25} 25$, $\log_{25} 5$, $\log_{25} 1$, $\log_{25} (1/5)$.

In exercises 7–12, find the given values of the function $g(x)$.

7. $g(x) = \log_5 x$; find $g(1)$, $g(5)$, $g(\sqrt{5})$, $g(0.2)$.
8. $g(x) = \log_3 x$; find $g(1)$, $g(3)$, $g(\sqrt{3})$, $g(\sqrt{27})$.
9. $g(x) = \log x$; find $g(1)$, $g(10)$, $g(\sqrt{10})$, $g(\sqrt{1000})$, $g(1/\sqrt{10})$.
10. $g(x) = \log_7 x$; find $g(1)$, $g(7)$, $g(\sqrt{7})$, $g(1/\sqrt{7})$.
11. $g(x) = \log_9 x$; find $g(9)$, $g(81)$, $g(3)$, $g(1)$, $g(1/3)$.
12. $g(x) = \log_{25} x$; find $g(25)$, $g(5)$, $g(1)$, $g(0.2)$.

In exercises 13 and 14, refer to the following chart. Values of 10^x were found on a mathematical calculator and rounded off.

x	0.10	0.15	0.20	0.25	0.30	0.35	0.40	0.45	0.50	0.55	0.60	0.65	0.70	0.75	0.80	0.85	0.90	0.95	1.00
10^x	1.26	1.41	1.58	1.78	2.00	2.24	2.51	2.82	3.16	3.55	3.98	4.47	5.01	5.62	6.31	7.08	7.94	8.91	10.00

13. Find $\log 1.58$, $\log 2.24$, $\log 5.01$, $\log 2.82$, $\log 7.08$.
14. Find $\log 1.41$, $\log 8.91$, $\log 3.98$, $\log 5.62$, $\log 2.51$.

GRAPHS OF LOGARITHMIC FUNCTIONS

To graph a logarithmic function with base b, we graph ordered pairs of the form (x, y) where $y = \log_b x$. By the definition of the logarithmic function, $x = b^y$. Therefore, we graph ordered pairs of the form (b^y, y). We see that we can compare the graph of the logarithmic function with the graph of the exponential function. That is, to graph the exponential function with base b, we graph ordered pairs of the form (y, b^y). So the graph of the logarithmic function is the graph of the exponential function with the coordinates interchanged.

Figure 8.4 depicts the graph of a typical exponential function, where we have assumed the base $b > 1$. But we have switched the x and y. That is, the vertical axis is the x-axis and the horizontal axis is the y-axis. So we have graphed $x = b^y$, which is the same function as $y = \log_b x$. Of course, we usually graph the x-axis horizontally and the y-axis vertically.

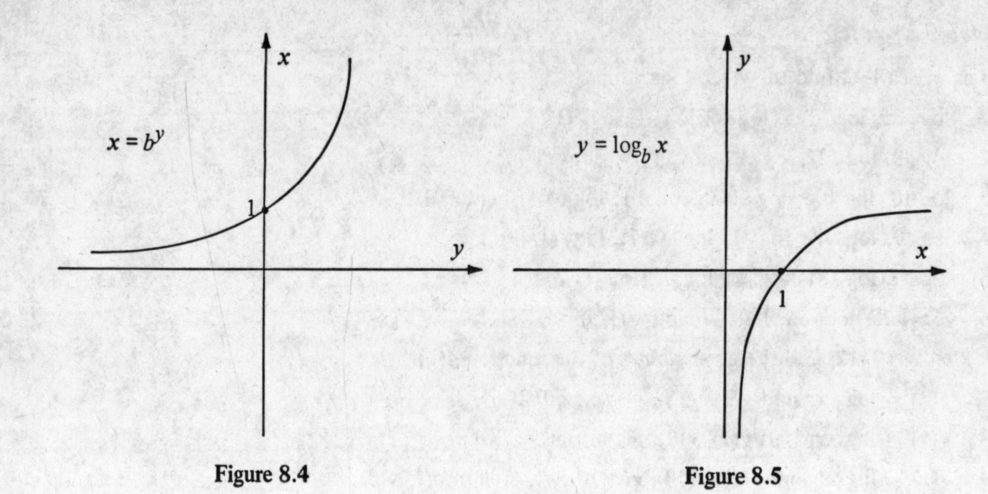

Figure 8.4

Figure 8.5

So we redraw the graph to do this. If we turn the graph in Figure 8.4 by 90 degrees and view it in a mirror, the x-axis is horizontal (and positive to the right) and the y-axis is vertical (and positive upward). We've sketched this graph in Figure 8.5. This is the graph of the logarithmic function for $b > 1$.

Exercise Set 8.2 (continued)

15. Sketch the graph of $g(x) = \log_2 x$ for $x > 0$.
16. Sketch the graph of $g(x) = \log_3 x$ for $x > 0$.
17. Use the following chart to graph $g(x) = \log x$ for $0.1 \leq x \leq 10$. The chart was made by rounding off entries from a table of common logarithms.

x	0.1	0.5	0.2	1.0	2.0	3.0	4.0	5.0	6.0	7.0	8.0	9.0	10.0
$\log x$	−1.0	−0.3	−0.7	0	0.3	0.5	0.6	0.7	0.8	0.8	0.9	0.9	1.0

18. Find $\log_b b$; $\log_b b^2$; $\log_b b^3$; $\log_b b^4$; $\log_b b^0$; $\log_b b^{-1}$; $\log_b b^{-2}$; $\log_b b^{-3}$. What is b^n? What is $\log_b b^x$?

LOGARITHMS COMPOSED WITH EXPONENTS

We now consider two properties of logarithmic functions when they are combined with exponential functions.

LOGARITHMS COMPOSED WITH EXPONENTS. If $b > 1$, then

$$\log_b b^x = x \qquad \text{and} \qquad b^{\log_b x} = x.$$

To prove the first property, let $y = b^x$. By the definition of logarithms, $\log_b y = x$; this statement is the first property.

To prove the second property, let $y = \log_b x$. By the definition of logarithms, $b^y = x$; this statement is the second property.

Example 2. Solve the equation $2^x = 10$ for x.

Solution. We take the logarithm with base 2 of each side of this equation:

$$2^x = 10$$
$$\log_2 2^x = \log_2 10$$
$$x = \log_2 10$$

We have obtained a solution, although the solution is expressed as a logarithm with base 2.

Example 3. Solve the equation $10^{x+1} = 3.14$ for x.

Solution. We take the common logarithm of each side of this equation:

$$10^{x+1} = 3.14$$
$$\log 10^{x+1} = \log 3.14$$
$$x + 1 = \log 3.14$$
$$x = -1 + \log 3.14$$

Exercise Set 8.2 (continued)

In exercises 19–30, solve each given equation; express the answer as a logarithm.

19. $2^x = 100$
20. $2^x = 0.1$
21. $3^x = 10$
22. $3^x = 0.1$
23. $5^x = 25$
24. $5^x = 0.2$
25. $10^{x-1} = 4$
26. $10^{x-2} = 3$
27. $10^{x+1} = 6$
28. $10^{x+2} = 5$
29. $10^{2x} = 6.28$
30. $10^{3x} = 3.33$

ND

section 3 • Properties of Logarithms

Logarithms were invented in the seventeenth century by John Napier (1550–1617) for the purpose of simplifying complicated computations.

Logarithms are useful in computations because they transform the operation of multiplication into addition, transform the operation of division into subtraction, and transform raising to powers into a product. The numerical operations of multiplication and division are frequently long and tedious, whereas addition and subtraction are relatively simple to perform. So the use of logarithms shortens many computations.

The use of logarithms in performing computations depends on three properties of the logarithmic function, which we will look at in this section. The first property pertains to transforming multiplication into addition, the second to transforming division into subtraction, and the third to transforming powers into a product.

THE LOGARITHM OF A PRODUCT IS THE SUM OF THE LOGARITHMS. If u and v are positive real numbers, then

$$\log_b uv = \log_b u + \log_b v.$$

To see why this property is true, we look at the definition of the logarithmic functions (Section 8.2). If we let $y = \log_b u$ and $z = \log_b v$, then $b^y = u$ and $b^z = v$. By the properties of the exponential functions, $uv = b^y b^z = b^{y+z}$. Therefore, $\log_b uv = y + z = \log_b u + \log_b v$.

Example 1. The logarithm of a product is the sum of the logarithms. We know that $\log_2 4 = 2$ and $\log_2 8 = 3$. Now $4 \cdot 8 = 32$ and $\log_2 32 = 5$. So in this case, we have checked that $\log_2 (4 \cdot 8) = \log_2 4 + \log_2 8$.

THE LOGARITHM OF A QUOTIENT IS THE DIFFERENCE OF THE LOGARITHMS. If u and v are positive real numbers, then

$$\log_b (u/v) = \log_b u - \log_b v.$$

The proof of this property again depends on the definition of logarithmic functions. If we let $y = \log_b u$ and $z = \log_b v$, then $b^y = u$ and $b^z = v$. By the properties of the exponential functions, $u/v = b^u/b^v = b^{u-v}$. Therefore, $\log (u/v) = u - v = \log_b u - \log_b v$.

Example 2. The logarithm of a quotient is the difference of the logarithms. We know that $\log_2 4 = 2$, $\log_2 8 = 3$, and $\log_2 2 = 1$. In this case, we have checked that $\log_2 (8/4) = \log_2 8 - \log_2 4$.

THE LOGARITHM OF A POWER IS THE POWER TIMES THE LOGARITHM. If n is a real number and x is a positive real number, then

$$\log_b x^n = n \log_b x.$$

To prove this property, let $y = \log_b x$. Then $b^y = x$ and $x^n = (b^y)^n = b^{ny}$. Therefore, $\log_b x^n = n \log_b x$.

Example 3. The logarithm of a power is the power times the logarithm. We know that $\log_2 4 = 2$ and $\log_2 16 = 4$ and $16 = 4^2$. In this special case, we have checked that $\log_2 4^2 = 2 \log_2 4$.

These three properties make logarithms useful for numerical computations. They tell how to deal with the logarithm of a product, quotient, or power. We point out, however, that many logarithmic expressions may arise for which we do not have equations. For example, there are no equations expressing

$$\log_b (u + v) \qquad \log_b (u - v)$$
$$(\log_b u)(\log_b v) \qquad \log_b u/\log_b v$$

It is important to realize that logarithms are appropriate only in situations where we take the logarithm of a product, quotient, or power.

Exercise Set 8.3

1. By finding log 10, log 100, and log 1000, check that log (10)(100) = log 10 + log 100.
2. By finding log 10, log 0.01, and log 0.1, check that log (10)(0.01) = log 10 + log 0.01.
3. By finding log 100, log 1000, and log 0.1, check that log 100/1000 = log 100 − log 1000.
4. By finding log 1000, log 10, and log 100, check that log 1000/10 = log 1000 − log 10.
5. By finding log 10 and log 1000, check that $\log 10^3 = 3 \log 10$.
6. By finding log 0.1 and log 100, check that $\log 0.1^{-2} = -2 \log 0.1$.

In exercises 7–10, use the fact that log 2 = 0.3010 and log 3 = 0.4771 (rounded off to four decimal places).

7. Is log (2 + 1) = log 2 + log 1?
8. Is log (2 − 1) = log 2 − log 1?
9. Is (log 3)(log 1) = log (3 · 1)?
10. Is log 3/log 1 = log (3/1)?

PRACTICE WITH LOGARITHMS

We gain practice using the properties of logarithms.

Example 4. Express the given logarithm in terms of logarithms of a single letter. We assume the letters represent positive real numbers.

a. $\log_b (x^2y^5) = \log_b x^2 + \log_b y^5$ (logarithm of a product)
$= 2 \log_b x + 5 \log_b y$ (logarithm of a power)

b. $\log_b \left(\frac{xy}{z^3}\right) = \log_b (xy) - \log_b z^3$ (logarithm of a quotient)
$= \log_b x + \log_b y - \log_b z^3$ (logarithm of a product)
$= \log_b x + \log_b y - 3 \log_b z$ (logarithm of a power)

c. $\log_b \left(\frac{\sqrt{x}}{y^3}\right) = \log_b \sqrt{x} - \log_b y^3$ (logarithm of a quotient)
$= \log_b x^{1/2} - \log_b y^3$ (rational exponents)
$= \frac{1}{2} \log_b x - 3 \log_b y$ (logarithm of a power)

d. $\log_b \sqrt{\dfrac{xy}{z}} = \log_b \left(\dfrac{xy}{z}\right)^{1/2}$ (rational exponents)

$= \dfrac{1}{2} \log_b \left(\dfrac{xy}{z}\right)$ (logarithm of a power)

$= \dfrac{1}{2} [\log_b (xy) - \log_b z]$ (logarithm of a quotient)

$= \dfrac{1}{2} [\log_b x + \log_b y - \log_b z]$ (logarithm of a product)

Exercise Set 8.3 (continued)

In exercises 11–30, express each given logarithm in terms of logarithms of a single letter. Assume that the letters represent positive real numbers.

11. $\log_b (xyz)$
12. $\log_b \left(\dfrac{xy}{z}\right)$
13. $\log_b \left(\dfrac{1}{x}\right)$
14. $\log_b \left(\dfrac{1}{xy}\right)$
15. $\log_b \left(\dfrac{1}{x^3}\right)$
16. $\log_b \left(\dfrac{x^2}{y^3}\right)$
17. $\log_b \left(\dfrac{x}{y}\right)^3$
18. $\log_b (\sqrt{x}\, y)$
19. $\log_b \sqrt{xy}$
20. $\log_b \sqrt{xyz}$
21. $\log_b \sqrt{\dfrac{xy}{z^3}}$
22. $\log_b \sqrt[3]{xy}$
23. $\log_b \left(\dfrac{xy}{wz}\right)$
24. $\log_b \left(\dfrac{x^2y^5}{wz}\right)$
25. $\log_b \dfrac{\sqrt{xy}}{w^2z^3}$
26. $\log_b \sqrt{\dfrac{xy}{wz}}$
27. $\log_b w \cdot \sqrt{\dfrac{x}{y}}$
28. $\log_b (bw\sqrt{x})$
29. $\log_b \sqrt{b\dfrac{x}{y}}$
30. $\log_b \dfrac{\sqrt{bx}}{y^2}$
31. Show that, if $x_1/x_2 = t_1/t_2$, then $\log_b x_1 - \log_b x_2 = \log_b t_1 - \log_b t_2$. This says that equal *ratios* give rise to equal *differences* of logarithms.
32. Show that, in a right triangle with sides x and y and hypotenuse z,

$$\log_b x = \frac{1}{2} [\log_b (z + y) + \log_b (z - y)],$$

$$\log_b y = \frac{1}{2} [\log_b (z + x) + \log_b (z - x)].$$

PRACTICE WITH COMMON LOGARITHMS

Example 5. Given that log 2.365 = 0.3738, find log 236.5 and log 0.2365.

Solution. $236.5 = 2.365 \times 100$ and $0.2365 = 2.365/10$

$$\log 236.5 = \log (2.365 \times 100) = \log 2.365 + \log 100$$
$$= 0.3738 + 2 = 2.3738$$

$$\log 0.2365 = \log 2.365/10 = \log 2.365 - \log 10$$
$$= 0.3738 - 1 = -0.6262$$

Example 6. Given that $\log 2 = 0.3010$ and $\log 3 = 0.4771$, find $\log 60$.

Solution.

$$\log 60 = \log 2 \cdot 3 \cdot 10 = \log 2 + \log 3 + \log 10 = 0.3010 + 0.4771 + 1 = 1.7781.$$

Exercise Set 8.3 (continued)

33. Given that $\log 3.652 = 0.5625$, find $\log 36.52$ and $\log 3652$.
34. Given that $\log 8.216 = 0.9147$, find $\log 821.6$ and $\log 82160$.
35. Given that $\log 3.14 = 0.4969$, find $\log 314$ and $\log (3.14)^2$.
36. Given that $\log 2.22 = 0.3464$, find $\log 222$ and $\log (2.22)^3$.
37. Given that $\log 5 = 0.6990$, find $\log 25$, $\log 125$ and $\log 0.2$.
38. Given that $\log 4 = 0.6021$, find $\log 16$, $\log 2$ and $\log 0.25$.

In exercises 39–60, given that $\log 2 = 0.3010$ and $\log 3 = 0.4771$, find each given common logarithm.

39. $\log 4$
40. $\log 8$
41. $\log 32$
42. $\log 24$
43. $\log 18$
44. $\log 200$
45. $\log 1200$
46. $\log 54{,}000$
47. $\log \frac{1}{2}$
48. $\log \frac{1}{3}$
49. $\log 0.05$
50. $\log (0.6666\ldots)$
51. $\log (2.25)$
52. $\log (1.125)$
53. $\log (6.75)$
54. $\log (0.075)$
55. $\log \sqrt{2}$
56. $\log \sqrt{3}$
57. $\log \sqrt{6}$
58. $\log \sqrt{1.5}$
59. $\log \sqrt{8}$
60. $\log \sqrt{12}$

section 4 • Use of Tables

To use logarithmic functions in computations we must use either an electronic calculator having a logarithm key or a table of logarithms. In this section we show how to use a table of common logarithms. Appendix D of this book is a table of common logarithms.

There are two aspects to using a table of logarithms:

1. Given a number, we find its logarithm.
2. Given the logarithm of a number, we find the number.

With both of these aspects we discuss linear interpolation.

FINDING THE LOGARITHM OF A NUMBER

Recall that any positive real number can be written in scientific notation (Section 3.2);

that is, we can write the number in the form

$$a \times 10^k,$$

where a is a number between 1 and 10 and k is an *integer* exponent (positive, negative, or 0). But if a number is written in scientific notation, then

$$\begin{aligned}\log(a \times 10^k) &= \log a + \log 10^k \\ &= \log a + k = k + \log a\end{aligned}$$

The significance of this is that we can find the logarithm of *any* number if we know the logarithms of numbers between 1 and 10, by writing the number in scientific notation. Common logarithm tables take advantage of this fact and list only the logarithms of numbers between 1 and 10. Since the numbers whose logarithms are listed in a common logarithm table are all between 1 and 10, the logarithms themselves turn out to be numbers between 0 and 1, as can be seen from the graph of the common logarithm function (Figure 8.6).

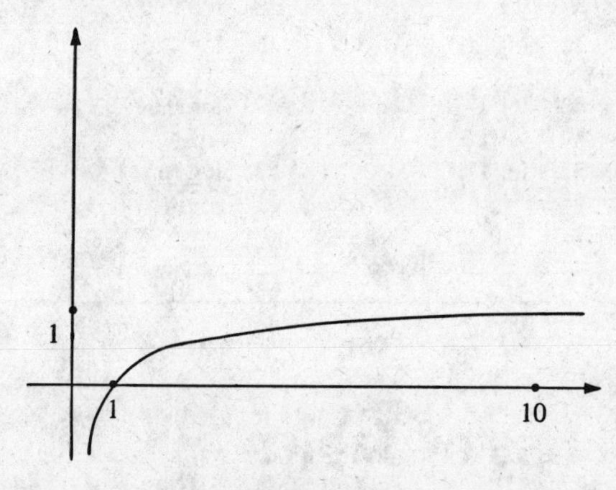

Figure 8.6

DEFINITION OF MANTISSA AND CHARACTERISTIC OF A COMMON LOGARITHM. If a number is written in scientific notation—that is, in the form $a \times 10^k$ with $1 \le a < 10$ and k and integer—then $\log(a \times 10^k) = k + \log a$. We call $\log a$ the *mantissa* of the logarithm, and we call the integer k the *characteristic* of the logarithm.

We use the table of common logarithms to find the mantissa, but we determine the characteristic ourselves.

Example 1. Find log 152.

Solution. scientific notation: $152 = 1.52 \times 10^2$
characteristic is 2
mantissa is $\log 1.52 = ?$

Turn to the table, Appendix D. Going down the left-hand column to 1.5 and then across to the column labelled 2, we find the entry 0.1818. Therefore, log 1.52 = 0.1818.

$$\log 152 = 2 + 0.1818 = 2.1818$$

Example 2. Find log 663,000.

Solution. scientific notation: $663{,}000 = 6.63 \times 10^5$
characteristic is 5
mantissa is $\log 6.63 = ?$
Turn to Appendix D. Going down the left-hand column to 6.6 and then across to the column labelled 3, we find the entry 0.8215. Therefore, log 6.63 = 0.8215.

$$\log 663{,}000 = 5 + 0.8215 = 5.8215$$

Example 3. Find log 0.000242

Solution. scientific notation: $0.000242 = 2.42 \times 10^{-4}$
characteristic is -4
mantissa is $\log 2.42 = 0.3838$
$\log 0.000242 = -4 + 0.3838$
We will find it convenient to write

$$\log 0.000242 = 6.3838 - 10.$$

The point is that the logarithm of this number is negative, but we write the logarithm in such a way as to keep the *mantissa* positive. This makes subsequent computations easier.

Example 4. Find log 0.0717

Solution. scientific notation: $0.0717 = 7.17 \times 10^{-2}$
characteristic is -2
mantissa is $\log 7.17 = 0.8555$

$$\log 0.0717 = -2 + 0.8555 = 8.8555 - 10$$

Exercise Set 8.4

In exercises 1–20, find the common logarithm of each given number.

1. 97,600
2. 43.9
3. 839
4. 8320
5. 0.0424
6. 0.522
7. 6660
8. 707
9. 0.0244
10. 0.00286

11. 825	12. 22.5	13. 2	14. 0.5	15. 0.0772
16. 0.624	17. 9890	18. 2780	19. 0.272	20. 0.247

FINDING THE NUMBER

We use the table of common logarithms to find N if we are given $\log N$. To do so, we break $\log N$ into its characteristic and mantissa.

Example 5. Find N if $\log N = 3.6395$.

Solution. characteristic is 3
mantissa is 0.6395

Turn to Appendix D and look through to find the entry 0.6395. This entry occurs in the row marked 4.3 and the column marked 6.

$$\log 4.36 = 0.6395$$

$$N = 4.36 \times 10^3 = 4{,}360$$

Example 6. Find N if $\log N = 9.8932 - 10$

Solution. characteristic is -1
mantissa is 0.8932

In the table of common logarithms, the entry 0.8932 occurs in the row marked 7.8 and the column marked 2. So

$$\log 7.82 = 0.8932$$

$$N = 7.82 \times 10^{-1} = 0.782$$

Exercise Set 8.4 (continued)

In exercises 21–32, find the number whose common logarithm is given.

21. 1.6405	22. 1.7686	23. 2.8971	24. 2.5453
25. 4.7482	26. 3.4955	27. 9.6693 − 10	28. 9.8376 − 10
29. 8.8202 − 10	30. 8.6848 − 10	31. 6.4362 − 10	32. 7.8681 − 10

INTERPOLATION—FINDING THE LOGARITHM

The table of Appendix D gives approximations to logarithms of certain numbers. Those numbers are all between 1 and 10 at intervals of 0.01, and the mantissas are accurate to four decimal places.

We can improve on the accuracy of these tables by the process of *linear interpolation.* The basic idea of linear interpolation is to obtain an approximation to a small portion of the logarithm function by a straight line.

Example 7. Find log 3.216

Solution. The number 3.216 falls between 3.210 and 3.220. From the table of common logarithms, we find

$$\log 3.210 = 0.5065 \quad \text{and} \quad \log 3.220 = 0.5079.$$

Now log 3.216 is the y-coordinate of the logarithm function above the x-axis. But instead of going all the way up to the curve, we approximate this by going up to the straight line instead (Figure 8.7). Actually, the straight line is a fairly good approximation to the logarithm function because the intervals involved are small (Figure 8.7 is not drawn to scale). Because we use a straight line for the approximation, we call this a *linear* interpolation.

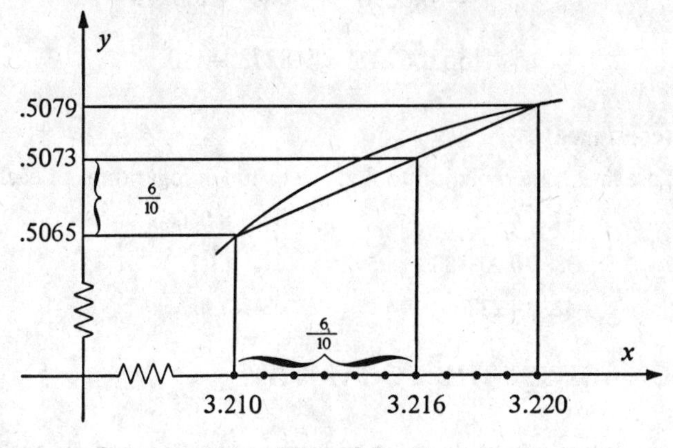

Figure 8.7

The advantage of linear interpolation is that, with a little practice, the interpolation can be done mentally. We reason as follows: The number 3.216 is 6/10 of the way between the values of the tables; that is, 3.216 is 6/10 of the way from 3.210 to 3.220. Since the interpolation is linear, the y-value is 6/10 of the way between the two logarithms (See Figure 8.7). So the interpolated value is 6/10 of the way from 0.5065 to 0.5079. Since the difference between these y-values is 0.0014, the interpolated value is

$$0.5065 + \frac{6}{10}(0.0014) = 0.5065 + 0.0008$$

$$= 0.5073.$$

So by interpolation, log 3.216 = 0.5073.

We round off an interpolation so that the mantissa has four decimal places—this is all that is permitted by the accuracy of the tables. So interpolation permits us to look up four-digit numbers but does not improve the number of digits in the mantissa.

Notice that linear interpolation depends on comparing certain *differences* on the x-axis and y-axis. We can think of interpolation in terms of using the *slope* of the line we have drawn in Figure 8.7. Of course, this slope changes, depending on what portion of the graph we are looking at.

Example 8. Find log 0.6727

Solution. scientific notation: $0.6727 = 6.727 \times 10^{-1}$
characteristic is -1
mantissa is log 6.727

$$\log 6.727 = 0.8274 + \frac{7}{10}(0.8280 - 0.8274)$$

$$= 0.8274 + \frac{7}{10}(.0006)$$

$$= 0.8274 + .0004 = 0.8278$$

$$\log 0.6727 = 9.8278 - 10$$

Exercise Set 8.4 (continued)

In exercises 33–44, use linear interpolation to find the common logarithms of each given number.

33. 21.37
34. 4762
35. 0.06666
36. 0.9215
37. 371.4
38. 0.008888
39. 61,210
40. 446,300
41. 0.5279
42. 4.222
43. 1.055
44. 1.014

INTERPOLATION—GIVEN THE LOGARITHM

We can also use linear interpolation to find the number when we are given its logarithm. Now we wish to find an extra decimal place in the number.

Example 9. Find N if $\log N = 0.6925$

Solution. We look through the table and find that this mantissa comes between the mantissas 0.6920 and 0.6928, and

$$\log 4.92 = 0.6920 \quad \text{and} \quad \log 4.93 = 0.6928.$$

Again, the interpolation is *linear* (Figure 8.8). The difference between the two mantissas is 0.0008, and so the mantissa 0.6925 is 5/8 of the way from 0.6920 to 0.6928. Since the interpolation is linear, we want the interpolation to be 5/8 of the way from 4.920 to 4.930. The difference between these numbers (on the x-axis) is 0.010, and so

$$N = 4.920 + \frac{5}{8}(0.010)$$

$$= 4.920 + 0.006$$

$$= 4.926.$$

We round off to *one* extra decimal place because the tables do not permit further accuracy.

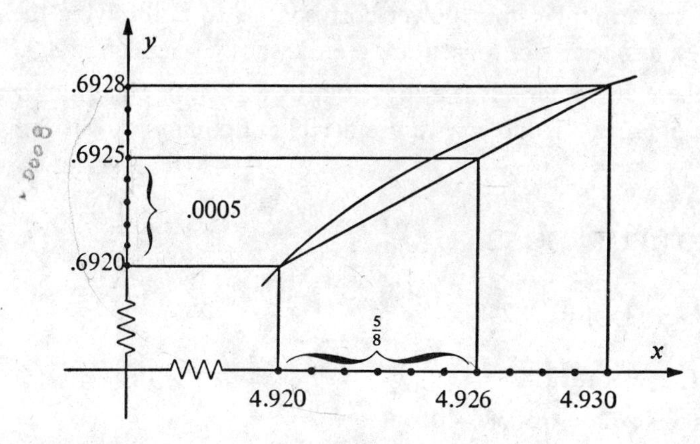

Figure 8.8

Example 10. Find N if $\log N = 10.7055$

Solution. The characteristic is 10 and the mantissa is 0.7055. This mantissa is the logarithm of

$$5.07 + \frac{5}{9}(0.010) = 5.076$$

$$N = 5.076 \times 10^{10} = 50{,}760{,}000{,}000$$

Exercise Set 8.4 (continued)

In exercises 45–56, use linear interpolation to find each number whose common logarithm is given.

45. 1.7368	46. 2.6110	47. 0.8764	48. 0.5180
49. 9.7915 − 10	50. 7.5458 − 10	51. 2.8428	52. 1.9728
53. 0.2000	54. 0.3000	55. 8.5656 − 10	56. 9.3205 − 10

section 5 • Computations with Logarithms

In this section we use common logarithms to perform arithmetic computations. Computation with logarithms involves three steps:

1. We must find the logarithm of each number in the computation. To do so, we use a table of common logarithms as discussed in Section 8.4.

2. We perform the computations using logarithms. Multiplication of numbers corresponds to addition of logarithms, division of numbers corresponds to subtraction of logarithms, and raising a number to a power corresponds to multiplying the logarithm by the power. These properties of logarithms were discussed in Section 8.3.
3. The result of these computations is a logarithm; we must find the number whose logarithm we have obtained. To do so, we use a table of common logarithms as discussed in Section 8.4.

ARITHMETIC COMPUTATIONS

Example 1. Find $N = (152)(0.000242)$.

Solution. $\log 152 = 2.1818$ (step 1)
$\log 0.000242 = 6.3838 - 10$ (step 1)

$$\begin{aligned} \log N &= \log 152 + \log 0.000242 && \text{(step 2)} \\ &= 2.1818 + 6.3838 - 10 \\ &= 8.5656 - 10 \\ N &= 0.03678 && \text{(step 3)} \end{aligned}$$

Note: The value of N we have found is an approximation. Using an electronic calculator, we find $N = 0.036784$.

Example 2. Find $N = 0.6727/3.216$.

Solution. $\log 0.6727 = 9.8278 - 10$ (step 1)
$\log 3.216 = 0.5073$ (step 1)

$$\begin{aligned} \log N &= \log 0.6727 - \log 3.216 && \text{(step 2)} \\ &= 9.8278 - 10 - 0.5073 \\ &= 9.3205 - 10 \\ N &= 0.2092 && \text{(step 3)} \end{aligned}$$

Note: This value of N is an approximation; using an electronic calculator, we obtain the approximation $N = 0.2091728$.

Exercise Set 8.5

In exercises 1–20, use logarithms (with interpolation) to approximate each given number.

1. (4830)(665,000)
2. (22.6)(5020)
3. (0.000625)(0.00737)
4. (17,300)(0.0279)

5. $\dfrac{97{,}600}{839}$

6. $\dfrac{43.9}{8320}$

7. $\dfrac{0.0000248}{0.00666}$

8. $\dfrac{0.00158}{7.42}$

9. $(58700)(9180)(0.0424)$

10. $(0.522)(7.20)(35.1)$

11. $\dfrac{(77.7)(0.0244)}{6660}$

12. $\dfrac{(0.555)(707)}{0.00286}$

13. $(4073)(226{,}600)$

14. $(0.9244)(0.0003668)$

15. $\dfrac{662.2}{0.3695}$

16. $\dfrac{0.04113}{7\,204}$

17. $(35.27)\dfrac{0.2021}{9.115}$

18. $\dfrac{(6727)(1.019)}{(4406)}$

19. $\dfrac{3{,}369{,}000}{(0.8295)(504.4)}$

20. $\dfrac{0.02222}{(9.103)(0.1781)}$

POWERS AND ROOTS

An important use of common logarithms is to approximate powers and roots of numbers.

Example 3. Find $N = 3890\,(1.019)^{26}$.

Solution.
$$\begin{aligned}\log N &= \log 3890 + \log (1.019)^{26} \\ &= \log 3890 + 26 \log 1.019 \\ &= 3.5899 + 26\,[0.0082] \\ &= 3.8031\end{aligned}$$

$$N = 6354$$

Note: Using an electronic calculator, we obtain $N = 6346$. Thus, because logarithms are approximations and because the calculator rounds off, we have agreement only in the first two digits.

Example 4. Find $N = \sqrt{0.6126/3.14}$.

Solution. First we write the square root using fractional exponents:

$$N = (0.6126/3.14)^{1/2}$$

$$\log N = \frac{1}{2} \log (0.6126/3.14)$$

$$= \frac{1}{2}\,[\log 0.6126 - \log 3.14]$$

$$= \frac{1}{2}[9.7872 - 10 - 0.4969]$$

$$= \frac{1}{2}[9.2903 - 10]$$

$$= 4.6451 - 5$$

$$N = 0.4417$$

Note: Using an electronic calculator we obtain $N = 0.4417$.

Exercise Set 8.5 (continued)

In exercises 21–48, use logarithms to approximate each given number.

21. $(1.055)^{10}$
22. $(1.055)^{25}$
23. $(1.014)^{40}$
24. $(1.014)^{100}$
25. $(1.016)^{8}$
26. $(1.01625)^{80}$
27. $(825)(1.017)^{11}$
28. $586.1(1.021)^{11}$
29. $211.9(1.008)^{26}$
30. $22.5(1.013)^{26}$
31. $(0.5)^{5/9}$
32. $(0.5)^{8/9}$
33. 2^{100}
34. 2^{1000}
35. $(5.07)^3(44.3)^5(33.1)^6$
36. $\dfrac{(440)^2(6.06)^3}{(0.0272)^3}$
37. $\sqrt{45.8}\,\dfrac{(32.5)^4}{(0.0772)^6}$
38. $\dfrac{(3380)^5(0.624)^2}{\sqrt{9090}}$
39. $\sqrt{\dfrac{(326)(623)}{88800}}$
40. $\sqrt{\dfrac{39.9}{(0.247)(7870)}}$
41. $\sqrt[3]{\dfrac{(9890)(0.0722)}{5.44}}$
42. $\left(\dfrac{(2780)(10.4)}{93.6}\right)^{3/2}$
43. $(2.027)^3\sqrt{308.8}$
44. $\sqrt{44.55}\,\sqrt[3]{55.44}$
45. $\sqrt{\dfrac{57.82}{6.137}}$
46. $\sqrt[3]{(4080)(5173)}$
47. $\sqrt{\dfrac{(471.2)(217.4)}{873.5}}$
48. $\left(\dfrac{707.2}{555.5}\right)^{5/2}$

SOLVING EQUATIONS

Frequently in applications of exponential functions we must solve an equation in which the variable is the exponent. We use common logarithms to do so.

Example 5. Solve $(1.019)^x = 2$.

Solution. We take the common logarithm of each side of this equation:

$$\log (1.019)^x = \log 2$$

$$x \log 1.019 = \log 2$$

$$x = \log 2/\log 1.019$$

$$= 0.3010/0.0082 = 36.7$$

The solution is 36.7.

Example 6. Solve $(1/2)^{x/4.5} = 0.1$

Solution. Taking the common logarithm of each side of the equation,

$$\log (1/2)^{x/4.5} = \log (0.1) = -1$$

$$(x/4.5) \log (1/2) = -1$$

$$-x \log 2 = -4.5$$

$$x = 4.5/0.3010$$

$$= 14.95$$

The solution is 14.95.

Exercise Set 8.5 (continued)

In exercises 49–64, solve each given equation.

49. $(1.008)^x = 2$
50. $(1.013)^x = 2$
51. $(1.017)^x = 2$
52. $(1.021)^x = 2$
53. $(1.016)^{4x} = 2$
54. $(1.018)^{4x} = 2$
55. $825\,(1.017)^x = 1000$
56. $586.1\,(1.021)^x = 1000$
57. $211.9\,(1.008)^x = 300$
58. $22.5\,(1.013)^x = 50$
59. $(1/2)^{x/15} = 0.1$
60. $(1/2)^{x/10.1} = 0.1$
61. $(1/2)^{x/5580} = 0.78$
62. $(1/2)^{x/5580} = 0.65$
63. $104.2\,(1.028)^x = 211.9\,(1.08)^x$
64. $68.2\,(1.033)^x = 252.1\,(1.009)^x$

section 6 • Applications

This section gives applications of exponential and logarthmic functions to exponential growth, exponential decay, and compound interest.

EXPONENTIAL GROWTH

A phenomenon displays exponential growth if it can be described using an exponential function with base greater than 1. Population growth is frequently described as exponential growth.

Example 1. In 1974 the population of the world was 3890 million and the annual rate of population increase was 1.9 percent. Thus the world population increases by 1.9 percent each year. If the population at a particular time is P_0, then the population one year later is

$$P_0 + P_0(0.019) = P_0(1 + 0.019) = P_0(1.019).$$

Since the population is multiplied by 1.019 each year, we have

$$P = 3890\,(1.019)^t,$$

where P is the population of the world (measured in millions) and t is the time (measured in years from 1974). That is, the population is 3890 times the exponential function with base 1.019.

a. What will the population be in the year 2000?

b. In what year will the population reach 5 billion?

c. In what year will the population be double what it was in 1974?

Solution.

a. The year 2000 occurs when $t = 2000 - 1974 = 26$. The population will be

$$P = 3890\,(1.019)^{26} = 3890\,(1.631) = 6346 \text{ million.}$$

b. The population will reach 5 billion when

$$P = 5000 = 3890\,(1.019)^t$$

$$\log 5000 = \log 3890 + t \log 1.019$$

$$t = (\log 5000 - \log 3890)/\log 1.019$$
$$= (3.6990 - 3.5899)/0.0082 = 13.3$$

The population will be 5 billion in 1974 + 13 = 1987.

c. The population will double when

$$3890\,(1.019)^t = 2\,(3890)$$
$$(1.019)^t = 2$$
$$t \log 1.019 = \log 2$$
$$t = \log 2/\log 1.019 = 0.3010/0.0082 = 36.7$$

So the population will be double in 1974 + 37 = 2011.

Table 8.1 gives the 1974 population (measured in millions) of various countries and the annual rate of population increase in 1974 (in percent).

Table 8.1

	Population	Rate increase
Brazil	104.2	2.8
Canada	22.5	1.3
China	825.0	1.7
France	52.5	0.8
India	586.1	2.1
Pakistan	68.2	3.3
Sweden	8.2	0.4
United States	211.9	0.8
U.S.S.R.	252.1	0.9
World total	3890.0	1.9

Exercise Set 8.6

1. a. Write the population P (in millions) of the United States in terms of an exponential function of the time t (in years from 1974).
   b. What will be the population of the United States in 2000?
   c. In what year will the population of the United States reach 300 million?
   d. In what year will the population of the United States be double what it was in 1974?

2. a. Write the population P (in millions) of Canada in terms of an exponential function of the time t (in years from 1974).
   b. What will be the population of Canada in 2000?
   c. In what year will the population of Canada reach 50 million?
   d. In what year will the population of Canada be double what it was in 1974?

3. a. Write the population P (in millions) of China in terms of an exponential function of the time t (in years from 1974).
   b. What will be the population of China in 1985?
   c. In what year will the population of China reach 1 billion?
   d. In what year will the population of China be double what it was in 1974?
4. a. Write the population P (in millions) of India in terms of an exponential function of the time t (in years from 1974).
   b. What will be the population of India in 1985?
   c. In what year will the population of India reach 1 billion?
   d. In what year will the population of India be double what it was in 1974?
5. In what year will the country with the highest rate of increase have a population double its size in 1974?
6. In what year will the country with the lowest rate of increase have a population double its size in 1974?
7. The rate of increase for Brazil is greater than the rate of increase for the United States. In what year will the population of Brazil equal the population of the United States?
8. The rate of increase for Pakistan is greater than the rate of increase for the U.S.S.R. In what year will the population of Pakistan equal the population of the U.S.S.R.?

RADIOACTIVE DECAY

An example of exponential decay is radioactive decay. Some chemical elements are radioactive, which means that they decay into other elements by giving off particles and energy. If Q is the quantity of the radioactive material present, then Q is an exponential function of the time t. The basic measure of radioactive decay is the *half-life*. This is the length of time it takes for half of the quantity to decay.

Table 8.2 lists the half-lives of some elements.

Table 8.2

Element	Half-life
Uranium 238	4.5×10^9 years
Carbon 14	5580 years
Nitrogen 13	10.1 minutes
Sodium 24	15.0 hours
Iodine 136	86 seconds
Uranium 239	23.5 minutes

Example 2. Uranium 238 is used to estimate the age of the earth and of meteorites. Suppose G_0 is the number of grams of uranium 238 present at the creation of the universe. Each 4.5 billion years this amount is halved. So after t billion years, the number of grams G remaining is

$$G = G_0(1/2)^{t/4.5}$$

a. If the universe was created 6 billion years ago, what percentage of the original amount of uranium 238 remains today?

b. After how many years will 10% of the original amount of uranium 238 remain?

Solution.

a. For $t = 6$ we find

$$G = G_0(1/2)^{6/4.5} = G_0(1/2)^{4/3}$$

Therefore, the ratio of the original amount still present is

$$G/G_0 = G_0(1/2)^{4/3}/G_0 = (1/2)^{4/3} = 0.397$$

We find that 39.7% remains today.

b. We wish to solve

$$G = 0.1\ G_0$$

$$G_0\ (1/2)^{t/4.5} = 0.1\ G_0$$

$$(1/2)^{t/4.5} = 0.1$$

$$(t/4.5) \log (1/2) = \log 0.1 = -1$$

$$t = -4.5/\log (1/2) = -4.5/(-\log 2)$$

$$= 4.5/0.3010 = 14.95$$

So 10% of the original amount will remain after 15 billion years.

Exercise Set 8.6 (continued)

9. The age of the earth is estimated by comparing the amount of uranium 238 found in rocks with the amount of lead (an element produced by the radioactive decay). The oldest rocks measured in this way have ages of about 2.5 billion years. What percentage of the original amount of uranium 238 in these rocks remains today?

10. Meteorites on the earth are found to have ages of 4 billion years. What percentage of the original amount of uranium 238 in these meteorites remains today?

11. A chemist starts with G_0 grams of sodium 24.
    a. Write the number of grams G that remain in terms of an exponential function of the time t (in hours).
    b. What percentage of the original amount remains after 24 hours?
    c. After how many hours will 10% of the original amount remain?

12. A chemist starts with G_0 grams of nitrogen 13.
    a. Write the number of grams G that remain in terms of an exponential function of the time t (in minutes).
    b. What percentage of the original amount remains after 1 hour?
    c. After how many minutes will 10% of the original amount remain?

13. Carbon 14 is used to date archaeological relics. The amount of carbon 14 found in a relic is compared with the amount of carbon 12 (an element produced by the radioactive decay). How old is the relic if it is determined that 78% of the original amount of carbon 14 still remains?

14. Carbon 14 is used to date an archaeological relic. How old is the relic if it is determined that 65% of the original amount of carbon 14 still remains?

COMPOUND INTEREST

We saw in Section 3.10 that, if a principal of P dollars is invested at an interest rate r compounded yearly, the amount A after t years is

$$A = P(1 + r)^t.$$

Thus A equals P times the exponential function of t with base $1 + r$. However, if a principal of P dollars is invested at an interest rate r compounded n times a year, the amount A after t years is

$$A = P\left(1 + \frac{r}{n}\right)^{nt}.$$

Example 3. In 1977 most banks gave 5.5% interest compounded quarterly on a passbook savings account. This means that 4 times a year the amount in the account is increased by $5.5/4 = 1.375$ percent. Therefore, $r = 0.055$, $n = 4$, and the amount A on a principal of \$1000 after a time of t years is

$$A = 1000\,(1.01375)^{4t}.$$

a. What is the amount after 4 years?

b. How many years does it take for the principal to double?

Solution.

a. For $t = 4$ we have

$$A = 1000\,(1.01375)^{4(4)} = 1000\,(1.01375)^{16}$$
$$= 1000\,(1.24421) = 1244.21$$

The amount is \$1244.21.

b. The principal will double when

$$A = 2P$$

$$1000\,(1.01375)^{4t} = 2\,(1000)$$

$$(1.01375)^{4t} = 2$$

$$4t \log 1.01375 = \log 2$$

$$t = \log 2/4 \log 1.01375$$

$$= 0.3010/4\ (0.0059) = 12.8$$

The principal will double after 13 years.

Exercise Set 8.6 (continued)

15. At the same time, a principal of \$1000 is placed in an account yielding 5.5% interest compounded yearly and another \$1000 is placed in an account yielding 5.5% interest compounded quarterly.
    a. What is the amount in each account after 1 year?
    b. What is the amount in each account after 10 years?
    c. What is the amount in each account after 25 years?
16. At the same time, a principal of \$1000 is placed in an account yielding 6% interest compounded yearly and another \$1000 is placed in an account yielding 6% interest compounded quarterly.
    a. What is the amount in each account after 1 year?
    b. What is the amount in each account after 10 years?
    c. What is the amount in each account after 25 years?
17. Some savings accounts yield 6.5% interest compounded quarterly if no withdrawal is made for at least 2 years. Suppose \$1000 principal is placed in such an account.
    a. Express the amount A in terms of an exponential function of the time t (in years).
    b. What is the amount after 2 years?
    c. What is the amount after 20 years?
    d. How many years does it take for the principal to double?
18. Some savings accounts yield 7.25% interest compounded quarterly if no withdrawal is made for at least 4 years. Suppose \$1000 principal is placed in such an account.
    a. Express the amount A in terms of an exponential function of the time t (in years).
    b. What is the amount after 4 years?
    c. What is the amount after 20 years?
    d. How many years does it take for the principal to double?
19. What principal must be invested at 6.5% interest compounded quarterly to have \$40,000 in 20 years?
20. What principal must be invested at 7.25% interest compounded quarterly to have \$40,000 in 20 years?

Review Exercises

1. Let $f(x) = 6^x$, and it is given that $f(0.1) = 1.2$
    a. Find $f(2)$, $f(1)$, $f(0)$, $f(-1)$, and $f(-2)$.
    b. Find $f(0.2)$, $f(0.3)$, and $f(0.9)$.
    c. Use logarithms to approximate $f(20)$ and $f(50)$.

2. Let $g(x) = 11^x$, and it is given that $g(0.1) = 1.3$
   a. Find $g(2)$, $g(1)$, $g(0)$, $g(-1)$, and $g(-2)$.
   b. Find $g(0.2)$, $g(0.3)$, and $g(0.9)$.
   c. Use logarithms to approximate $g(20)$ and $g(50)$.
3. Suppose $f(x) = b^x$, and it is given that $f(2) = 5$. Find $f(4)$, $f(6)$, $f(-2)$, $f(1)$, and $f(3)$.
4. Suppose $f(x) = b^x$, and it is given that $f(2) = 3$. Find $f(4)$, $f(6)$, $f(-2)$, $f(1)$, and $f(3)$.

5. Find $\log_6 6$, $\log_6 36$, $\log_6 1$, $\log_6 (1/36)$, $\log_6 \sqrt{6}$
6. Find $\log_{11} 11$, $\log_{11} 121$, $\log_{11} 1$, $\log_{11} (1/11)$, $\log_{11} \sqrt{11}$

In exercises 7–18, express each given logarithm in terms of logarithms of a single letter. Assume that the letters represent positive real numbers.

7. $\log x^2y$ 8. $\log xy^2$ 9. $\log x^{-2}$ 10. $\log y^{-3}$
11. $\log (x/y^2)$ 12. $\log (x^2/y)$ 13. $\log (x^3/y^2)$ 14. $\log (x^2/y^3)$
15 $\log x^{3/2}$ 16. $\log y^{2/3}$ 17. $\log \sqrt{x}$ 18. $\log \sqrt{xy}$

19. Given $\log 4 = 0.6021$, find
   a. $\log 40$ b. $\log 0.4$ c. $\log 400$ d. $\log 0.004$
   e. $\log 16$ f. $\log 2$ g. $\log (1/4)$ h. $\log 0.25$
20. Given $\log 9 = 0.9542$, find
   a. $\log 90$ b. $\log 900$ c. $\log 0.9$ d. $\log 81$
   e. $\log 8.1$ f. $\log 3$ g. $\log 27$ h. $\log (1/9)$
21. Given $\log 5 = 0.6990$ and $6 = 0.7782$, find
   a. $\log 30$ b. $\log 3$ c. $\log 15$ d. $\log 1.5$
   e. $\log \sqrt{5}$ f. $\log \sqrt{6}$ g. $\log 0.2$ h. $\log 12$
22. Given $\log 3 = 0.4771$ and $\log 7 = 0.8451$, find
   a. $\log 21$ b. $\log 9$ c. $\log 63$ d. $\log 49$
   e. $\log \sqrt{3}$ f. $\log \sqrt{7}$ g. $\log 3/7$ h. $\log 7/3$

In exercises 23–34, perform each operation using logarithms and interpolation.

23. $(3.724)(0.022)(588.6)$ 24. $(978.4)(0.3362)(75.2)$ 25. $\dfrac{(66.02)(3720)}{(0.444)(7333)}$

26. $\dfrac{(0.145)(2.492)}{(99.20)(6.321)}$ 27. $\dfrac{1.327}{(6215)(299)}$ 28. $\dfrac{75.24}{(107)(326300)}$

29. $(4522)^7\,(5422)^6$ 30. $\dfrac{(0.00922)^{10}}{(0.0366)^{12}}$ 31. $\sqrt{0.000621}\,(37.77)^2$

32. $\dfrac{(0.03333)}{\sqrt{55.21}}$ 33. $\sqrt{\dfrac{0.01782}{88.24}}$ 34. $\sqrt{(65280)(90200)}$

In exercises 35–40, use common logarithms to solve each given equation.

35. $(1.1)^x = 2$ 36. $(1.09)^x = 2$ 37. $(1.9)^x = 3$
38. $(2.1)^x = 3$ 39. $(1/2)^{x/3} = 4.2$ 40. $(1/2)^{x/4} = 1.6$

41. In 1974 the population of France was 52.5 million and the annual rate of population increase was 0.8% per year.
    a. Write the population P (in millions) of France in terms of an exponential function of the time t (in years from 1974).
    b. What will be the population of France in 2000?
    c. In what year will the population of France be double what it was in 1974?
42. The half-life of uranium 239 is 23.5 minutes. Suppose a chemist starts with G_0 grams of uranium 239.
    a. Write the number of grams G that remain in terms of an exponential function of the time t (in minutes).
    b. What percentage of the original amount remains after 47 minutes?
    c. After how many minutes will 10% of the original amount remain?
43. A principal of $1000 is placed in an account yielding 5% interest compounded yearly.
    a. Express the amount A in terms of an exponential function of the time t (in years).
    b. What is the amount after 50 years?
    c. How many years does it take for the principal to double?
44. A principal of $1000 is placed in an account yielding 6% interest compounded quarterly.
    a. Express the amount A in terms of an exponential function of the time t (in years).
    b. What is the amount after 50 years?
    c. After how many years will the amount be $10,000?

chapter 9

Systems of Equations and Inequalities

If we graph two equations in the same coordinate system, we have a *system of equations*. The points at which these graphs intersect are called the *simultaneous solution* of the system of equations. We consider systems of two linear equations in two variables in Section 9.1 and systems of two quadratic equations in two variables in Section 9.3.

If we graph two or more inequalities in the same coordinate system, we say that we have a *system of inequalities*. The points at which these graphs intersect are called the *simultaneous solution* of the system of inequalities. We consider systems of linear inequalities in Section 9.2.

In Section 9.4 we consider systems of linear equations in more than two variables. Although we cannot graph such systems in the Cartesian coordinate system, we will develop a technique for solving such systems of equations.

In Section 9.5 we study matrices and determinants. We use determinants in Section 9.6 to study Cramer's Rule, which provides another method for solving systems of linear equations.

In Section 9.7 we consider applications of systems of equations and inequalities to rate problems, geometry problems, diet problems, mixture problems, and historical problems.

section 1 • Linear Equations

Suppose we have two linear equations in two variables x and y; say

$$a_1x + b_1y = c_1 \qquad \text{and} \qquad a_2x + b_2y = c_2.$$

The graph of each equation in the Cartesian coordinate system is a straight line. We say that we have a *system of linear equations* and the point at which the graphs intersect is the *simultaneous solution.* We will describe two methods for solving a system of linear equations—the substitution method and the linear combination method—and we will examine the simultaneous solution.

SUBSTITUTION METHOD

To solve a system of two linear equations by the substitution method, we (1) solve for a variable x or y in one of the equations; (2) substitute this expression for x or y in the second equation; (3) solve the resulting linear equation in one variable; (4) find the corresponding value of the other variable; and (5) check the resulting solution in both equations.

Example 1. Solve

$$\begin{aligned} 3x + 2y &= 13 \\ 4x - 3y &= 6. \end{aligned}$$

Solution. Solving for y in the first equation (step 1), we obtain

$$y = \frac{13 - 3x}{2}.$$

Substituting this expression for y in the second equation (step 2), we have

$$4x - 3\frac{13 - 3x}{2} = 6.$$

This is a linear equation in one variable x, which we proceed to solve (step 3).

$$\begin{aligned} 8x - 39 + 9x &= 12 \\ 17x &= 51 \\ x &= 3 \end{aligned}$$

Now we find the corresponding value of y (step 4).

$$y = \frac{13 - 3(3)}{2} = 2.$$

We have obtained the solution $x = 3$ and $y = 2$. We can write this solution as the ordered pair (3, 2). We check the ordered pair (3, 2) in each equation (step 5):

$$3(3) + 2(2) = 13 \quad \text{and} \quad 4(3) - 3(2) = 6.$$

The simultaneous solution is (3, 2).

Note: The graph of each equation of this system is a straight line in the Cartesian coordinate system (see Figure 9.1). We have found that these lines intersect at the point

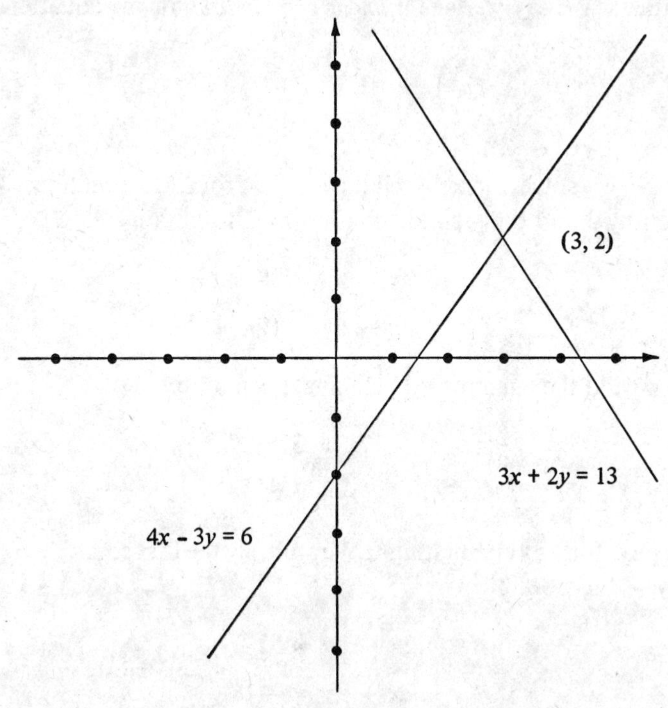

Figure 9.1

(3, 2). The equations $3x + 2y = 13$ and $y = (13 - 3x)/2$ are equivalent, so they have the same graph. When we substitute this value of y into the second equation, we are finding the point along the line $4x - 3y = 6$ where the lines intersect.

Exercise Set 9.1

In exercises 1–10, use the substitution method to solve each system of equations.

1. $x + y = 12$
   $3x + 2y = 34$

2. $x + y = 6$
   $2x + 3y = 15$

3. $4x + y = 6$
   $3x - y = 1$

4. $3x + 2y = 34$
   $x + 3y = 30$

5. $3x - 2y = 6$
   $4x + 5y = -38$

6. $5x + 7y = -20$
   $7x + 5y = -28$

7. $2x - 3y = -1$
   $3x + 5y = -11$

8. $4x + 2y = -6$
   $3x - 5y = 2$

9. $3x - 5y = 4$
   $2x - 4y = 3$

10. $3x + 3y = 3$
    $5x - 4y = 2$

LINEAR COMBINATION METHOD

The linear combination method of solving a system of linear equations again depends on reducing the problem to solving a linear equation in just one variable. We eliminate a variable from the equations by multiplying each equation by an appropriate constant and adding the equations. We say we take a *linear combination* of the equations.

Example 2. Solve

$$3x + 2y = 13$$
$$4x - 3y = 6.$$

Solution. First we solve for x by eliminating y from the equations. We multiply the first equation by 3 and the second equation by 2 to obtain

$$9x + 6y = 39$$
$$8x - 6y = 12.$$

Now when we add the equations we eliminate y and obtain

$$17x = 51$$
$$x = 3$$

Second, we solve for y by eliminating x. We multiply the first equation by 4 and the second equation by -3 to obtain

$$12x + 8y = 52$$
$$-12x + 9y = -18.$$

Adding the equations eliminates x, and we obtain

$$17y = 34$$
$$y = 2$$

The simultaneous solution is (3, 2). This is the same system of equations we solved in Example 1.

Example 3. Solve $2x + 3y = 2$
$4x + 4y = 2$

Solution. First we solve for x by eliminating y from the equations. To do so, we multiply the first equation by 4 and the second equation by -3 to obtain

$$8x + 12y = 8$$
$$-12x - 12y = -6$$

Adding these equations, we obtain

$$-4x = 2$$
$$x = -1/2$$

To eliminate x from the equations, we can multiply the first equation by 2 and the second equation by -1:

$$4x + 6y = 4$$
$$-4x - 4y = -2$$

Adding these equations we find

$$2y = 2$$
$$y = 1$$

Now we check the ordered pair $(-1/2, 1)$ in both equations:

$$2(-1/2) + 3(1) = 2 \quad \text{and} \quad 4(-1/2) + 4(1) = 2.$$

Therefore, the simultaneous solution is $(-1/2, 1)$.

Exercise Set 9.1 (continued)

In exercises 11–20, use the linear combination method to solve each system of equations.

11. $2x + 3y = 7$
    $3x + 2y = 8$

12. $4x + 3y = 10$
    $3x + 4y = 11$

13. $4x + 3y = 32$
    $4x + 5y = 40$

14. $x + y = 12$
    $3x + 2y = 30$

15. $2x - 3y = 13$
    $4x + 3y = -1$

16. $3x + 5y = 45$
    $5x + 4y = 62$

17. $4x - 3y = 23$
    $7x + 2y = 4$

18. $4x - 2y = 6$
    $10x - 5y = 15$

19. $4x + 9y = 5$
    $2x - 3y = 0$

20. $8x + 4y = -1$
    $2x - 2y = -1$

CLASSIFICATION OF SOLUTIONS

You can use either the substitution method or the linear combination method to solve a system of linear equations; choose the method that seems easier for a given systems of equations.

So far we have considered systems of linear equations which possess a single ordered pair as solution. That is, the graphs of the equations intersect in just one point. However, this possibility is just one of three that can occur, as Figure 9.2 illustrates. A second possibility is that the graphs are parallel lines, and a third possibility is that the graphs are the

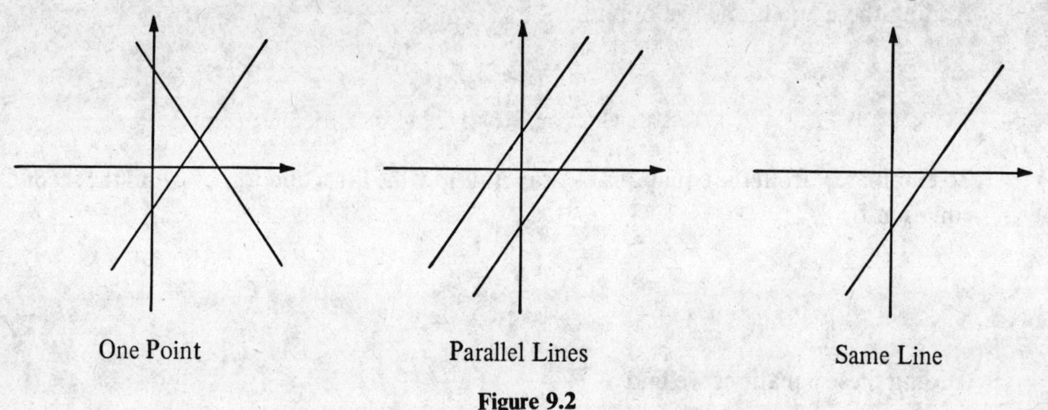

Figure 9.2

same straight line. We classify systems of linear equations according to these three possibilities.

DEFINITION OF SIMULTANEOUS, INCONSISTENT, AND DEPENDENT EQUATIONS. A system of linear equations in two variables is:

1. *Simultaneous* if the solution consists of one ordered pair; that is, if the graphs intersect in one point.
2. *Inconsistent* if there is no solution; that is, if the graphs are parallel lines (and thus do not intersect).
3. *Dependent* if the solution contains infinitely many ordered pairs; that is, if the graphs are the same line.

Example 4. Solve

$$3x - 4y = 6$$
$$-6x + 8y = 5.$$

Solution. We use linear combinations to eliminate y. Multiply the first equation by 2 and the second equation by 1.

$$6x - 8y = 12$$
$$-6x + 8y = 5.$$

When we add the equations, we obtain

$$0 = 17.$$

It looks as if something has gone wrong. But all that has happened is that the lines are parallel. In fact, if we check, each line has slope 3/4. So the system of equations is inconsistent. That it is inconsistent is reflected in the fact that we obtained a contradiction, namely $0 = 17$, which indicates that there is no solution.

Example 5. Solve $12x - 2y = 3$
$36x - 6y = 9.$

Solution. If we solve for y in the first equation, we obtain

$$y = \frac{12x - 3}{2}.$$

Substituting this expression for y in the second equation, we find

$$36x - 6\frac{12x - 3}{2} = 9$$

$$36x - 36x + 9 = 9.$$

This equation is an identity, which indicates that the two equations are equivalent. In fact, each equation has slope 6 and y-intercept $-3/2$. So the two equations have the same graph; they are dependent.

Exercise Set 9.1 (continued)

In exercises 21–30, classify each given system of equations as simultaneous, inconsistent, or dependent.

21. $x + 2y = 2$
$3x + 6y = 3$

22. $3x + y = 5$
$6x + 2y = 4$

23. $x + y = 5$
$x - y = 3$

24. $x + y = 400$
$x - y = 300$

25. $3x - 2y = 1$
$9x - 6y = 3$

26. $4x - 6y = 8$
$2x - 3y = 4$

27. $3x - 9y = 4$
$4x - 12y = -3$

28. $6x + 15y = -2$
$4x + 10y = 5$

29. $8x + 6y = 4$
$12x + 9y = 6$

30. $6x - 9y = 12$
$8x - 12y = 16$

EQUATIONS WITH DECIMAL COEFFICIENTS

Frequently in applications we must solve a system of linear equations with decimal coefficients. We can use either the substitution method or the linear combination method to solve such a system of equations.

Example 6. Solve $\quad 0.3x - 0.4y = 1.25$
$\quad\quad\quad\quad\quad\quad\quad 0.4x + 0.3y = 1.50$

Solution. Using the linear combination method, we multiply the first equation by 3 and the second equation by 4 to obtain

$$0.9x - 1.2y = 3.75$$
$$1.6x + 1.2y = 6.00$$

When we add these equations we obtain

$$2.5x = 9.75$$
$$x = 3.9$$

Substituting this value of x in the second equation, we find

$$0.4\,(3.9) + 0.3y = 1.50$$
$$0.3y = -0.06$$
$$y = -0.2$$

The simultaneous solution is $(3.9, -0.2)$.

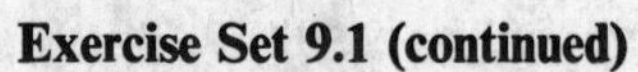

Exercise Set 9.1 (continued)

In exercises 31–36, solve each system of equations; if necessary, round off the solution to one decimal place.

31. $0.6x - 0.8y = 0.08$
    $0.8x + 0.6y = 1.44$

32. $1.2x + 1.6y = 2.16$
    $1.6x - 1.2y = -1.12$

33. $x + y = 14$
    $0.4x + 0.75y = 0.5(14)$

34. $x + y = 20$
    $0.15x + 0.55y = 0.3(20)$

35. $x + y = 10$
    $0.12x + 0.28y = 0.24(10)$

36. $x + y = 20$
    $0.15x + 0.24y = 0.18(20)$

CALCULATOR EXERCISES: In exercises 37–42, use an electronic calculator to solve the system of equations; if necessary, round off the solution to two decimal places.

37. $0.39x - 0.92y = 0.45$
    $0.92x + 0.39y = 1.22$

38. $0.91x + 0.42y = 3.15$
    $0.42x - 0.91y = 1.06$

39. $0.26x + 1.08y = 2.03$
    $1.22x + 0.47y = 2.58$

40. $0.92x + 2.05y = 3.33$
    $1.48x + 0.44y = 2.88$

41. $3.41x - 4.22y = 1.44$
    $5.54x + 3.66y = 8.16$

42. $2.54x - 3.14y = 0.88$
    $4.29x + 2.91y = 6.71$

section 2 • Linear Inequalities

Now we will graph systems of linear inequalities in two variables x and y. The solution of such a system of inequalities is the region in the Cartesian coordinate system that satisfies every inequality of the system. Recall that we have graphed linear inequalities in x and y in Section 7.6.

TWO LINEAR INEQUALITIES

Example 1. Graph $x + y \leq 4$
$x - y \geq 3$

Solution. On the same coordinate system we graph the linear inequality $x + y \leq 4$ and also the linear inequality $x - y \geq 3$. The graph of the system of inequalities is the region that satisfies both inequalities; that is, the double-hatched region of Figure 9.3.

Example 2. Graph $2x + 3y \leq 6$
$x - 2y \leq 4$

Solution. On the same coordinate system we graph the linear inequality $2x + 3y \leq 6$ and also the linear inequality $x - 2y \leq 4$. The graph of the system of inequalities is the region that satisfies both inequalities; that is, the double-hatched region of Figure 9.4.

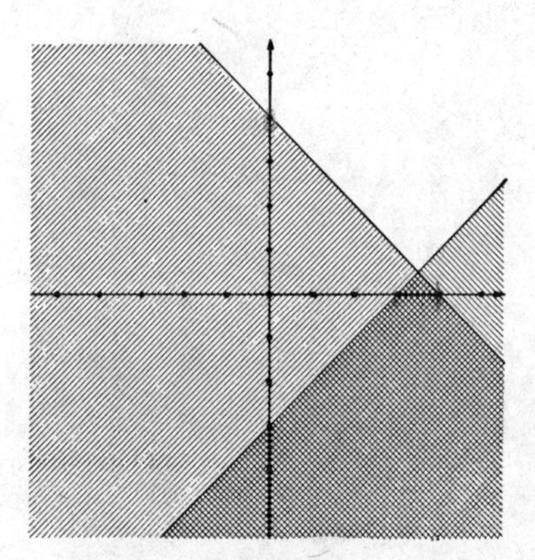

Figure 9.3

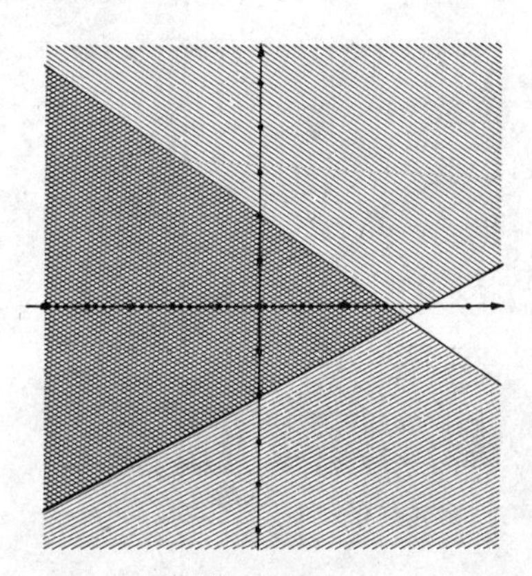

Figure 9.4

Exercise Set 9.2

In exercises 1–10, graph each given system of inequalities.

1. $x + y \leq 4$
   $x - y \leq 4$
2. $x + y \geq 3$
   $x - y \geq 3$
3. $x + y \geq 2$
   $x - y \leq 3$
4. $x + y \leq 1$
   $x - y \geq 4$
5. $3x + 2y \leq 6$
   $x \leq y$
6. $4x - 3y \leq 12$
   $x \leq y$

7. $x + 4y \geq 4$
   $2x - y \leq 4$

8. $2x - 3y \geq 6$
   $3x + y \leq 6$

9. $4x - 3y \leq 12$
   $3x + y \geq 3$

10. $5x + 2y \geq 10$
    $2x - 5y \leq 10$

FIRST QUADRANT INEQUALITIES

In many applications of linear inequalities, the two variables x and y cannot be negative. In this case the inequalities $x \geq 0$ and $y \geq 0$ are part of the system of inequalities. These two inequalities confine the graph to the first quadrant of the Cartesian coordinate system.

Example 3. Graph
$$3x + 2y \geq 20$$
$$3x + 5y \geq 36$$
$$x \geq 0$$
$$y \geq 0$$

Solution. The two inequalities $x \geq 0$ and $y \geq 0$ confine the graph to the first quadrant of the Cartesian coordinate system. We graph each inequality $3x + 2y \geq 20$ and $3x + 5y \geq 36$ and find the region in the first quadrant that satisfies both inequalities. This region is graphed in Figure 9.5.

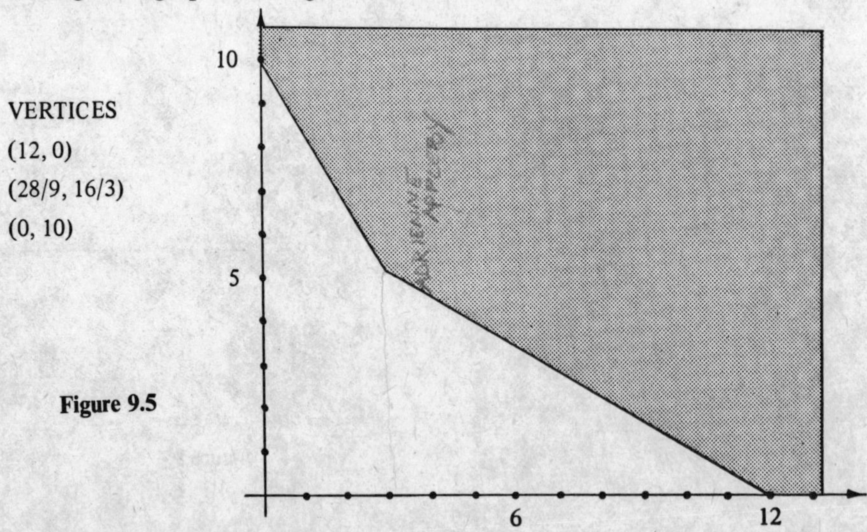

Figure 9.5

Note: Figure 9.5 indicates the three *vertices* of this graph. We find these vertices by solving the linear *equations* in pairs. For example, to find the vertex (28/9, 16/3) we solve the system of equations

$$3x + 2y = 20$$
$$3x + 5y = 36$$

because these equations represent the border of the graph.

Exercise Set 9.2 (continued)

In exercises 11–18, graph each system of inequalities and find the vertices of the graph.

11. $x + y \geq 12$
$2x + 3y \geq 30$
$x \geq 0$
$y \geq 0$

12. $x + y \geq 12$
$x + 3y \geq 21$
$x \geq 0$
$y \geq 0$

13. $2x + 2y \geq 24$
$6x + 4y \geq 60$
$x \geq 0$
$y \geq 0$

14. $3x + 2y \geq 15$
$2x + 3y \geq 15$
$x \geq 0$
$y \geq 0$

15. $x + y \geq 12$
$2x + 3y \geq 30$
$4x + 5y \leq 60$
$x \geq 0$
$y \geq 0$

16. $x + y \geq 12$
$x + 3y \geq 21$
$x + y \leq 18$
$x \geq 0$
$y \geq 0$

17. $2x + 2y \geq 24$
$6x + 4y \geq 60$
$4x + 5y \leq 60$
$x \geq 0$
$y \geq 0$

18. $3x + 2y \geq 15$
$2x + 3y \geq 15$
$4x + 5y \leq 40$
$x \geq 0$
$y \geq 0$

section 3 • Quadratic Equations

In this section we consider systems of two equations in two variables x and y. In Section 9.1 we considered systems in which each equation was linear; now we will consider systems in which one or both of the equations is quadratic. We will see when to use the substitution method and when to use the linear combination method.

A LINEAR WITH A QUADRATIC EQUATION

We can use the substitution method to solve a system of equations in x and y in which one equation is linear and the other is quadratic. We solve for either x or y in the linear equation and substitute this expression in the quadratic equation. We then solve the resulting quadratic equation in one variable.

Example 1. Solve

$$2x + y - 8 = 0$$
$$y = 3x^2 - 2x - 4.$$

Solution. Solving for y in the linear equation,

$$y = -2x + 8.$$

Substituting this expression for y in the second equation, we obtain

$$-2x + 8 = 3x^2 - 2x - 4.$$
$$0 = 3x^2 - 12$$
$$0 = x^2 - 4$$
$$0 = (x - 2)(x + 2)$$

$$x = 2 \quad \text{or} \quad x = -2.$$

Using these two values of x, we find the corresponding values of y:

$$y = -2(2) + 8 = 4 \quad \text{and} \quad y = -2(-2) + 8 = 12.$$

Since $(2, 4)$ and $(-2, 12)$ check in both equations, the simultaneous solution is

$$(2, 4) \quad \text{and} \quad (-2, 12).$$

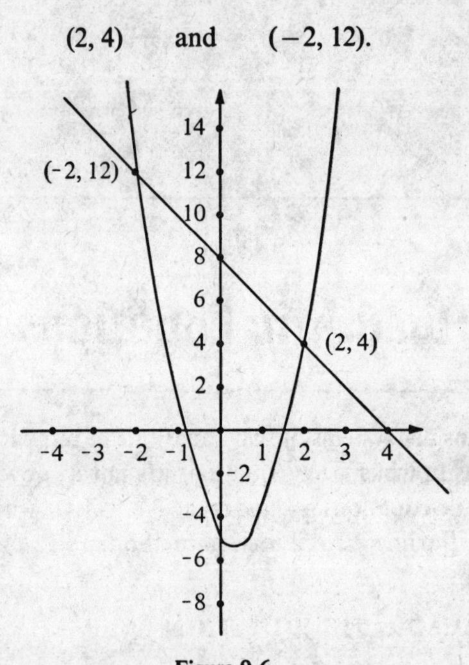

Figure 9.6

Note: The graph of these equations in the same coordinate system is shown in Figure 9.6. The graph of the linear equation is a straight line, and the graph of the quadratic equation is a parabola. These graphs intersect in two points, $(2, 4)$ and $(-2, 12)$.

Example 2. Solve
$$x - y = 6$$
$$x^2 + y^2 = 16.$$

Solution. Solving for y in the linear equation, we obtain

$$y = x - 6.$$

Substituting this expression for y in the second equation,

$$x^2 + (x - 6)^2 = 16$$
$$x^2 + x^2 - 12x + 36 = 16$$
$$2x^2 - 12x + 20 = 0$$
$$x^2 - 6x + 10 = 0$$

Discriminant is $(-6)^2 - 4(1)(10) = -4$. Since the discriminant is negative, this quadratic equation does not possess real number solutions. In this chapter we are interested only in real number solutions; therefore we say that the system of equations is *inconsistent.*

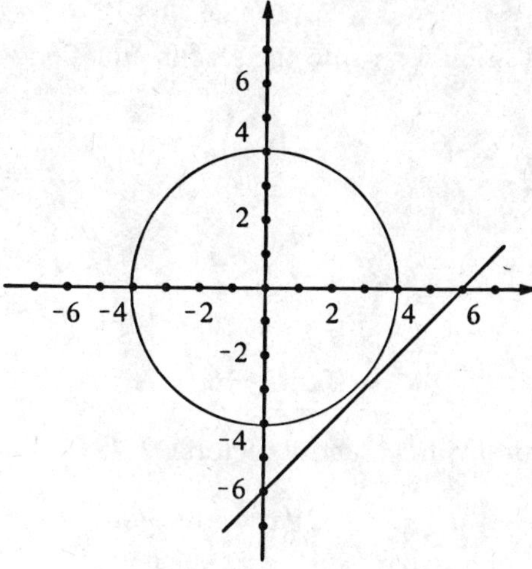

Figure 9.7

Note: Figure 9.7 is the graph of these two equations in the same coordinate system. The graph of the linear equation is a straight line, and the graph of the quadratic equation is a circle. These graphs do not intersect, indicating that the system of equations is inconsistent.

Exercise Set 9.3

In exercises 1–12, solve each given system of equations or specify that it is inconsistent.

1. $3x - y - 4 = 0$
   $y = -3x^2 + 6x + 2$
2. $4x - y - 1 = 0$
   $y = x^2 + 2x$
3. $y = -x^2 - 6x - 4$
   $2x + y - 1 = 0$
4. $4x + y = -1$
   $y = x^2 - 2x$
5. $2x + 2y = 7$
   $x^2 + y^2 = \frac{25}{4}$
6. $x + y = 4$
   $x^2 + y^2 = 10x$
7. $x - y = 4$
   $x^2 + y^2 = 6$
8. $2x + 2y = 28$
   $x^2 + y^2 = 100$
9. $x + y = 18$
   $x^2 = y^2 + 36$
10. $x + y = 1$
    $x^2 = y^2 + 3$
11. $y = x^2 - 4x + 2$
    $y = -x^2 + 4x + 2$
12. $y = x^2 - 4x + 4$
    $y = x^2 + 6x + 9$

FURTHER SUBSTITUTIONS

We consider further systems of equations that can be solved by the method of substitution.

Example 3. Solve $xy = 4$

$$x^2 + y^2 = 10.$$

Solution. Solving for y in the first equation, we obtain

$$y = \frac{4}{x} \qquad \text{(for } x \neq 0\text{).}$$

Substituting this expression for y into the second equation we have

$$x^2 + \left(\frac{4}{x}\right)^2 = 10$$

$$x^2 + \frac{16}{x^2} = 10$$

$$x^4 + 16 = 10x^2$$

$$x^4 - 10x^2 + 16 = 0.$$

This equation is quadratic in x^2, and it factors:

$$(x^2 - 2)(x^2 - 8) = 0$$

$$x^2 = 2 \quad \text{or} \quad x^2 = 8$$

$$x = \sqrt{2}, \quad x = -\sqrt{2}, \quad x = 2\sqrt{2}, \quad \text{or} \quad x = -2\sqrt{2}$$

Using these four values of x to find the corresponding values of y, we obtain the simultaneous solution

$$(\sqrt{2}, 2\sqrt{2}), \quad (-\sqrt{2}, -2\sqrt{2}), \quad (2\sqrt{2}, \sqrt{2}), \quad (-2\sqrt{2}, -\sqrt{2}).$$

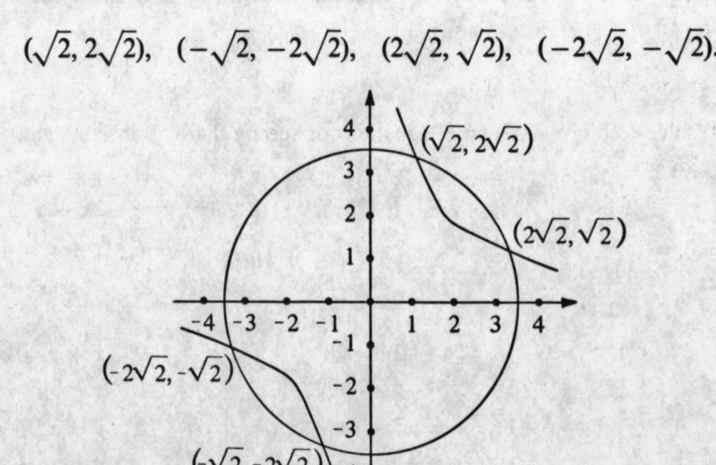

Figure 9.8

Note: Figure 9.8 shows the graphs of these two equations in the same coordinate system; and we see that the graphs actually intersect in four points.

Exercise Set 9.3 (continued)

In exercises 13–24, solve each given system of equations or specify that it is inconsistent.

13. $x^2 + y^2 = 100$
 $xy = 48$

14. $x^2 + y^2 = 25/4$
 $xy = 6$

15. $2x + 2y = 22$
 $xy = 28$

16. $2x + 2y = 19$
 $xy = 12$

17. $2x + 2y = 32$
 $xy = 48$

18. $2x + 2y = 13$
 $xy = 10$

19. $xy = 300$
 $(x + 10)(y - 1) = 300$

20. $xy = 60$
 $(x + 5)(y - 2) = 60$

21. $2x + 2y = 4$
 $xy = 2(x + 4)(y + 4)$

22. $x - y = 1$
 $xy = 2(x + 1)(y + 1)$

23. $2x - 1 = 2y$
 $\frac{1}{x} + \frac{1}{y} = 3$

24. $x - 2 = y$
 $\frac{1}{x} + \frac{1}{y} = \frac{5}{12}$

TWO QUADRATIC EQUATIONS

If we have a system of two quadratic equations in x and y, we can try to use the linear combination method to solve the system.

Example 4. Solve $\quad x^2 + y^2 = 8y - 15$
$\quad x^2 + y^2 = 6x + 7.$

Solution. Multiplying the first equation by 1, multiplying the second equation by -1, and adding the equations, we obtain

$$0 = -6x + 8y - 22.$$

Solving for y in this linear equation,

$$y = \frac{3x + 11}{4}.$$

We substitute this expression for y in the second equation:

$$x^2 + \left(\frac{3x + 11}{4}\right)^2 = 6x + 7$$

$$x^2 + \frac{9x^2 + 66x + 121}{16} = 6x + 7$$

$$16x^2 + 9x^2 + 66x + 121 = 96x + 112$$

$$25x^2 - 30x + 9 = 0$$
$$(5x - 3)^2 = 0$$

$$x = \frac{3}{5} \quad \text{and} \quad y = \frac{3(3/5) + 11}{4} = \frac{16}{5}$$

The simultaneous solution is (3/5, 16/5).

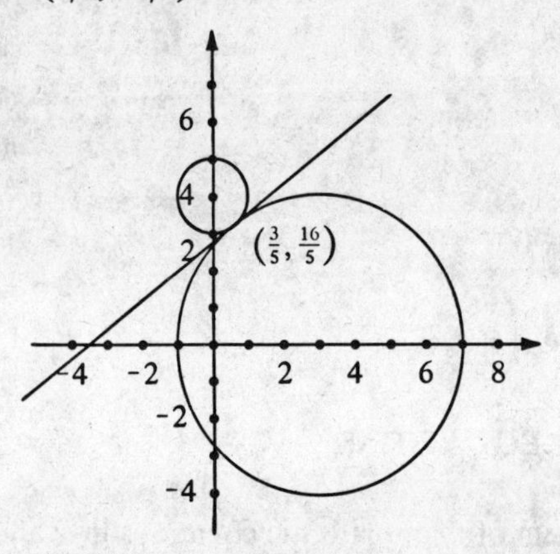

Figure 9.9

Note: Figure 9.9 shows the graphs of these two equations in the same coordinate system. The graph of each equation is a circle. We see that the circles intersect in one point; that is, the circles are tangent at (3/5, 16/5). The linear equation we obtained in the solution passes through the same point; in fact, it is tangent to each circle.

Exercise Set 9.3 (continued)

In exercises 25–30, solve each given system of equations or specify that it is inconsistent.

25. $x^2 + y^2 = 8$
 $x^2 + y^2 = 2x + 2y$

26. $x^2 + y^2 = 36$
 $x^2 + y^2 = 2x + 2y$

27. $x^2 + y^2 = 32$
 $x^2 + y^2 = 4x + 4y$

28. $x^2 + y^2 = 225$
 $x^2 + y^2 = 6x - 6y + 207$

29. $x^2 + y^2 = 225$
 $(x + 3)^2 + (y - 3)^2 = 225$

30. $x^2 + y^2 = 9$
 $(x - 1)^2 + (y - 1)^2 = 1$

section 4 • Systems in *n* Variables

We have seen how to solve a system of two linear equations in two variables (Section 9.1). Sometimes situations arise in which we must solve a system of three linear equations in three variables, four linear equations in four variables, and so on. Finding the simultaneous solution set of a system of linear equations is important in mathematics and in applications, and it has been studied extensively.

We now examine the situation in which we have equations in more than two variables. We will confine our attention to linear equations.

DEFINITION OF A LINEAR EQUATION IN n VARIABLES. A *linear equation* in n variables $x_1, x_2, \ldots x_n$ is an equation equivalent to one of the form

$$a_1x_1 + a_2x_2 + \cdots + a_nx_n = c,$$

where $a_1, a_2, \ldots, a_n$ are constants which are not all 0, and c is a constant.

We will consider the case in which we have a system of n linear equations in n variables; that is, we have as many equations as variables. The cases in which there are not the same number of equations and variables present added complications which we will not discuss.

In this section, we discuss two methods of solving a system of n linear equations in n variables. The methods are the same as those discussed in Section 9.1; both involve the elimination of variables. The first method involves the elimination of variables by solving for one variable in terms of the others. The second method involves the elimination of variables by using linear combinations.

SUBSTITUTION METHOD

Example 1. Solve

$$\begin{aligned} 2x + 3y - z &= 1 \\ x - 4y + z &= 0 \\ 3x - 2y - 2z &= -2. \end{aligned}$$

Solution. We will solve these equations by solving for one variable in terms of the others. This can be done in several ways. We first solve for z in terms of x and y. Using the second equation, we obtain

$$z = -x + 4y.$$

We substitute this expression for z into the first and third equations:

$$\begin{aligned} 2x + 3y - (-x + 4y) &= 1 \\ 3x - 2y - 2(-x + 4y) &= -2. \end{aligned}$$

Simplifying these two equations, we obtain

$$\begin{aligned} 3x - y &= 1 \\ 5x - 10y &= -2. \end{aligned}$$

We now have a system of two linear equations in the two variables x and y. We proceed to solve these. Solving for y, we obtain

$$y = 3x - 1.$$

Thus,

$$5x - 10(3x - 1) = -2$$
$$-25x = -12$$
$$x = 12/25$$
$$y = 3(12/25) - 1 = 11/25$$
$$z = -(12/25) + 4(11/25) = 32/25$$

You should check this solution in each of the three equations.

Exercise Set 9.4

In exercises 1–10, solve each given system of equations.

1. $3x + y + z = 5$
   $2x - 2y - 3z = -5$
   $x + 4y + 2z = 3$

2. $x + y - 2z = 9$
   $2x + 2y - z = 6$
   $2x - y - z = -3$

3. $x + y - z = -1$
   $x - y + z = -1$
   $-x + y + z = -1$

4. $x + y + z = 5$
   $3x + y - 3z = 5$
   $x + 3y - 3z = 5$

5. $x + 23 = 2y$
   $y + 23 = 3z$
   $z + 23 = 4x$

6. $3x + 4y + z = 2$
   $2x - 2y - 3z = -2$
   $x - 2y - z = 0$

7. $x - y + 3z = 15$
   $2x + 2y - z = -23$
   $3x - y + 2z = 8$

8. $2x - 3y + 2z = -47$
   $x + 2y + 2z = 3$
   $3x - y - z = 16$

9. $w + x + y = 27$
   $x + y + z = 31$
   $w + y + z = 34$
   $w + x + z = 37$

10. $w + x + y = 1$
    $x + y + z = 1$
    $w + y + z = 1$
    $w + x + z = 1$

LINEAR COMBINATION METHOD

Example 2. Solve

$$4x - y + 3z = -4$$
$$2x + 2y - 7z = -3$$
$$5x + 2y - 2z = -1.$$

Solution. We use the method of linear combinations to eliminate a variable. We can choose to eliminate any variable, so we will choose first to eliminate y. We multiply the first equation by 2 and the second by 1 to obtain

$$8x - 2y + 6z = -8$$
$$2x + 2y - 7z = -3$$

Adding these equations gives us

$$(*) \qquad 10x - z = -11$$

Now, multiplying the second equation by 1 and the third equation by -1, we obtain

$$2x + 2y - 7z = -3$$
$$-5x - 2y + 2z = 1$$

Adding these equations, we have

$$(*) \qquad -3x - 5z = -2$$

The two equations marked (*) form a system of two linear equations in two variables x and z. We now solve this system of equations. Multiplying the first (*) equation by 5 and the second (*) equation by -1, we obtain

$$50x - 5z = -55$$
$$3x + 5z = 2$$

Adding, we find

$$53x = -53$$
$$x = -1$$

Now we continue to find $z = 1$ and $y = 3$. We check this answer in each equation, and we have the solution $x = -1$, $y = 3$, $z = 1$.

Exercise Set 9.4 (continued)

In exercises 11–18, use the linear combination method to solve each given system of equations.

11. $3x - y + 2z = 5$
    $2x + 3y + z = 1$
    $5x + y + 4z = 8$
12. $5x + 3y - z = 3$
    $3x + y + 5z = -3$
    $x - 5y - z = -1$
13. $2y + 3z = 1$
    $2x + 4z = 6$
    $3x + 4y = -1$
14. $4x + 5y = -3$
    $2x + z = 2$
    $3y - 2z = -9$
15. $3x - 5y + 2z = 0$
    $x + 2y - 3z = 0$
    $2x - 2y + z = 4$
16. $3x - y + 2z = 1$
    $2x + 3y + 3z = -2$
    $3x - 2y - 3z = 4$
17. $w + x + y + z = 2$
    $w - x + y - z = -2$
    $w + x - y - z = 2$
    $w - x - y + z = 2$
18. $w + x + y + z = 0$
    $w - x + y - z = 2$
    $w + x - y - z = 0$
    $w - x - y + z = 0$

CLASSIFICATION OF SOLUTIONS

Just as with a system of two equations in two variables, we classify larger systems of linear equations as simultaneous, inconsistent, and dependent.

DEFINITION OF SIMULTANEOUS, INCONSISTENT, AND DEPENDENT EQUATIONS. A system of linear equations is *simultaneous* if it has a single solution; it is *inconsistent* if it has no solution; it is dependent if it has infinitely many solutions.

Example 3. Solve

$$\begin{aligned} 3x + y + 2z &= 2 \\ -2x - 3y + z &= -6 \\ x - y + 2z &= -2. \end{aligned}$$

Solution. We solve the equations by linear combinations, first eliminating z. If we multiply the first equation by 1, multiply the second equation by -2, and add, we obtain

$$7x + 7y = 14 \quad \text{or} \quad x + y = 2.$$

If we multiply the first equation by 1, multiply the third equation by -1, and add, we obtain

$$2x + 2y = 4, \quad \text{or} \quad x + y = 2.$$

We have obtained the same equation. If we multiply the second equation by -2, multiply the third equation by 1, and add, we obtain

$$5x + 5y = 10, \quad \text{or} \quad x + y = 2.$$

This system is dependent. We will show that there are infinitely many solutions. We have $y = 2 - x$ and, from the second equation,

$$z = 2x + 3y - 6 = 2x + 3(2 - x) - 6 = -x.$$

You can check that the infinite set of solutions where $y = 2 - x$ and $z = -x$ satisfies each equation. Therefore, the system of equations is dependent.

Exercise Set 9.4 (continued)

In exercises 19–24, classify each given system of equations as simultaneous, dependent, or inconsistent.

19. $\begin{aligned} x + 2y + z &= 1 \\ x \qquad + z &= 1 \\ x - 2y + z &= 1 \end{aligned}$

20. $\begin{aligned} 5x + 2y + 3z &= 1 \\ 3x \qquad + 5z &= -4 \\ 3x + 2y - 2z &= 5 \end{aligned}$

21. $\begin{aligned} 2x + 3y + 4z &= 1 \\ 4x + 9y + 16z &= 1 \\ 8x + 27y + 64z &= 1 \end{aligned}$

22. $\begin{aligned} 2x - 2y + z &= 0 \\ 3x + 3y + 6z &= 4 \\ 6x + 2y + z &= 0 \end{aligned}$

23. $\begin{aligned} x - y + 5z &= 1 \\ 2x + 5y - 4z &= -1 \\ 3x + 4y + z &= 4 \end{aligned}$

24. $\begin{aligned} 2x + y + 3z &= 5 \\ 2x - y + 2z &= 3 \\ 6x - 7y + 4z &= 5 \end{aligned}$

section 5 • Determinants

Now we define matrices and show how to find the determinant of a matrix.

MATRICES

DEFINITION OF A MATRIX. A *matrix* is a rectangular array of numbers. A matrix is an m by n matrix if it has m rows and n columns. A matrix is a *square matrix* if it has the same number of rows as columns. We identify an *entry* in a matrix by specifying its row and column position. A standard notation for matrices is to enclose the array of numbers with square brackets.

Example 1. $\begin{bmatrix} 1 & 2 & 1 \\ -2 & -1 & -2 \\ 4 & -3 & 6 \end{bmatrix}$

is a 3 by 3 matrix, and so it is a square matrix. The entry in the first row and first column is 1; the entry in the third row and first column is 4; the entry in the second row and third column is -2; and so on.

Given a matrix, we will consider the matrix obtained by omitting a row or a column. For the matrix of Example 1, if we omit the first row and first column we obtain the 2 by 2 matrix

$$\begin{bmatrix} -1 & -2 \\ -3 & 6 \end{bmatrix}$$

If we omit the first row and second column, we obtain the 2 by 2 matrix

$$\begin{bmatrix} -2 & -2 \\ 4 & 6 \end{bmatrix}$$

If we omit the first row and third column, we obtain the 2 by 2 matrix

$$\begin{bmatrix} -2 & -1 \\ 4 & -3 \end{bmatrix}$$

Exercise Set 9.5

In exercises 1–6, do the following for the given 3 by 3 matrix: a. specify the entry in the first row and first column; b. write the matrix obtained by omitting the first row and first column; c. specify the entry in the first row and third column; d. write the matrix obtained by omitting the first row and third column; e. specify the entry in the second row and second column; f. write the matrix obtained by omitting the second row and second column.

1. $\begin{bmatrix} 3 & 1 & 1 \\ 2 & -2 & -3 \\ 1 & 4 & 2 \end{bmatrix}$
2. $\begin{bmatrix} 5 & 1 & 1 \\ -5 & -2 & -3 \\ 3 & 4 & 2 \end{bmatrix}$
3. $\begin{bmatrix} 1 & -1 & 5 \\ 2 & 5 & -4 \\ 3 & 4 & 1 \end{bmatrix}$
4. $\begin{bmatrix} 1 & -1 & 1 \\ -1 & 1 & -1 \\ 1 & -1 & 1 \end{bmatrix}$
5. $\begin{bmatrix} 2 & -3 & 0 \\ 0 & 2 & -3 \\ -3 & 0 & 2 \end{bmatrix}$
6. $\begin{bmatrix} 2 & 4 & 0 \\ 0 & 5 & -3 \\ -3 & -3 & 2 \end{bmatrix}$

TWO BY TWO DETERMINANTS

Now we define the determinant of a 2 by 2 matrix. The determinant is a number associated with the matrix.

DETERMINANT OF A 2 BY 2 MATRIX. For a 2 by 2 matrix, the determinant is

$$\det \begin{bmatrix} a & b \\ c & d \end{bmatrix} = ad - bc.$$

Example 2. Determinants.

a. $\det \begin{bmatrix} 1 & -2 \\ 3 & 6 \end{bmatrix} = (1)(6) - (-2)(3) = 12.$

b. $\det \begin{bmatrix} 4 & 0 \\ -1 & -2 \end{bmatrix} = 4(-2) - (0)(-1) = -8.$

c. $\det \begin{bmatrix} 2 & -6 \\ 3 & -9 \end{bmatrix} = 2(-9) - (-6)(3) = 0.$

Exercise Set 9.5 (continued)

In exercises 7–18, find the determinant of each given 2 by 2 matrix.

7. $\begin{bmatrix} 1 & 1 \\ 3 & 2 \end{bmatrix}$
8. $\begin{bmatrix} 12 & 1 \\ 34 & 2 \end{bmatrix}$
9. $\begin{bmatrix} 1 & 12 \\ 3 & 34 \end{bmatrix}$
10. $\begin{bmatrix} 3 & -2 \\ 4 & 5 \end{bmatrix}$
11. $\begin{bmatrix} 5 & -15 \\ 7 & -21 \end{bmatrix}$
12. $\begin{bmatrix} 2 & -15 \\ -3 & -21 \end{bmatrix}$
13. $\begin{bmatrix} 4 & -3 \\ 7 & 2 \end{bmatrix}$
14. $\begin{bmatrix} 23 & -3 \\ 4 & 2 \end{bmatrix}$

15. $\begin{bmatrix} 4 & 23 \\ 7 & 4 \end{bmatrix}$ 16. $\begin{bmatrix} 4 & -2 \\ 10 & -5 \end{bmatrix}$ 17. $\begin{bmatrix} 6 & -2 \\ 15 & -5 \end{bmatrix}$ 18. $\begin{bmatrix} 4 & 6 \\ 10 & 15 \end{bmatrix}$

DETERMINANT OF A SQUARE MATRIX

Now we discuss the determinants of larger square matrices. We will illustrate the concepts involved with 3 by 3 matrices, but the concepts apply to any n by n matrix.

The method we use to evaluate the determinant of a matrix is called *expansion by minors.* A loose description of expansion by minors is that we select any row or any column in the matrix and expand along the row or down the column. We multiply each entry in the row or column by either $+1$ or -1 and then by the *minor* of that entry. So we now define the minor of an entry.

DEFINITION OF THE MINOR OF AN ENTRY IN A MATRIX. The *minor* of an entry in an n by n matrix is the determinant of the $(n - 1)$ by $(n - 1)$ matrix obtained by omitting the row and column in which the entry occurs.

Example 3. Consider the 3 by 3 matrix

$$\begin{bmatrix} 1 & 2 & 1 \\ -2 & -1 & -2 \\ 4 & -3 & 6 \end{bmatrix}$$

a. The minor of the entry in the first row and first column is

$$\det \begin{bmatrix} -1 & -2 \\ -3 & 6 \end{bmatrix} = (-1)(6) - (-2)(-3) = -12.$$

b. The minor of the entry in the third row and first column is

$$\det \begin{bmatrix} 2 & 1 \\ -1 & -2 \end{bmatrix} = 2(-2) - (1)(-1) = -3.$$

c. The minor of the entry in the second row and second column is

$$\det \begin{bmatrix} 1 & 1 \\ 4 & 6 \end{bmatrix} = (1)(6) - (1)(4) = 2.$$

Besides the minors of entries in a matrix, we consider an array of plus and minus signs associated with a matrix. The plus and minus signs alternate along each row and column, always starting with a plus in the first row and first column. This array is shown below for a 3 by 3 matrix and a 4 by 4 matrix. The array of plus and minus signs is independent of the fact that the entries in the matrix are themselves positive or negative.

$$\begin{bmatrix} + & - & + \\ - & + & - \\ + & - & + \end{bmatrix} \qquad \begin{bmatrix} + & - & + & - \\ - & + & - & + \\ + & - & + & - \\ - & + & - & + \end{bmatrix}$$

Now we describe how to evaluate the determinant of a square matrix. We evaluate the determinant along any row of the matrix or down any column of the matrix. So we start by picking a row or column. For each entry in the row or column, we form the product of the entry, a $+1$ or -1 from the array of signs, and the minor of that entry. The determinant of the matrix is the sum of these products.

Example 3 (continued). Evaluate

$$\det \begin{bmatrix} 1 & 2 & 1 \\ -2 & -1 & -2 \\ 4 & -3 & 6 \end{bmatrix}$$

Solution.

a. If we evaluate along the first row, we obtain

$$1(1)\det\begin{bmatrix} -1 & -2 \\ -3 & 6 \end{bmatrix} + 2(-1)\det\begin{bmatrix} -2 & -2 \\ 4 & 6 \end{bmatrix} + 1(1)\det\begin{bmatrix} -2 & -1 \\ 4 & -3 \end{bmatrix}$$
$$= 1[(-1)(6) - (-2)(-3)] - 2[(-2)(6) - (-2)(4)] + 1[(-2)(-3) - (-1)(4)]$$
$$= 1(-12) - 2(-4) + 1(10) = 6.$$

b. If we evaluate down the first column, we obtain

$$1(1)\det\begin{bmatrix} -1 & -2 \\ -3 & 6 \end{bmatrix} - 2(-1)\det\begin{bmatrix} 2 & 1 \\ -3 & 6 \end{bmatrix} + 4(1)\det\begin{bmatrix} 2 & 1 \\ -1 & -2 \end{bmatrix}$$
$$= 1[(-1)(6) - (-2)(-3)] + 2[(2)(6) - (1)(-3)] + 4[(2)(-2) - (1)(-1)]$$
$$= 1(-12) + 2(15) + 4(-3) = 6.$$

c. If we evaluate along the third row, we obtain

$$4(1)\det\begin{bmatrix} 2 & 1 \\ -1 & -2 \end{bmatrix} - 3(-1)\det\begin{bmatrix} 1 & 1 \\ -2 & -2 \end{bmatrix} + 6(1)\det\begin{bmatrix} 1 & 2 \\ -2 & -1 \end{bmatrix}$$

$$= 4[(2)(-2) - (1)(-1)] + 3[(1)(-2) - (1)(-2)] + 6[(1)(-1) - (2)(-2)]$$
$$= 4(-3) + 3(0) + 6(3) = 6.$$

Example 3 illustrates the fact that we obtain the same result in each case. The basic property of expansion by minors is that we obtain the same answer for the determinant no matter which row or column we choose.

Expansion by minors can be used to evaluate the determinant of a 4 by 4 matrix. The minors of each entry are now determinants of 3 by 3 matrices.

Exercise Set 9.5 (continued)

In exercises 19–34, find the determinant of each given square matrix.

19. $\begin{bmatrix} 3 & 1 & 1 \\ 2 & -2 & -3 \\ 1 & 4 & 2 \end{bmatrix}$

20. $\begin{bmatrix} 5 & 1 & 1 \\ -5 & -2 & -3 \\ 3 & 4 & 2 \end{bmatrix}$

21. $\begin{bmatrix} 1 & -1 & 5 \\ 2 & 5 & -4 \\ 3 & 4 & 1 \end{bmatrix}$

22. $\begin{bmatrix} 1 & -1 & 1 \\ -1 & 1 & -1 \\ 1 & -1 & 1 \end{bmatrix}$

23. $\begin{bmatrix} 2 & -3 & 0 \\ 0 & 2 & -3 \\ -3 & 0 & 2 \end{bmatrix}$

24. $\begin{bmatrix} 2 & 4 & 0 \\ 0 & 5 & -3 \\ -3 & -3 & 2 \end{bmatrix}$

25. $\begin{bmatrix} 0 & 2 & 3 \\ 2 & 0 & 4 \\ 3 & 4 & 0 \end{bmatrix}$

26. $\begin{bmatrix} 0 & 2 & 0 \\ 2 & 0 & 1 \\ 3 & 4 & -1 \end{bmatrix}$

27. $\begin{bmatrix} 5 & 2 & 3 \\ 3 & 0 & 5 \\ 2 & 2 & -2 \end{bmatrix}$

28. $\begin{bmatrix} 1 & 2 & 3 \\ -4 & 0 & 5 \\ 5 & 2 & -2 \end{bmatrix}$

29. $\begin{bmatrix} 2 & 1 & 3 \\ 2 & -1 & 2 \\ 6 & -7 & 4 \end{bmatrix}$

30. $\begin{bmatrix} 2 & 5 & 3 \\ 2 & 3 & 2 \\ 6 & 5 & 4 \end{bmatrix}$

31. $\begin{bmatrix} 1 & 1 & 1 & 0 \\ 0 & 1 & 1 & 1 \\ 1 & 0 & 1 & 1 \\ 1 & 1 & 0 & 1 \end{bmatrix}$

32. $\begin{bmatrix} 1 & 1 & 1 & 0 \\ 1 & 1 & 1 & 1 \\ 1 & 0 & 1 & 1 \\ 1 & 1 & 0 & 1 \end{bmatrix}$

33. $\begin{bmatrix} 0 & 1 & 1 & 1 \\ 2 & -1 & 1 & -1 \\ 0 & 1 & -1 & -1 \\ 0 & -1 & -1 & 1 \end{bmatrix}$

34. $\begin{bmatrix} 1 & 2 & -1 & 1 \\ 3 & 1 & -1 & 1 \\ 2 & 2 & 1 & -2 \\ 1 & 3 & 3 & 2 \end{bmatrix}$

section 6 • Cramer's Rule

In this section we show how to use determinants to solve a system of n linear equations in n variables. The method we use is called *Cramer's Rule.*

TWO EQUATIONS IN TWO VARIABLES

We start with the case of two linear equations in two variables, say

$$a_1x + b_1y = c_1$$
$$a_2x + b_2y = c_2.$$

We eliminate y from these equations by the method of linear combinations (Section 9.1). We multiply the first equation by b_2 and the second equation by $-b_1$.

$$a_1b_2x + b_1b_2y = c_1b_2$$
$$-a_2b_1x - b_1b_2y = -c_2b_1.$$

Adding, we obtain

$$(a_1b_2 - a_2b_1)x = c_1b_2 - c_2b_1.$$

Now it is possible to have $a_1b_2 - a_2b_1 = 0$. If this happens, however, then the equations are either inconsistent or dependent. Under the assumption that $a_1b_2 - a_2b_1 \neq 0$, we obtain the unique solution for x,

$$x = \frac{c_1b_2 - c_2b_1}{a_1b_2 - a_2b_1}.$$

Now we eliminate x from the equations by multiplying the first equation by $-a_2$, multiplying the second equation a_1, and adding. We obtain

$$(a_1b_2 - a_2b_1)y = a_1c_2 - a_2c_1.$$

Again we obtain the coefficient $a_1b_2 - a_2b_1$. Assuming that $a_1b_2 - a_2b_1 \neq 0$, we obtain the unique solution for y,

$$y = \frac{a_1c_2 - a_2c_1}{a_1b_2 - a_2b_1}.$$

The significance of this derivation is that we can write both x and y as a quotient of two determinants. The determinant in the denominator is the same in each case; it is the determinant of the matrix of the coefficients of x and y from the equations. We let

$$D = \det\begin{bmatrix} a_1 & b_1 \\ a_2 & b_2 \end{bmatrix}, \quad D_x = \det\begin{bmatrix} c_1 & b_1 \\ c_2 & b_2 \end{bmatrix}, \quad D_y = \det\begin{bmatrix} a_1 & c_1 \\ a_2 & c_2 \end{bmatrix}.$$

Now D_x is the numerator for x and is the determinant of the matrix obtained by replacing the first column of the coefficient matrix by the constants. Also, D_y, the numerator for y, is the determinant of the matrix obtained by replacing the second column of the coefficient matrix by the constants. We have assumed $D \neq 0$; if $D = 0$, the system is either inconsistent or dependent. This is Cramer's Rule for a system of two linear equations in two variables.

Example 1. Solve
$$\begin{aligned} 7x + 2y &= 6 \\ 3x - y &= -2. \end{aligned}$$

Solution. We use Cramer's Rule:

$$D = \det\begin{bmatrix} 7 & 2 \\ 3 & -1 \end{bmatrix} = -13, \qquad D_x = \det\begin{bmatrix} 6 & 2 \\ -2 & -1 \end{bmatrix} = -2,$$

$$D_y = \det\begin{bmatrix} 7 & 6 \\ 3 & -2 \end{bmatrix} = -32.$$

$$x = \frac{D_x}{D} = \frac{-2}{-13} = \frac{2}{13} \text{ and } y = \frac{D_y}{D} = \frac{-32}{-13} = \frac{32}{13}.$$

The simultaneous solution is (2/13, 32/13).

Exercise Set 9.6

In exercises 1–8, use Cramer's Rule to solve each given system of equations.

1. $3x - 4y = 1$
   $4x + 3y = 0$
2. $3x - 4y = 0$
   $4x + 3y = 1$
3. $3x - 2y = 11$
   $2x + 3y = 3$
4. $3x - 2y = 7$
   $2x + 3y = -9$
5. $3x + y = 8$
   $2x - 3y = -2$
6. $x + 2y = 7$
   $3x - y = 0$
7. $3x - 6y = -4$
   $x - y = -1$
8. $4x - 2y = 2$
   $2x - 5y = 3$

N EQUATIONS IN N VARIABLES

Cramer's Rule can be extended to the solution of n linear equations in n variables. Suppose, for example, that we have three linear equations in three variables, say x, y, and z. Then

$$x = \frac{D_x}{D}, \qquad y = \frac{D_y}{D}, \qquad z = \frac{D_z}{D}.$$

Here D is the determinant of the matrix of the coefficients of x, y, and z. D_x is the determinant of the matrix obtained by replacing the column of coefficients of x by the constants; D_y is the determinant of the matrix obtained by replacing the column of coefficients of y by the constants; D_z is the determinant of the matrix obtained by replacing the column of coefficients of z by the constants. We have assumed here that $D \neq 0$; if $D = 0$, the system is either inconsistent or dependent.

Example 2. Solve

$$\begin{aligned} 2x + 3y - z &= 1 \\ x - 4y + z &= 0 \\ 3x - 2y - 2z &= -2. \end{aligned}$$

Solution. Using Cramer's Rule, we have

$$D = \det\begin{bmatrix} 2 & 3 & -1 \\ 1 & -4 & 1 \\ 3 & -2 & -2 \end{bmatrix} = 25, \quad D_x = \det\begin{bmatrix} 1 & 3 & -1 \\ 0 & -4 & 1 \\ -2 & -2 & -2 \end{bmatrix} = 12,$$

$$D_y = \det\begin{bmatrix} 2 & 1 & -1 \\ 1 & 0 & 1 \\ 3 & -2 & -2 \end{bmatrix} = 11, \quad D_z = \det\begin{bmatrix} 2 & 3 & 1 \\ 1 & -4 & 0 \\ 3 & -2 & -2 \end{bmatrix} = 32.$$

$$x = \frac{D_x}{D} = \frac{12}{25}, y = \frac{D_y}{D} = \frac{11}{25}, z = \frac{D_z}{D} = \frac{32}{25}.$$

Exercise Set 9.6 (continued)

In exercises 9–16, use Cramer's Rule to solve each given system of equations.

9. $\begin{aligned} x + y \quad &= 1 \\ x \quad + z &= 2 \\ y + z &= 3 \end{aligned}$

10. $\begin{aligned} x + y \quad &= -1 \\ x \quad + z &= -2 \\ y + z &= -3 \end{aligned}$

11. $\begin{aligned} -x + y + z &= 1 \\ x - y + z &= 1 \\ x + y - z &= 1 \end{aligned}$

12. $\begin{aligned} -x + y + z &= -1 \\ x - y + z &= -1 \\ x + y + z &= -1 \end{aligned}$

13. $\begin{aligned} x + y + z &= 1 \\ 2x + 3y + 4z &= 1 \\ 4x + 9y + 16z &= 1 \end{aligned}$

14. $\begin{aligned} x + y + z &= 1 \\ x + 2y + 3z &= 1 \\ x + 4y + 9z &= 1 \end{aligned}$

15. $\begin{aligned} x + 2y + 4z &= 1 \\ x + 3y + 9z &= 2 \\ x + 4y + 16z &= 3 \end{aligned}$

16. $\begin{aligned} x + y + z &= 1 \\ x + 2y + 4z &= 2 \\ x + 3y + 9z &= 4 \end{aligned}$

In exercises 17–22, find the determinant of each given matrix.

17. $\begin{bmatrix} -a & b & c \\ a & -b & c \\ a & b & -c \end{bmatrix}$

18. $\begin{bmatrix} 0 & b & c \\ a & 0 & c \\ a & b & 0 \end{bmatrix}$

19. $\begin{bmatrix} 0 & b & a \\ a & 0 & b \\ a & b & c \end{bmatrix}$

20. $\begin{bmatrix} 0 & a & c \\ a & b & c \\ a & c & 0 \end{bmatrix}$

21. $\begin{bmatrix} 1 & 1 & 1 \\ a & b & c \\ a^2 & b^2 & c^2 \end{bmatrix}$

22. $\begin{bmatrix} a & b & c \\ a^2 & b^2 & c^2 \\ a^3 & b^3 & c^3 \end{bmatrix}$

Frequently determinants are useful for solving systems of linear equations where the coefficients or constants are letters and not specific numbers. In exercises 23–32, use Cramer's Rule to solve each system of equations.

23. $ax - by = a - b$
$bx + ay = a + b$

24. $ax - by = a + b$
$bx + ay = a - b$

25. $a(x + y) + b(x - y) = 2a$
$a(x - y) + b(x + y) = 2b$

26. $a\,x + b\,y = 1$
$a^2x + b^2y = 1$

27. $-ax + by + cz = 1$
$ax - by + cz = 1$
$ax + by - cz = 1$

28. $by + cz = a$
$ax + cz = b$
$ax + by = c$

29. $x + y + z = 1$
$ax + b\,y + c\,z = d$
$a^2x + b^2y + c^2z = d^2$

30. $a\,x + b\,y + c\,z = d$
$a^2x + b^2y + c^2z = d^2$
$a^3x + b^3y + c^3z = d^3$

31. $x + ay + a^2z = 1$
$x + by + b^2z = 1$
$x + cy + c^2z = 1$

32. $x + ay + a^2z = a^3$
$x + by + b^2z = b^3$
$x + cy + c^2z = c^3$

---

section 7 • Applications

We give applications of systems of equations and inequalities.

RATE PROBLEMS

Example 1. An airplane flies 1200 miles with the wind in 3 hours and then flies 800 miles against the wind in 2.5 hours. Find the speed of the plane in still air and the speed of the wind.

Solution. We wish to solve for two speeds, so we set up a system of two equations in two variables.

Let x denote the speed of the plane in still air and let y denote the speed of the wind. Then the speed of the plane with the wind is $x + y$, and it is also 1200/3. So we have the equation

$$x + y = \frac{1200}{3} = 400.$$

The speed of the plane against the wind is $x - y$, and is also 800/2.5. We have the second equation

$$x - y = \frac{800}{2.5} = 320.$$

When we add these two equations, we obtain

$$2x = 720$$
$$x = 360$$
$$y = 400 - 360 = 40$$

So the speed of the plane in still air is 360 mph, and the speed of the wind is 40 mph.

Exercise Set 9.7

1. A boat travels 30 miles downstream on a river in 6 hours and makes the return trip in 10 hours. Find the speed of the boat in still water and the speed of the water.
2. A plane goes 1200 miles with the wind in 3 hours and makes the return trip in 4 hours. Find the speed of the plane in still air and the speed of the wind.
3. A plane goes 260 kilometers with the wind in 2 hours and then goes 250 kilometers against the wind in 2.5 hours. Find the speed of the plane in still air and the speed of the wind.
4. A boat travels 50 kilometers downstream on a river in 2 hours and then travels 27 kilometers upstream in 1.5 hours. Find the speed of the boat in still water and the speed of the water.
5. When a train's speed was increased by 10 mph, the time for a 300-mile run was cut by 1 hour. Find the original speed and time of the train.
6. Two boats cover the same 60-mile course. One boat has a speed 5 mph faster than the other and covers the course in 2 hours less. Find the rate and time of the faster boat.
7. A tank is filled by two pipes which together fill it in 20 minutes. The larger pipe alone can fill the tank in 30 minutes less than the smaller pipe alone. How long does it take each pipe alone to fill the tank?
8. A pool is filled by two pipes which together fill it in 2 hours and 24 minutes. The larger pipe alone can fill the pool in 2 hours less than the smaller pipe alone. How long does it take each pipe alone to fill the pool?

GEOMETRY PROBLEMS

Example 2. A rectangle has perimeter 32 meters and area 28 square meters. Find the sides of the rectangle.

Solution. Let x denote the length of one pair of opposite sides and let y denote the length of the other pair of opposite sides. To solve for these two variables we look for a system of two equations in x and y. Because we know the perimeter, we have the equation

$$2x + 2y = 32.$$

Because we know the area, we have the equation

$$xy = 28.$$

Solving for y in the first equation, we find

$$y = \frac{32 - 2x}{2} = 16 - x$$

We substitute this expression for y in the second equation:

$$x(16 - x) = 28$$

$$16x - x^2 = 28$$

$$x^2 - 16x + 28 = 0$$

$$(x - 14)(x - 2) = 0$$

$$x = 14 \quad \text{or} \quad x = 2$$

Using $x = 14$ we obtain $y = 2$, and using $x = 2$ we obtain $y = 14$. Therefore, the rectangle is 14 meters by 2 meters.

Exercise Set 9.7 (continued)

9. Find the sides of a rectangle if the perimeter is 32 meters and the area is 48 square meters.
10. Find the sides of a rectangle if the perimeter is 22 centimeters and the area is 28 square centimeters.
11. Find the sides of a rectangle if the perimeter is 13 kilometers and the area is 10 square kilometers.
12. Find the sides of a rectangle if the perimeter is 19 meters and the area is 12 square meters.
13. Find the sides of a right triangle if the hypotenuse is 10 meters and the area is 24 square meters.
14. Find the sides of a right triangle if the hypotenuse is 2.5 centimeters and the area is 3 square centimeters.
15. The perimeter of a rectangle is 7 and the diagonal is 5/2. Find the length and width of the rectangle.
16. The perimeter of a rectangle is 28 and the diagonal is 10. Find the length and width of the rectangle.
17. A rectangle has diagonal 13. If the width is increased by 4, the diagonal is 15. Find the length and width of the rectangle.
18. A rectangle has diagonal 15. If the width is increased by 3 and the length is decreased by 3, the diagonal is still 15. Find the length and width of the rectangle.

DIET PROBLEMS

Example 3. We prepare an animal feed by mixing two basic foodstuffs, a less expensive foodstuff and a more expensive foodstuff. To insure the health and growth of the animal, the resulting mixture must contain at least 20 grams of protein and 36 grams of carbohydrates. If each unit of the less expensive foodstuff contains 3 grams of protein and 3 grams of carbohydrates and each unit of the more expensive foodstuff contains 2 grams of protein and 5 grams of carbohydrates, graph the region of the plane that satisfies these requirements.

Solution. Suppose we mix x units of the less expensive foodstuff with y units of the more expensive foodstuff. Then the mixture contains $3x + 2y$ grams of protein and $3x + 5y$

grams of carbohydrates. So the requirements are expressed by the system of inequalities:

$$\begin{aligned} \text{proteins:} \quad & 3x + 2y \geq 20 \\ \text{carbohydrates:} \quad & 3x + 5y \geq 36 \\ & x \geq 0 \\ & y \geq 0 \end{aligned}$$

The inequalities $x \geq 0$ and $y \geq 0$ say that we cannot mix a negative amount of foodstuffs. The graph of this system of inequalities is Figure 9.10.

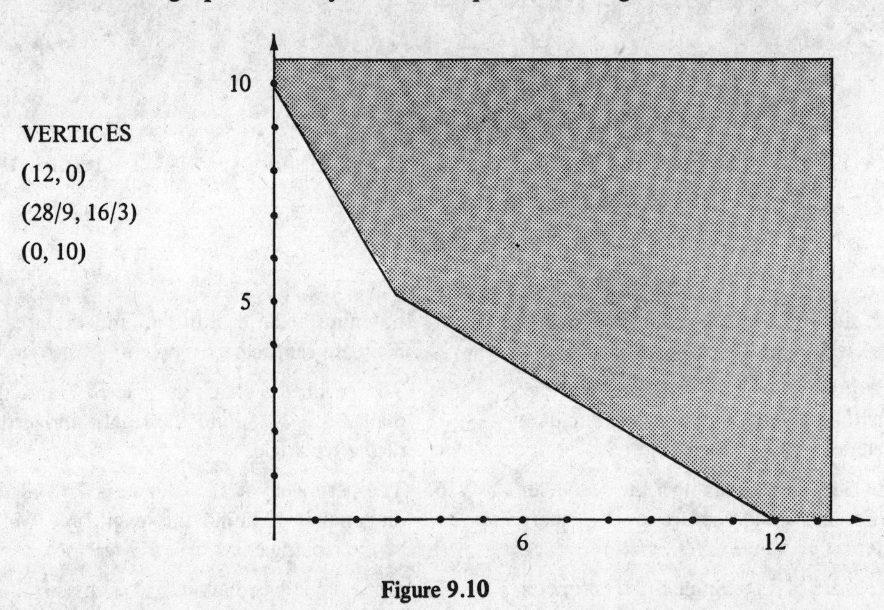

Figure 9.10

Exercise Set 9.7 (continued)

19. We prepare an animal feed by mixing a less expensive foodstuff with a more expensive foodstuff. The mixture must contain at least 24 grams of protein and 60 grams of carbohydrates. Graph the region of the plane satisfying these requirements if each unit of the less expensive foodstuff contains 2 grams of protein and 6 grams of carbohydrates and each unit of the more expensive foodstuff contains 2 grams of protein and 4 grams of carbohydrates. (Let x denote the number of units of the less expensive foodstuff and y denote the number of units of the more expensive foodstuff.)

20. We prepare an animal feed by mixing a less expensive foodstuff with a more expensive foodstuff. The mixture must contain at least 15 grams of protein and 15 grams of fats. Graph the region of the plane that satisfies these requirements if each unit of the less expensive foodstuff contains 3 grams of protein and 2 grams of fats and each unit of the more expensive foodstuff contains 2 grams of protein and 3 grams of fats. (Let x denote the number of units of the less expensive foodstuff and y denote the number of units of the more expensive foodstuff.)

21. A food company plans to come out with a drink consisting of a mixture of orange juice and pineapple juice. To advertise the nutritional value of the drink, the management requires each can to contain at least 12 milligrams of Vitamin A and 30 mg of Vitamin C. Graph the region of the plane that satisfies these requirements if each fluid ounce of orange juice contains 1 mg of Vitamin A and 2 mg of Vitamin C and each fluid ounce of pineapple juice contains 1 mg of Vitamin A and 3 mg of Vitamin C. (Let x denote the number of fluid ounces of orange juice and y the number of fluid ounces of pineapple juice.)
22. A food company plans to come out with a drink consisting of a mixture of orange juice and pineapple juice. To advertise the nutritional value of the drink, the management requires each can to contain at least 12 milligrams of Vitamin A and 21 mg of Vitamin D. Graph the region of the plane satisfying these requirements if each fluid ounce of orange juice contains 1 mg of Vitamin A and 1 mg of Vitamin D and each fluid ounce of pineapple juice contains 1 mg of Vitamin A and 3 mg of Vitamin D. (Let x denote the number of fluid ounces of orange juice and y the number of fluid ounces of pineapple juice.)
23. Suppose that in exercise 19, in addition to the protein and carbohydrate requirements, the mixture must cost at most \$60. Graph the region of the plane satisfying all these requirements if each unit of the less expensive foodstuff costs \$4 and each unit of the more expensive foodstuff costs \$5.
24. Suppose that in exercise 20, in addition to the protein and fats requirements, the mixture must cost at most \$40. Graph the region of the plane that satisfies all these requirements if each unit of the less expensive foodstuff costs \$4 and each unit of the more expensive foodstuff costs \$5.
25. Suppose that in exercise 21, in addition to the Vitamin A and Vitamin C requirements, the mixture must cost at most \$0.60. Graph the region of the plane that satisfies all these requirements if each fluid ounce of orange juice costs \$0.04 and each fluid ounce of pineapple juce costs \$0.05.
26. Suppose that in exercise 22, in addition to the Vitamin A and Vitamin D requirements, the mixture must cost at most \$0.54. Graph the region of the plane that satisfies all these requirements if each fluid ounce of orange juice costs \$0.03 and each fluid ounce of pineapple juice costs \$0.03.

MIXTURE PROBLEMS

Example 4. We mix a solution containing 35% alcohol with a solution containing 65% alcohol to obtain a solution containing 40% alcohol. How many liters of each solution do we mix to obtain 20 liters of the 40% solution?

Solution. Let x denote the number of liters of the 35% alcohol solution in the mixture, and let y denote the number of liters of the 65% alcohol solution in the mixture. We must find two equations in x and y. Since $x + y$ is the total number of liters in the mixture, we have

$$x + y = 20.$$

The number of liters of alcohol in the mixture is $0.35x + 0.65y$ which is equal to $0.40(20)$. So the second equation is

$$0.35x + 0.65y = 0.40(20) = 8.$$

Solving for y in the first equation, we have

$$y = 20 - x.$$

Substituting this expression for y in the second equation, we obtain

$$\begin{aligned} 0.35x + 0.65(20 - x) &= 8 \\ 0.35x + 13 - 0.65x &= 8 \\ -0.3x &= -5 \\ x &= 50/3 \\ y &= 20 - (50/3) = 10/3 \end{aligned}$$

Therefore, we mix 50/3 liters of the 35% solution with 10/3 liters of the 65% solution.

Exercise Set 9.7 (continued)

27. We mix a solution containing 40% alcohol with a solution containing 75% alcohol. How many liters of each solution do we mix to obtain 14 liters of a mixture containing 50% alcohol?

28. We mix a solution containing 15% alcohol with a solution containing 55% alcohol. How many liters of each solution do we mix to obtain 20 liters of a mixture containing 30% alcohol?

29. We mix a solution containing 12% salt with a solution containing 28% salt. How many liters of each solution do we mix to obtain 10 liters of a mixture containing 24% salt?

30. We mix a solution containing 15% salt with a solution containing 24% salt. How many liters of each solution do we mix to obtain 20 liters of a mixture containing 18% salt?

Exercises 31–34 pertain to the octane of a gasoline. A good gasoline must burn smoothly in an engine without detonating (called *knocking*). Isooctane in gasoline prevents knocking, while normal heptane is prone to knocking. The higher the octane of the gasoline, the larger the percentage of isooctane it contains. For example, an octane of 90 means that the gasoline is equivalent to a mixture of 90% isooctane by volume to 10% normal heptane by volume.

31. We blend a gasoline with octane 95 with a gasoline with octane 85. How many liters of each gasoline must we blend to obtain 10 liters of a gasoline with octane 90?

32. We blend a gasoline with octane 94 with a gasoline with octane 86. How many liters of each gasoline must we blend to obtain 10 liters of a gasoline with octane 90?

33. We blend a gasoline with octane 100 with a gasoline with octane 90. How many liters of each gasoline must we blend to obtain 20 liters of a gasoline with octane 95?

34. We blend a gasoline with octane 98 with a gasoline with octane 92. How many liters of each gasoline must we blend to obtain 20 liters of a gasoline with octane 95?

HISTORICAL EXERCISES

Exercise Set 9.7 (continued)

Exercise 35 is from the *Greek Anthology*, a collection of problems assembled for Greek students about A.D. 500.

35. Make a crown of gold, copper, tin, and iron weighing 60 minae. Gold and copper together shall be two thirds of it, gold and tin three-fourths of it, gold and iron three-fifths of it, and copper and tin shall be one-fourth of it. Find the weights of gold, copper, tin, and iron required.

Exercise 36 is attributed to Euclid.

36. A mule and an ass went on their way with burdens of wineskins; weighed down by its load, the ass was bitterly groaning. The mule, hearing its complaints, addressed it: "Friend, why do you complain? If you give me one measure of wine, then I will carry twice your burden; but if you take one measure from me, then our loads will be equal." Tell me the measure they bore, good sir, geometry's master.

Exercises 37 and 38 are from the *Liber Abaci* of Leonard Fibonacci (1202).

37. There are four men. The first, second, and third together have 27 denarii; the second, third, and fourth together have 31 denarii; the third, fourth, and first together have 34 denarii; the fourth, first, and second together have 37 denarii. How many denarii does each man have?

38. Three men have denarii, and they find a purse containing 23 denarii. The first man said to the second, "If I take this purse, I will have twice as much as you." The second man said to the third, "If I take this purse, I will have three times as much as you. And the third man said to the first, "If I take this purse, I will have four times as much as you." How much did each man have?

Exercise 39 is from the Hindu mathematician Brahmagupta (ca. 630).

39. When a bamboo 18 cubits high was broken by the wind, its top touched the ground 6 cubits from the root. Tell the lengths of the segments of the bamboo.

Exercise 40 is adopted from Chuquet's *Triparty en la science des nombres* (1484).

40. Two wine merchants enter Paris, one of them with 64 casks of wine and the other with 20. Since they do not have enough cash to pay the customs duties, the first pays 5 casks of wine and 40 francs, and the second pays 2 casks of wine and receives 40 francs in change. Find the price of each cask of wine and the duty on each cask.

Review Exercises

In exercises 1–10, solve each given system of linear equations.

1. $x + y = 1$
   $x - y = 5$

2. $x + y = 3$
   $x - y = -5$

3. $2x + 3y = 2$
   $3x - 2y = -10$

4. $3x + 4y = 21$
   $4x - 3y = 3$

5. $5x - 3y = 2$
   $4x - y = -4$

6. $6x - 2y = -12$
   $5x - 3y = -14$

7. $3x + 9y = 4$
   $4x + 12y = 7$

8. $4x - 8y = 12$
   $5x - 10y = 15$

9. $3x + 5y = 4$
   $3x + y = 3$

10. $5x + y = -2$
    $3x + 7y = 2$

In exercises 11–16, graph each given system of inequalities.

11. $x + y \leq 2$
    $x - y \leq 2$

12. $2x + y \leq 4$
    $x \leq y$

13. $3x + y \leq 3$
    $x + 3y \geq 3$

14. $2x + 3y \geq 6$
    $3x - 2y \leq 6$

15. $2x + 3y \leq 12$
    $4x + 3y \leq 18$
    $x \geq 0$
    $y \geq 0$

16. $3x + 4y \geq 25$
    $x + y \geq 7$
    $x \geq 0$
    $y \geq 0$

In exercises 17–26, solve each system of equations.

17. $x + y = 10$
    $y = x^2 + 2x$

18. $x - y = -4$
    $y = x^2 - 2x$

19. $x + y = 2$
    $x^2 + y^2 = 2$

20. $x - y = 1$
    $x^2 + y^2 = 13$

21. $x^2 + y^2 = 5$
    $xy = 2$

22. $x^2 + y^2 = 2$
    $xy = 1$

23. $2x + 2y = 10$
    $xy = 6$

24. $2x + 2y = 16$
    $xy = 15$

25. $x^2 + y^2 = 4$
    $x^2 + y^2 = 2x - 2y$

26. $x^2 + y^2 = 4$
    $x^2 + y^2 = 2y - 2x$

In exercises 27–32, solve each system of linear equations.

27. $x + y + z = 6$
    $x - y + z = 2$
    $x + y - z = 4$

28. $-x + y + z = 2$
    $x - y + z = 2$
    $x + y - z = 2$

29. $x + y + z = 1$
    $x - y + z = 3$
    $x - y - z = 1$

30. $x - y - z = 5$
    $x + y - z = 3$
    $x - y + z = 1$

31. $x + y + z = 1$
    $2x + 2y + z = 1$
    $4x + 4y + z = 1$

32. $x - y + z = 2$
    $2x - y + 2z = 6$
    $4x - y + 4z = 14$

In exercises 33–40, find the determinant of each given matrix.

33. $\begin{bmatrix} 4 & 7 \\ 2 & 3 \end{bmatrix}$

34. $\begin{bmatrix} 3 & -2 \\ 3 & 4 \end{bmatrix}$

35. $\begin{bmatrix} 6 & 3 \\ 8 & 4 \end{bmatrix}$

36. $\begin{bmatrix} -2 & -3 \\ 6 & 9 \end{bmatrix}$

37. $\begin{bmatrix} 1 & -3 & 2 \\ -1 & 3 & 0 \\ 2 & 4 & -1 \end{bmatrix}$

38. $\begin{bmatrix} -2 & -3 & 0 \\ 2 & -1 & 4 \\ 3 & 5 & -1 \end{bmatrix}$

39. $\begin{bmatrix} 1 & 4 & 5 \\ 2 & 3 & 5 \\ -1 & 5 & 4 \end{bmatrix}$

40. $\begin{bmatrix} 2 & 1 & 3 \\ -2 & 5 & 3 \\ -1 & -2 & -3 \end{bmatrix}$

41. A boat travels 80 kilometers downstream in 10 hours and makes the return trip upstream in 16 hours. Find the speed of the boat in still water and the speed of the water.
42. Find the sides of a rectangle which has perimeter 22 meters and area 24 square meters.
43. We mix a solution containing 10% alcohol with a solution containing 60% alcohol to obtain a solution containing 40% alcohol. How many liters of each solution do we mix to obtain 20 liters of the 40% solution?

Exercise 44 is adopted from the Hindu mathematician Mahāvīra from about A.D. 850.

44. The mixed price of 9 citrons and 7 apples is 107, and the mixed price of 7 citrons and 9 apples is 101. O you arithmetician, tell me the price of a single citron and apple.

appendix A • The Metric System

Measurement of length

Basic unit is the meter.

1 meter	=		3.28 feet
1 kilometer	=	1000 meters =	0.621 miles
1 centimeter	=	1/100 meter =	0.3937 inches
1 millimeter	=	1/1000 meter =	0.039 inches

Measurement of weight

Basic unit is the gram.

1 gram	=		0.0352 ounces
1 kilogram	=	1000 grams =	2.2 pounds
1 milligram	=	1/1000 gram =	0.0000352 ounces

Measurement of volume

Basic unit is the liter.

1 liter	=		1.06 liquid quarts
1 kiloliter	=	1000 liters =	264 gallons
1 centiliter	=	1/100 liter =	0.338 fluid ounces

appendix B • Rounding Off

In this book we round off numbers written in decimal notation or scientific notation to n decimal places. To do so, we examine the digit in the $n + 1^{st}$ decimal place. If this digit is less than 5, we write 0 for every digit after the nth decimal place. If this digit is 5 or more, we increase the digit in the nth place by 1 and write 0 for every digit after the nth decimal place.

Example. Rounding off.

a. To round off 1.33 to the first decimal place, we write 1.3000 · · · or simply 1.3.
b. To round off 3.14159 to the second decimal place, we write 3.14000 · · · or simply 3.14.
c. To round off 2.6666 to the third decimal place, we write 2.667000 · · · or simply 2.667.
d. To round off 5.4545 to the third decimal place, we write 5.455000 · · · or simply 5.455.

appendix C • Squares and Square Roots

x	x^2	$\sqrt{x}$	x	x^2	$\sqrt{x}$	x	x^2	$\sqrt{x}$
1	1	1.000	35	1225	5.916	68	4624	8.246
2	4	1.414	36	1296	6.000	69	4761	8.307
3	9	1.732	37	1369	6.083	70	4900	8.367
4	16	2.000	38	1444	6.164	71	5041	8.426
5	25	2.236	39	1521	6.245	72	5184	8.485
6	36	2.449	40	1600	6.325	73	5329	8.544
7	49	2.646	41	1681	6.403	74	5476	8.602
8	64	2.828	42	1764	6.481	75	5625	8.660
9	81	3.000	43	1849	6.557	76	5776	8.718
10	100	3.162	44	1936	6.633	77	5929	8.775
11	121	3.317	45	2025	6.708	78	6084	8.832
12	144	3.464	46	2116	6.782	79	6241	8.888
13	169	3.606	47	2209	6.856	80	6400	8.944
14	196	3.742	48	2304	6.928	81	6561	9.000
15	225	3.873	49	2401	7.000	82	6724	9.055
16	256	4.000	50	2500	7.071	83	6889	9.110
17	289	4.123	51	2601	7.141	84	7056	9.165
18	324	4.243	52	2704	7.211	85	7225	9.220
19	361	4.359	53	2809	7.280	86	7396	9.274
20	400	4.472	54	2916	7.348	87	7569	9.327
21	441	4.583	55	3025	7.416	88	7744	9.381
22	484	4.690	56	3136	7.483	89	7921	9.434
23	529	4.796	57	3249	7.550	90	8100	9.487
24	576	4.899	58	3364	7.616	91	8281	9.539
25	625	5.000	59	3481	7.681	92	8464	9.592
26	676	5.099	60	3600	7.746	93	8649	9.644
27	729	5.196	61	3721	7.810	94	8836	9.695
28	784	5.292	62	3844	7.874	95	9025	9.747
29	841	5.385	63	3969	7.937	96	9216	9.798
30	900	5.477	64	4096	8.000	97	9409	9.849
31	961	5.568	65	4225	8.062	98	9604	9.899
32	1024	5.657	66	4356	8.124	99	9801	9.950
33	1089	5.745	67	4489	8.185	100	10,000	10.000
34	1156	5.831						

appendix D • Common Logarithms

N	0	1	2	3	4	5	6	7	8	9
10	0000	0043	0086	0128	0170	0212	0253	0294	0334	0374
11	0414	0453	0492	0531	0569	0607	0645	0682	0719	0755
12	0792	0828	0864	0899	0934	0969	1004	1038	1072	1106
13	1139	1173	1206	1239	1271	1303	1335	1367	1399	1430
14	1461	1492	1523	1553	1583	1614	1644	1673	1703	1732
15	1761	1790	1818	1847	1875	1903	1931	1959	1987	2014
16	2041	2068	2095	2122	2148	2175	2201	2227	2253	2279
17	2304	2330	2355	2380	2405	2430	2455	2480	2504	2529
18	2553	2577	2601	2625	2648	2672	2695	2718	2742	2765
19	2788	2810	2833	2856	2878	2900	2923	2945	2967	2989
20	3010	3032	3054	3075	3096	3118	3139	3160	3181	3201
21	3222	3243	3263	3284	3304	3324	3345	3365	3385	3404
22	3424	3444	3464	3483	3502	3522	3541	3560	3579	3598
23	3617	3636	3655	3674	3692	3711	3729	3747	3766	3784
24	3802	3820	3838	3856	3874	3892	3909	3927	3945	3962
25	3979	3997	4014	4031	4048	4065	4082	4099	4116	4133
26	4150	4166	4183	4200	4216	4232	4249	4265	4281	4298
27	4314	4330	4346	4362	4378	4393	4409	4425	4440	4456
28	4472	4487	4502	4518	4533	4548	4564	4579	4594	5609
29	4624	4639	4654	4669	4683	4698	4713	4728	4742	4757
30	4771	4786	4800	4814	4829	4843	4857	4871	4886	4900
31	4914	4928	4942	4955	4969	4983	4997	5011	5024	5038
32	5051	5065	5079	5092	5105	5119	5132	5145	5159	5172
33	5185	5198	5211	5224	5237	5250	5263	5276	5289	5302
34	5315	5328	5340	5353	5366	5378	5391	5403	5416	5428
35	5441	5453	5465	5478	5490	5502	5514	5527	5539	5551
36	5563	5575	5587	5599	5611	5623	5635	5647	5658	5670
37	5682	5694	5705	5717	5729	5740	5752	5763	5775	5786
38	5798	5809	5821	5832	5843	5855	5866	5877	5888	5899
39	5911	5922	5933	5944	5955	5966	5977	5988	5999	6010
40	6021	6031	6042	6053	6064	6075	6085	6096	6107	6117
41	6128	6138	6149	6160	6170	6180	6191	6201	6212	6222
42	6232	6243	6253	6263	6274	6284	6294	6304	6314	6325
43	6335	6345	6355	6365	6375	6385	6395	6405	6415	6425
44	6435	6444	6454	6464	6474	6484	6493	6503	6513	6522
45	6532	6542	6551	6561	6571	6580	6590	6599	6609	6618
46	6628	6637	6646	6656	6665	6675	6684	6693	6702	6712
47	6712	6730	6739	6749	6758	6767	6776	6785	6794	6803
48	6812	6821	6830	6839	6848	6857	6866	6875	6884	6893
49	6902	6911	6920	6928	6937	6946	6955	6964	6972	6981
50	6990	6998	7007	7016	7024	7033	7042	7050	7059	7067
51	7076	7084	7093	7101	7110	7118	7126	7135	7143	7152
52	7160	7168	7177	7185	7193	7202	7210	7218	7226	7235
53	7243	7251	7259	7267	7275	7284	7292	7300	7308	7316
54	7324	7332	7340	7348	7356	7364	7372	7380	7388	7396

N	0	1	2	3	4	5	6	7	8	9
55	7404	7412	7419	7427	7435	7443	7451	7459	7466	7474
56	7482	7490	7497	7505	7513	7520	7528	7536	7543	7551
57	7559	7566	7574	7582	7589	7597	7604	7612	7619	7627
58	7634	7642	7649	7657	7664	7672	7679	7686	7694	7701
59	7709	7716	7723	7731	7738	7745	7752	7760	7767	7774
60	7782	7789	7796	7803	7810	7818	7825	7832	7839	7846
61	7853	7860	7868	7875	7882	7889	7896	7903	7910	7917
62	7924	7931	7938	7945	7952	7959	7966	7973	7980	7987
63	7993	8000	8007	8014	8021	8028	8035	8041	8048	8055
64	8062	8069	8075	8082	8089	9096	8102	8109	8116	8122
65	8129	8136	8142	8149	8156	8162	8169	8176	8182	8189
66	8195	8202	8209	8215	8222	8228	8235	8241	8248	8254
67	8261	8267	8274	8280	8287	8293	8299	8306	8312	8319
68	8325	8331	8338	8344	8351	8357	8363	8370	8376	8382
69	8388	8395	8401	8407	8414	8420	8426	8432	8439	8445
70	8451	8457	8463	8470	8476	8482	8488	8494	8500	8506
71	8513	8519	8525	8531	8537	8543	8549	8555	8561	8567
72	8573	8579	8585	8591	8597	8603	8609	8615	8621	8627
73	8633	8639	8645	8651	8657	8663	8669	8675	8681	8686
74	8692	8698	8704	8710	8716	8722	8727	8733	8739	8745
75	8751	8756	8762	8768	8774	8779	8785	8791	8797	8802
76	8808	8814	8820	8825	8831	8837	8842	8848	8854	8859
77	8865	8871	8876	8882	8887	8893	8899	8904	8910	8915
78	8921	8927	8932	8938	8943	8949	8954	8960	8965	8971
79	8976	8982	8987	8993	8998	9004	9009	9015	9020	9025
80	9031	9036	9042	9047	9053	9058	9063	9069	9074	9079
81	9085	9090	9096	9101	9106	9112	9117	9122	9128	9133
82	9138	9143	9149	9154	9159	9165	9170	9175	9180	9186
83	9191	9196	9201	9206	9212	9217	9222	9227	9232	9238
84	9243	9248	9253	9258	9263	9269	9274	9279	9284	9289
85	9294	9299	9304	9309	9315	9320	9325	9330	9335	9340
86	9345	9350	9355	9360	9365	9370	9375	9380	9385	9390
87	9395	9400	9405	9410	9415	9420	9425	9430	9435	9440
88	9445	9450	9455	9460	9465	9469	9474	9479	9484	9489
89	9494	9499	9504	9509	9513	9518	9523	9528	9533	9538
90	9542	9547	9552	9557	9562	9566	9571	9576	9581	9586
91	9590	9595	9600	9605	9609	9614	9619	9624	9628	9633
92	9638	9643	9647	9652	9657	9661	9666	9671	9675	9680
93	9685	9689	9694	9699	9703	9708	9713	9717	9722	9727
94	9731	9736	9741	9745	9750	9754	9759	9763	9768	9773
95	9777	9782	9786	9791	9795	9890	9805	9809	9814	9818
96	9823	9827	9832	9836	9841	9845	9850	9854	9859	9863
97	9868	9872	9877	9881	9886	9890	9894	9899	9903	9908
98	9912	9917	9921	9926	9930	9934	9939	9943	9948	9952
99	9956	9961	9965	9969	9974	9978	9983	9987	9991	9996

Answers to Odd-Numbered Exercises

Chapter 1 • Review

Exercise Set 1.1

1. 0.25 (terminating) **3.** 0.75 (terminating) **5.** 0.666 . . . (nonterminating)
7. 0.8333 . . . (nonterminating) **9.** 0.3 (terminating) **11.** 0.08333 . . . (nonterminating)

13.
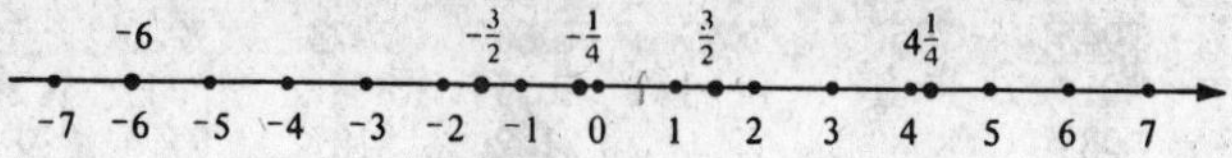

15. $-3 < x < 2$
17. $-4 < x \le 2$ **19.** $x \le 3$ **21.** $x < -\pi$ or $x > \pi$ **23.** $x \le -2$ or $x \ge -1$
25. $x \le 2\pi$ **27.** $-4 < x < -2$ or $x \ge 1$ **29.** $-3 < x \le -1$ or $1 \le x < 3$

31.
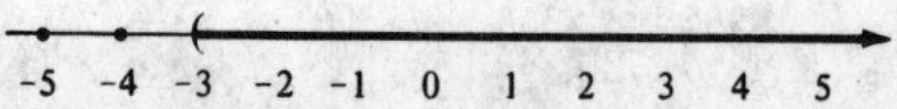

33.
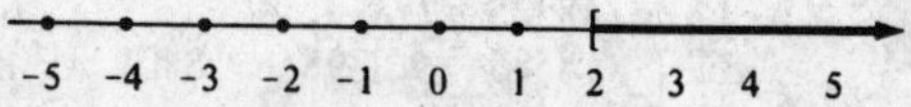

35.
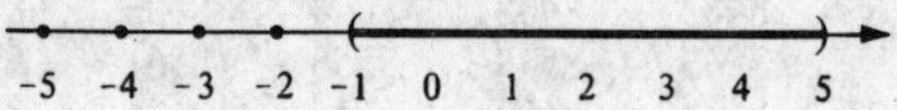

37.
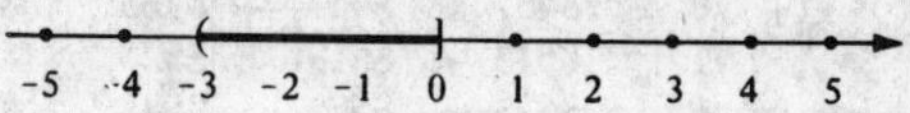

39.
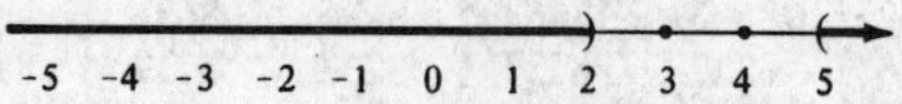

41.
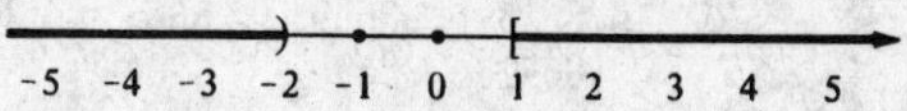

43.
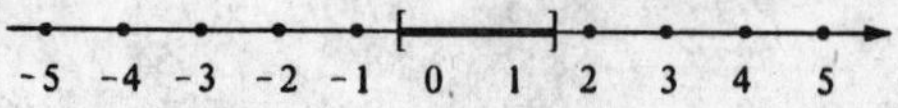

45.
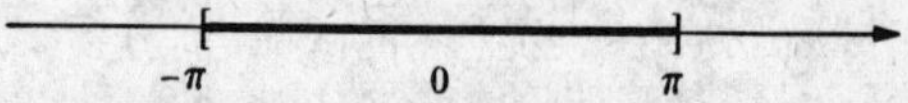

Exercise Set 1.2

1. $2 + 4 = 6 = 4 + 2$; $3 + 6 = 9 = 6 + 3$; $1 + 9 = 10 = 9 + 1$ **3.** $-0 = 0$
5. a. $5 - 2 = 3$ and $2 - 5 = -3$; also, $3 - 7 = -4$ and $7 - 3 = 4$; also, $9 - 4 = 5$ and $4 - 9 = -5$ b. Subtraction is not commutative.
7. $(3 \cdot 2)4 = (6)4 = 24$ and $3(2 \cdot 4) = 3(8) = 24$; also, $(2 \cdot 5)6 = (10)6 = 60$ and $2(5 \cdot 6) = 2(30) = 60$; also, $(4 \cdot 7)2 = (28)2 = 56$ and $4(7 \cdot 2) = 4(14) = 56$
9. a. 35 b. -35 c. -35 d. 35 **11.** a. 8 b. -8 c. -8 d. 8
13. -6 **15.** -42 **17.** -6 **19.** -6 **21.** $12x$ **23.** $4x$ **25.** $7x - y$
27. $-x + 3y$ **29.** $3a + 3b - 8c$ **31.** $-a + 5b - 8c$ **33.** $27xy$ **35.** $-36abc$
37. On the author's calculator, $6 \times 1 \div 6 = 6$ and $1 \div 6 \times 6 = 0.9999996$
39. On the calculator, $(1.00001)(1.00001) = 1.00002$; by hand, $(1.00001)(1.00001) = 1.0000200001$; the calculator truncates

Exercise Set 1.3

1. 26 **3.** 0 **5.** 56 **7.** -31 **9.** -11 **11.** 12 **13.** -5 **15.** -1
17. -4 **19.** -30 **21.** -3 **23.** -4 **25.** $8xy + 10xz$ **27.** $31ab$ **29.** $5ab$

31. $-2ab$ **33.** $25ax$ **35.** $-6ab$ **37.** $-10abc$ **39.** 23; 35; 15; 11
41. 24; 36; 16; 12 **43.** 6; 30; 0; 0 **45.** 0; 27; -8; -9 **47.** 0.5; 0.2; undefined; -1
49. $A = 72$; $A = 8$; $A = 15.5$ **51.** $C = -160/9$; $C = -260/9$; $C = 0$; $C = 55/3$; $C = 340/9$; $C = 37$; $C = 100$ **53.** $A = 73.2336$; $A = 8.008161$; $A = 15.150802$
55. $C = -17.20555$; $C = -1.0777777$; $C = 24.48444$; $C = 49.75888$; $C = -24.45611$; $C = -34.07555$
57. a. $ab - c$ b. $(a - b)c$ c. $a + b - c$ d. $a - b + c$

Exercise Set 1.4

1. 8 **3.** 81 **5.** 64 **7.** 216 **9.** 1 **11.** 1 **13.** -8 **15.** -8 **17.** 25
19. 625 **21.** x^3y^2 **23.** ax^4y^2 **25.** $(abc)^2$ **27.** $(ab)^4c^2$ **29.** x^2/y^4 **31.** x^4
33. y^6 **35.** a^3b^4 **37.** x^5y^5 **39.** a^6 **41.** a^9 **43.** a^7 **45.** t^{10} **47.** a^3/b^4
49. a^7/b^4 **51.** $A = 3.14$; $A = 200.96$; $A = 7850$; $A = 125{,}600$
53. $s = 19.6$; $s = 490$; $s = 1960$; $s = 196{,}000$ **55.** $E = 20$; $E = 4$; $E = 1$; $E = 0.05$
57. $V = 9.42$; $V = 169.56$; $V = 392.5$; $V = 100.48$ **59.** $A = 5.828$; $A = 197.359$; $A = 7470.167$; $A = 133{,}588.46$ **61.** $s = 17.663$; $s = 565.233$; $s = 1753.811$; $s = 204{,}363.44$
63. $(xy)^2 = 1.0000178$ and $x^2y^2 = 0.999998$; also, $(xy)^2 = 1.0000358$ and $x^2y^2 = 0.999998$; the calculator truncates the products, causing a discrepancy.

Review Exercises for Chapter 1

1. 0.75 (terminating) **3.** 0.1666 . . . (nonterminating)

5. -5 -4 -3 -2 -1 0 1 2 3 4 5 **7.** -5 -4 -3 -2 -1 0 1 2 3 4 5

9. -5 -4 -3 -2 -1 0 1 2 3 4 5 **11.** -5 -4 -3 -2 -1 0 1 2 3 4 5

13. $6x + 3y$ **15.** $4x + y$ **17.** $24xy$ **19.** $6xy$ **21.** $-30xy$ **23.** $19xy$ **25.** 0
27. $24xy$ **29.** $-47xy$ **31.** $-xy$ **33.** 100 **35.** 25 **37.** 1 **39.** 32 **41.** x^2
43. r^4 **45.** a^3b **47.** s^2r^2 **49.** x^5 **51.** a^5b^3 **53.** a^5b^5 **55.** a^4
57. $A = 26$; $A = 360$; $A = 500$; $A = 1600$ **59.** $c = 75.36$; $c = 188.4$; $c = 565.2$; $c = 6280$

Chapter 2 • Algebraic Expressions

Exercise Set 2.1

1. $6x + 72$ **3.** $5x^2y + 15xz$ **5.** $-12x^2 + 36y$ **7.** $4x^3 + 3x^2$ **9.** $x^2y - xy^2$
11. $3a^2 + 3ab + 3a^2b$ **13.** $x^2y^2z + x^2yz^2 + xy^2z^2$ **15.** $-4x^3y + 4x^2y^2 + 4x^2y^3$
17. $-x^3y + x^2y^2 - xy^3$ **19.** $xy + 3x + y + 3$ **21.** $xy - 5x - 3y + 15$
23. $ab - 3a - 5b + 15$ **25.** $ab + a - b - 1$ **27.** $12xy + 6x + 2y + 1$
29. $14xy + 7x + 10y + 5$ **31.** $a^3 - a^2b + ab^2 - b^3$ **33.** $x^3 + 2x^2 + x + 2$
35. $a^2 + 2ab - a^2b + b^2 - ab^2$ **37.** $x^2 + 2xy + y^2 - 1$ **39.** $a^4 - 2a^2b^2 + b^4$
41. $a^6 - a^5 - a^4 + a^2 + a - 1$ **43.** $a^2 - b^2$ **45.** $x^2 + ax + bx + ab$ **47.** $a^3 - b^3$

Exercise Set 2.2

1. $x^2 + 2x + 1$ **3.** $x^2 + 10x + 25$ **5.** $x^2y^2 + 12xy + 36$ **7.** $x^4 + 2x^2y^2 + y^4$
9. $x^2 - 4$ **11.** $4x^2 - 25y^2$ **13.** $x^4 - y^4$ **15.** $x^2y^2 - 100$ **17.** $x^2 + 8x + 15$
19. $x^2 + x - 42$ **21.** $x^2 + 8x - 48$ **23.** $x^2 + 3xy + 2y^2$ **25.** $6x^2 + 32x + 32$
27. $10x^2 - 81x + 45$ **29.** $15x^2 + 8x + 1$ **31.** $9x^2 + 33x + 28$ **33.** $6x^2 + ax - a^2$
35. $6x^2 + 5axy + a^2y^2$ **37.** $x^3 + 3x^2 + 3x + 1$ **39.** $x^3y^3 + 9x^2y^2 + 27xy + 27$
41. $a^3 - 8$ **43.** $x^3 - 27y^3$ **45.** $x^3 + 125$ **47.** $x^6 + 27$

Exercise Set 2.3

1. $5x(y + 5)$ **3.** $3a(1 - 3x)$ **5.** $ab(a^2 + b)$ **7.** $xy(x + y - 1)$ **9.** $(x + 7)^2$
11. $(a - 3)^2$ **13.** $(xy + 5)^2$ **15.** $(2x + 3)^2$ **17.** $(x + 8)(x - 8)$
19. $(2x + 15y)(2x - 15y)$ **21.** $(6ab + 7)(6ab - 7)$ **23.** $(2x + 7)(2x - 7)$
25. $xy(x + y)(x - y)$ **27.** $x^2y(x + y)(x - y)$ **29.** $6(x - 2)^2$ **31.** $(x^3 + y)^2$
33. $xy(x^3 + y)^2$ **35.** $(x^3 + y^2)(x^3 - y^2)$ **37.** $(x^4 + y^2)(x^2 + y)(x^2 - y)$
39. $(x^2 + 1)^2$ **41.** $(x + 2y + 3)(x - 2y + 3)$ **43.** $(x + y + 2)(x - y + 2)$
45. $(2x + y + 1)(2x - y - 1)$ **47.** $(x + y + 4)(x - y + 2)$ **49.** 25.1447
51. 257.439 **53.** 360.0324 **55.** 1434.8544

Exercise Set 2.4

1. $(x + 4)(x - 1)$ **3.** $(x - 3)(x - 4)$ **5.** $(x + 6)(x - 4)$ **7.** Irreducible
9. $(x + 1)(x - 1)$ **11.** Irreducible **13.** Irreducible **15.** Irreducible
17. $(x + 3)(x + 7)$ **19.** Irreducible **21.** $(x - 3)(2x + 1)$ **23.** $(x + 4)(3x + 2)$
25. $(x - 2)(5x - 3)$ **27.** $(2x + 1)(2x + 3)$ **29.** $(x - 1)(2x + 1)$ **31.** $(2x - 1)^2$
33. $(x + 6)(4x - 3)$ **35.** $(y + 1)(4y - 1)$ **37.** $(3x + 1)(5x + 3)$
39. $(4x - 3)(5x + 6)$ **41.** $2(x + 2)(5x - 4)$ **43.** Irreducible **45.** $(x^2 + 3)(x^2 + 7)$
47. $(x + 1)(x - 1)(3x^2 - 2)$

Exercise Set 2.5

1. $(x + 2)(y + 3)$ **3.** $(x + 2)(y - 3)$ **5.** $(r + s)(r + t)$ **7.** $(ax - y)(bx + z)$
9. $(x + 2)^3$ **11.** $(x^2 + 1)^3$ **13.** $(z - 3)(z^2 + 3z + 9)$
15. $(a - 5b)(a^2 + 5ab + 25b^2)$ **17.** $ab(a - b)(a^2 + ab + b^2)$ **19.** $(z + 3)(z^2 - 3z + 9)$
21. $x(x + 1)(x^2 - x + 1)$ **23.** $3ab(a + 2)(a^2 - 2a + 4)$ **25.** $(x^2 + x + 1)(x^2 - x + 1)$
27. $(x^2 + 2xy - y^2)(x^2 - 2xy - y^2)$ **29.** $(x^2 + y^2)(x^4 - x^2y^2 + y^4)$
31. $(x + 1)(x^2 + 1)$ **33.** $(3x + 7)(3x - 7)$ **35.** $(x - 6)^2$ **37.** $x(x + 4)(x - 3)$
39. $(x - 4)(2x + 3)$ **41.** $(2x + 1)(2x + 3)$ **43.** $x(2x + 7)(2x - 7)$
45. $(x^2 + 1)(x + 2)$ **47.** $(x^2 + 4)(x + 1)(x - 1)$

Exercise Set 2.6

1. Equal **3.** Not equal **5.** Equal **7.** Not equal **9.** Equal **11.** 6/7 **13.** 6/7
15. a/b **17.** x/y **19.** $\dfrac{xy}{x^2 - y^2}$ **21.** $a + b$ **23.** $\dfrac{1}{x - y}$ **25.** $x - y$ **27.** 6/21
29. 49/35 **31.** x^3y/xy^2 **33.** x^2yz^2/xyz **35.** $\dfrac{a^2 + 2ab + b^2}{a^2 - b^2}$ **37.** $\dfrac{a^2b + ab^2}{a^2 - b^2}$

39. $\dfrac{x^2 - 2x - 3}{x^2 + 3x + 2}$ **41.** $x = 15$ **43.** $x = 1/2$ **45.** $x = 1250/9$ **47.** $x = 400/3$

49. $x = 5000/3$ **51.** $x = 1500$ **53.** $x = 1000$ **55.** $x = 750$ **57.** $x = 180$

59. $x = 275/3$

Exercise Set 2.7

1. 2 **3.** $\dfrac{a + b}{x}$ **5.** 2 **7.** $\dfrac{1}{a + 1}$ **9.** 25/3 **11.** $\dfrac{x}{x + 3}$ **13.** $\dfrac{x(2x + 1)}{x + 1}$

15. $\dfrac{z(1 + x^2y^2)}{xy}$ **17.** 13/14 **19.** $\dfrac{x^2 + y^2}{xy}$ **21.** $\dfrac{b^2 - a^2}{a^2b^2}$ **23.** $\dfrac{2x}{(x + 1)(x - 1)}$

25. $\dfrac{6(x + 2)}{x(x + 6)}$ **27.** $\dfrac{6x}{(x + 1)(x - 1)}$ **29.** 13/8 **31.** $\dfrac{a(a + b)}{b^2}$ **33.** $\dfrac{1 + b^2}{abc}$

35. $\dfrac{y^2}{(y + 1)(y + 2)}$ **37.** $\dfrac{15x}{(x - 1)(3x + 2)}$ **39.** $\dfrac{7x^2 - 6}{(x + 1)(2x + 1)}$ **41.** $\dfrac{x - 5}{(x + 1)(x - 1)}$

43. $\dfrac{2x + 1}{(x + 1)(x + 2)}$ **45.** $\dfrac{x^2(x + 2)}{(x + 1)^2}$ **47.** $\dfrac{2}{x - 1}$

Exercise Set 2.8

1. 2/15 **3.** 5/6 **5.** $\dfrac{x^2}{x^2 - 1}$ **7.** $\dfrac{ax + ay}{bx - by}$ **9.** $\dfrac{x^2 + 1}{(x - 1)^2}$ **11.** $\dfrac{y(x + y)}{2x(x - y)}$

13. $\dfrac{2x^4}{y(x + 2)}$ **15.** $\dfrac{x + 4}{x - 1}$ **17.** $\dfrac{x - 3}{3x + 1}$ **19.** $\dfrac{2x + y}{x + y}$ **21.** 35/12 **23.** 7/15

25. a^2/xyz^3 **27.** x^3/y^2 **29.** $3/2x$ **31.** $a^2 - ab + b^2$

33. $\dfrac{x^2 + 1}{(x + 2)(x - 1)}$ **35.** $\dfrac{y}{xy + 3}$ **37.** $\dfrac{x - 3}{x - 4}$ **39.** $\dfrac{x + 1}{4x + 1}$ **41.** $\dfrac{x + 2}{x + 1}$

43. $\dfrac{(2x + 3)(x - 1)}{2(x + 3)(x + 1)}$ **45.** $\dfrac{x + 1}{x + 3}$ **47.** $\dfrac{x + 5}{x + 1}$

Exercise Set 2.9

1. $\dfrac{3x - 1}{3x - 2}$ **3.** $\dfrac{6x + 5}{3(2x + 1)}$ **5.** $\dfrac{1 + x}{1 - 3x}$ **7.** $\dfrac{x}{1 - x}$ **9.** $xy(x - y)$

11. $\dfrac{2(a^2 + b^2)}{(a + b)^2(a - b)^2}$ **13.** $\dfrac{x^2 + y^2 + z^2}{yz + xz + xy}$ **15.** 2 **17.** $\dfrac{3130}{601}$ **19.** $\dfrac{3x + 2}{2x + 1}$

1. S = 360.0 square centimeters, V = 515.8 cubic centimeters

3. S = 305.5 square centimeters, V = 385.8 cubic centimeters

5. S = 1434.9 square centimeters, V = 4165.3 cubic centimeters

7. \$6.54 **9.** \$4001.60 **11.** \$13.46; \$17.96 **13.** \$57.42; \$110.88 **15.** 139 bears

17. 1667 fish **19.** 3 tablets **21.** 4 tablets **23.** 3 cc **25.** 0.8 cc **27.** 320,000 *u.*

29. 180 milligrams **31.** 312 milligrams

Review Exercises for Chapter 2

1. $a^2bx + ab^2y$ **3.** $a^2b^2 + a^2bc + ab^2c$ **5.** $x^2 + 6x + 8$ **7.** $x^2 - 2x - 8$

9. $x^2 - 6x + 8$ **11.** $20x$ **13.** $2x^2 - 20$ **15.** 27 **17.** $x^2 + x + 7$

19. $x^2 + 21x + 13$ **21.** $x^2 + 108$ **23.** $10x$ **25.** $(a - b)(x - c)$ **27.** $y(2x + y)$

29. $xy(x + y - xy)$ **31.** $2\pi r(r + h)$ **33.** $(x + 4)^2$ **35.** $(2x - 1)^2$

37. $(2x + 3)(2x - 3)$ **39.** $(x^2 + 4)(x + 2)(x - 2)$ **41.** $(x^2 - 2)^2$

43. $(x + 1)(x + 5)$ **45.** $(x - 2)(x - 4)$ **47.** $(x - 4)(x - 5)$ **49.** Irreducible

51. $(x + 5)(x - 3)$ **53.** $(x - 5)(x + 1)$ **55.** $(x - 4)(x + 7)$ **57.** $(x + 1)(x - 28)$

59. $(x + 1)(2x + 1)$ **61.** $(x - 3)(2x - 1)$ **63.** Irreducible **65.** $(x - 3)(2x + 1)$

67. $(x - 1)(2x + 3)$ **69.** $(x^2 + 2)(x^2 + 6)$ **71.** $(x^2 + 7)(x + 1)(x - 1)$

73. $(x + 1)(y + 1)$ **75.** $(x + 1)^3$ **77.** $(x - 1)(x^2 + x + 1)$ **79.** $(x + 1)(x^2 - x + 1)$

81. $x = 10/3$ **83.** $x = 32/3$ **85.** $x = 1000/41$ **87.** $x = 500$ **89.** $x = 7/2000$

91. $x = 360{,}000$ **93.** $x = 3$ **95.** $x = 6/125$ **97.** $\dfrac{1}{x - 1}$ **99.** $\dfrac{2x}{x - 1}$ **101.** $\dfrac{x + y}{xy}$

103. $\dfrac{2x}{(x + 1)(x - 1)}$ **105.** $\dfrac{a - 1}{a^2}$ **107.** $\dfrac{a^2 + c^2}{abc}$ **109.** $\dfrac{x^2}{x^2 - 1}$ **111.** ac/b

113. ac/x **115.** $\dfrac{2x + 1}{2(x + 1)}$ **117.** $x + 1$ **119.** $\dfrac{x - 3}{x + 1}$ **121.** $\dfrac{3}{x - 2}$ **123.** $\dfrac{3}{x - 3}$

125. $\dfrac{x^2}{x + 3}$ **127.** $\dfrac{2(x^2 + ax + a^2)}{(x - a)(x + a)}$

129. $S = 251.2$ square centimeters, $V = 301.44$ cubic centimeters **131.** \$12.38; \$14.96

133. 3 cc.

Chapter 3 • Exponents, Roots, and Complex Numbers

Exercise Set 3.1

1. a. $10^1 = 10$; $10^2 = 100$; $10^3 = 1000$; $10^4 = 10{,}000$; $10^5 = 100{,}000$; $10^6 = 1{,}000{,}000$
b. 10^n is 1 followed by n 0's.

3. a. $(-1)^1 = -1$; $(-1)^2 = 1$; $(-1)^3 = -1$; $(-1)^4 = 1$; $(-1)^5 = -1$; $(-1)^6 = 1$
b. $(-1)^n = 1$ if n is even and $(-1)^n = -1$ if n is odd.

5. $(1.08)^1 = 1.08$; $(1.08)^2 = 1.1664$; $(1.08)^3 = 1.259712$; $(1.08)^4 = 1.360489$; $(1.08)^5 = 1.469328$

7. $(1.01)^1 = 1.01$; $(1.01)^2 = 1.0201$; $(1.01)^3 = 1.030301$; $(1.01)^4 = 1.040604$; $(1.01)^5 = 1.051010$

9. 1; $1/4 = 0.25$; $1/16 = 0.0625$; $1/64 = 0.015625$; $1/256 = 0.00390625$

11. 1; $1/8 = 0.125$; $1/64 = 0.015625$; $1/512 = 0.00195$

13. a. $10^0 = 1$; $10^{-1} = 0.1$; $10^{-2} = 0.01$; $10^{-3} = 0.001$; $10^{-4} = 0.0001$; $10^{-5} = 0.00001$; $10^{-6} = 0.000001$ b. 10^n is a decimal with $-1 - n$ zeros preceding 1.

15. $(1.08)^{-1} = 0.925926$; $(1.08)^{-2} = 0.857339$; $(1.08)^{-3} = 0.793832$; $(1.08)^{-4} = 0.735030$

17. $(1.01)^{-1} = 0.990099$; $(1.01)^{-2} = 0.980296$; $(1.01)^{-3} = 0.970590$; $(1.01)^{-4} = 0.960980$

19. $(1/2)^0 = 1$; $(1/2)^{-1} = 2$; $(1/2)^{-2} = 4$; $(1/2)^{-3} = 8$

21. a. Finally 2^n will exceed the capacity of the calculator.
b. Finally 2^n will be very close to 0.

23. $(1.06)^{10} = 1.790848$; $(1.06)^{20} = 3.207135$; $(1.06)^{-5} = 0.747258$; $(1.06)^{-18} = 0.350344$; $(1.06)^{-20} = 0.311805$

25. $(1.08)^{10} = 2.158925$; $(1.08)^{20} = 4.660957$; $(1.08)^{-5} = 0.680583$; $(1.08)^{-18} = 0.250249$; $(1.08)^{-20} = 0.214548$

27. 75.296061; 1773.324424; 2216.7358; 39,497.25; 4,341,775.7

29. x^{12} **31.** b^{21} **33.** x^3 **35.** b^3/a^4 **37.** a^3 **39.** y^5 **41.** $1/ab^3$

43. b^{10}/a^{10} **45.** a^7/b^2 **47.** a^{16}/b^3x^6 **49.** a^{12}/b^8 **51.** $576a^{12}b^7$ **53.** x^7

55. x^{5n+3} **57.** $1/a^{n+2}$ **59.** y^{n^2}

Exercise Set 3.2

1. 133,400 **3.** 0.00000000529 **5.** 31,000,000,000,000 **7.** 0.0136 **9.** 0.0000437
11. 203,200,000 **13.** 41,000,000,000,000 **15.** 250,000,000 **17.** 0.0353
19. 0.0000004 **21.** 2.52×10^5 **23.** 2.4×10^6 **25.** 2.12×10^{-2} **27.** 6.6×10^{-5}
29. 4.51×10^{-9} **31.** 1.86×10^5 **33.** 5.98×10^{27} **35.** 1.66×10^{-24}
37. 1.61×10^0 **39.** 6.378×10^8 **41.** 2.352×10^{10} **43.** 2.88×10^3
45. 2×10^{-6} **47.** 5×10^2 **49.** 8×10^{30} **51.** 9.4424×10^8 **53.** 1.6455×10^{26}
55. 8.7870×10^{-17} **57.** 8.378×10^{13} **59.** 1.1163×10^{41} **61.** 6.696×10^8 miles
63. 5.870×10^{12} miles **65.** 2.547×10^{13} miles **67.** 1.824×10^3 times
69. 1.199×10^{57} atoms

Exercise Set 3.3

1. 1 **3.** −1 **5.** Not a real number **7.** 9 **9.** −9 **11.** 8
13. Not a real number **15.** −8 **17.** 6 **19.** −6 **21.** 2 **23.** −2 **25.** 5
27. 4 **29.** 3 **31.** 1 **33.** 2 **35.** 0 **37.** 1.58 **39.** 3.57 **41.** 4.92
43. 9.39 **45.** 30.96 **47.** 87.97 **49.** 1.53 **51.** 2.57 **53.** 0.810 **55.** 0.569
57. 0.957 **59.** 0.927

Exercise Set 3.4

1. One example is $a = 25$ and $b = 16$. **3.** a^2 **5.** a^4 **7.** a^3b^6 **9.** x^2
11. $ab\sqrt{b}$ **13.** $2b^3\sqrt{5a}$ **15.** $9ab^3$ **17.** $2\sqrt{5}; 2 + \sqrt{5}$ **19.** $2\sqrt{6}; -1 - \sqrt{6}$
21. $4\sqrt{2}; -2 + 2\sqrt{2}$ **23.** $3\sqrt{2}; (1 + \sqrt{2})/2$ **25.** $3\sqrt{5}; (1 - \sqrt{5})/2$ **27.** 38; 4
29. $2\sqrt{34}; (-1 - \sqrt{34})/2$ **31.** $4\sqrt{11}; -1 + \sqrt{11}$ **33.** $20\sqrt{2}$; 28.28
35. $20\sqrt{6}$; 48.98 **37.** $16\sqrt{2}$; 22.624 **39.** 35 **41.** $\sqrt{3}/3$ **43.** $4\sqrt[3]{9}/3$
45. $-\sqrt{5} - \sqrt{6}$ **47.** $-2 + \sqrt{3}$ **49.** 3.0728×10^4 **51.** 5.8634×10^3
53. 7.5354×10^{-6} **55.** 4.3344×10^3 **57.** 2.5734×10^{-8} **59.** 1.5279×10^{-8}

Exercise Set 3.5

1. 3 **3.** 9 **5.** $2\sqrt{3}$ **7.** 3 **9.** −3 **11.** 2 **13.** 5 **15.** 2 **17.** 64
19. 243 **21.** 4 **23.** 16 **25.** 1/64 **27.** 1/27 **29.** 1/2 **31.** −1
33. $\sqrt[5]{a^4}$ **35.** $1/x\sqrt[3]{x}$ **37.** $\sqrt[3]{x^2}/\sqrt[3]{y^2}$ **39.** $1/2\sqrt{x+1}$ **41.** $(\sqrt{x} + 1)/2$
43. $\sqrt{x^2 + 1}/x\sqrt{x}$ **45.** $a^{3/2}$ **47.** $a^{2/3}$ **49.** $a^{1/m}b^{-1/n}$ **51.** $x^2(x^2 + 1)^{1/2}$
53. $x(x + 1)^{-3/2}$ **55.** $x^{1/2}(x - 1)^{n/2}$

Exercise Set 3.6

1. $(2^2)^{1/2} = 4^{1/2} = \sqrt{4} = 2$ and $2^{2(1/2)} = 2^1 = 2$; they are equal.
3. $(64^{1/2})(64^{1/3}) = (8)(4) = 32$ and $64^{(1/2)+(1/3)} = 64^{5/6} = (\sqrt[6]{64})^5 = 2^5 = 32$; they are equal.
5. $\sqrt[4]{x}$ **7.** $\sqrt[8]{x}$ **9.** $\sqrt{x}$ **11.** $\sqrt[5]{x}$ **13.** $(\sqrt[6]{a})^5$ **15.** $(\sqrt[12]{a})^7$ **17.** $\sqrt[12]{a}$
19. $(\sqrt[15]{ab})^8$ **21.** $x^{9/2}$ **23.** x **25.** $a^{-7}b^{1/4}$ **27.** $a^{-7/6}$ **29.** $x^2y^{-3/5}$
31. $x^{11/3}y^{2/3}$ **33.** $a^{19/3}b^{-14/3}$ **35.** $b^{-1/2}$ **37.** $x + 1$ **39.** $xy^{-1/2} + 1$
41. $(x^{1/2}y^{1/2} + x)y^{-1}$ **43.** $x^{1/2}y^{-3/2}(x + y)$ **45.** $x - y$ **47.** $x + 2x^{1/2}y^{1/2} + y$

Exercise Set 3.7

1. $3i$ **3.** $4i$ **5.** $\sqrt{7}i$ **7.** $\sqrt{10}i$ **9.** $2\sqrt{3}i$ **11.** $3\sqrt{2}i$ **13.** $2\sqrt{10}i$
15. $20\sqrt{2}i$ **17.** Real part is 1, imaginary part is $2i$; conjugate is $1 - 2i$
19. Real part is 1, imaginary part is $-2i$; conjugate is $1 + 2i$
21. Real part is 2, imaginary part is i; conjugate is $2 - i$
23. Real part is 4, imaginary part is $-3i$; conjugate is $4 + 3i$
25. Real part is 3, imaginary part is i; conjugate is $3 - i$
27. Real part is 1, imaginary part is $-i$; and conjugate is $1 + i$
29. Real part is 1, imaginary part is $-\sqrt{2}i$; conjugate is $1 + \sqrt{2}i$
31. Real part is 3, imaginary part is $\sqrt{3}i$; conjugate is $3 - \sqrt{3}i$
33. $D = -4, x = 1 + i$ **35.** $D = -4, x = -2 + i$ **37.** $D = 0, x = -2$
39. $D = 25, x = 1/2$ **41.** $D = 20, x = -1 + \sqrt{5}$ **43.** $D = -12, x = (-1 + \sqrt{3}i)/2$

Exercise Set 3.8

1. $3 + 4i$; $-1 - 2i$ **3.** $5 - i$; $-1 - 3i$ **5.** $9 - 5i$; $3 - i$ **7.** 8; $6i$ **9.** 2; $-4i$
11. 4; $2 + 6i$ **13.** $7 - i$; $3 - 7i$ **15.** $10 + 2i$; $-4i$ **17.** $-5 + 15i$ **19.** $2 + 4i$
21. $8 + i$ **23.** $1 - 3i$ **25.** $8 - 15i$ **27.** 5 **29.** 13 **31.** 5 **33.** 2
35. $3 + i$ **37.** $(11 + 2i)/5$ **39.** $(2 + i)/5$ **41.** i

Exercise Set 3.9

1. 62 centimeters **3.** 7 centimeters **5.** 78.74 meters **7.** a. 697.1 square centimeters
b. 961 square centimeters **9.** 8.787×10^{-17} square centimeters;
1.406×10^{-15} square centimeters; 7.117×10^{-15} square centimeters
11. 1.412×10^{33} cubic centimeters **13.** 1.646×10^{26} cubic centimeters
15. 1.149×10^{34} cubic centimeters **17.** 1.528×10^{-8} **19.** 2.574×10^{-8}
21. a. 750 feet b. The distance is quadrupled. **23.** a. π b. The area is quadrupled.
25. $\frac{1}{2}g = 15.96$ **27.** Yes, because $I = 24.95n$; the constant of variation is 24.95. **29.** No
31. 1.55×10^{15} kilometers **33.** a. 3.1×10^{13} kilometers b. 1.33 parsecs
35. 6.9950 times greater **37.** 0.4307 times less **39.** Mercury, 88 days; Venus, 225 days;
Earth, 365 days; Mars, 687 days Jupiter, 4334 days; Saturn, 10,773 days; Uranus, 30,729 days;
Neptune, 60,287 days; Pluto, 90,675 days
41. 1 year $3180; 2 years $3370.80; 3 years $3573.04; 4 years $3787.43; 5 years $4014.67;
10 years $5372.54; 20 years $9621.40 **43.** $2518.86 Calculator Exercise: $7006.88
45. 1 year $10,800; 2 years $11,664; 3 years $12,597.12; 4 years $13,604.89; 5 years $14,693.28;
10 years $21,589.25; 20 years $46,609.57 **47.** $6805.84 Calculator Exercise: $4290.97

Review Exercises for Chapter 3

1. 1; 2; 4; 8; 16; 32; 64; 1/2; 1/4; 1/8; 1/16 **3.** 1; 16; 256; 1/16; 1/256; 2; 4; 8; 1/2; 1/4; 1/8
5. 1; 8; 64; 1/8; 1/64; 2; 1/2; 4; 1/4; 16 **7.** 0.0366 **9.** 81,400 **11.** 0.00556
13. 6.48×10^{3} **15.** 4.3×10^{-3} **17.** 5.16×10^{1} **19.** 3.6×10^{20} **21.** 1.2×10^{-1}
23. 8×10^{6} **25.** 5×10^{3} **27.** 1.8×10^{-9} **29.** 20 **31.** 25 **33.** 10 **35.** 5

37. $2\sqrt{2}$; $2 + \sqrt{2}$ **39.** $2\sqrt{3}$; $1 - \sqrt{3}$ **41.** $4\sqrt{2}$; $1 - \sqrt{2}$ **43.** $3\sqrt{2}$; $(3 + \sqrt{2})/2$
45. $2\sqrt{2}$ **47.** $\sqrt{6}/6$ **49.** $\sqrt{3} + \sqrt{2}$ **51.** 10 **53.** 2 **55.** 1/10 **57.** 1/2
59. 1000 **61.** 4 **63.** a^{10} **65.** a^4 **67.** a^{12} **69.** a^{-15} **71.** $x^3y^{-9}z^6$
73. $x^{-1}y^{-8}$ **75.** a^{-1} **77.** 1 **79.** ab^3 **81.** a^3b^4 **83.** $x^{1/6}$ **85.** $a^{1/6}$
87. a^{-3} **89.** a^4 **91.** a^2 **93.** $x^{-4n}y^{-2m}$ **95.** x^{4n+4} **97.** a^{2n} **99.** $2i$
101. $\sqrt{6}\,i$ **103.** $2\sqrt{3}\,i$ **105.** $D = -4$, $x = -1 + i$ **107.** $D = -4$, $x = 3 + i$
109. $D = -36$, $x = (-1 + 3i)/2$ **111.** a. $4 + 3i$ b. $2 + i$ c. $1 + 5i$ d. $(5 - i)/2$
113. a. 4 b. $2i$ c. 5 d. $(3 + 4i)/5$ **115.** ~~1256 square millimeters~~ 314 mm²
117. 9.23×10^{29} cubic centimeters **119.** 1 year \$1050; 2 years \$1102.50

Chapter 4 • Linear Equations and Inequalities

Exercise Set 4.1

1. 8, −2 **3.** 1, 2, 3, 4, 5, 6, 7, 1/2, 3/2 **5.** −4 **7.** None of these **9.** −3/2
11. 6 and 6; equivalent **13.** The set of x so that $x \geq -2$ and the set of x so that $x \leq -2$; not equivalent **15.** −1 and 1, −1; not equivalent
17. 5 **19.** 4 **21.** The set of numbers x so that $x > 10$
23. The set of numbers x so that $x \leq 3$ **25.** 4 **27.** The set of numbers x so that $x > -5$
29. 6 **31.** −4 **33.** −4 **35.** 4 **37.** No solution **39.** 5
41. The set of numbers x such that $x < 5$ **43.** The set of numbers x such that $x \geq 7$
45. The set of numbers x such that $x < -10$ **47.** The set of numbers x such that $x \geq 4$
49. The set of numbers x such that $x \leq 1/3$ **51.** The set of numbers x such that $x \leq -1/2$

Exercise Set 4.2

1. 1 **3.** 1 **5.** 7 **7.** 1/2 **9.** −1/3 **11.** 2/5 **13.** No solution **15.** 29/2
17. 15/4 **19.** The set of numbers x such that $x > 3$ **21.** The set of numbers y such that $y \geq 1/2$
23. The set of numbers x such that $x < -1/2$ **25.** The set of numbers x such that $x \leq -2$
27. The set of numbers x such that $x < -3$ **29.** The set of numbers x such that $x \geq -1$
31. The set of numbers t such that $t < 1/2$ **33.** The set of numbers x such that $x > 5$
35. The set of numbers x such that $x > -1/4$ **37.** 15 **39.** −1/2 **41.** 0
43. 6 **45.** 5 **47.** 4

Exercise Set 4.3

1. 2/3 **3.** 6/5 **5.** −20 **7.** 0 **9.** No solution **11.** 12
13. −4 **15.** 2/3 **17.** 0 **19.** No solution **21.** 0 **23.** No solution
25. The set of numbers x such that $3/4 \leq x < 13/4$ **27.** The set of numbers x such that $1 \leq x \leq 9$
29. The set of numbers x such that $-6 \leq x \leq 0$ **31.** The set of numbers x such that $-8 \leq x \leq -7$
33. The set of numbers x such that $-5/4 < x < 3/4$ **35.** 0.056 **37.** 3
39. The set of numbers x such that $x \leq 10$ **41.** The set of numbers x such that $x \leq 9.33$
43. 15 **45.** −3.89 **47.** 3.09 **49.** 22.16 **51.** −0.32

Exercise Set 4.4

1. $r = c/2\pi$ **3.** $M = Fr^2/Gm$ **5.** $n = \dfrac{l - a + d}{d}$ **7.** $n = \dfrac{P + F}{p - c}$ **9.** $r = \dfrac{S - a}{S}$

11. $r = \dfrac{ab}{a + b}$ **13.** $t = \dfrac{r - d}{rd}$ **15.** $r = \dfrac{S - a}{S - t}$ **17.** $h = \dfrac{V}{\pi r^2}$ **19.** $x = \dfrac{v}{v + b - 1}$

21. $y = -\dfrac{3}{2}x - 6$ **23.** $y = 3x + \dfrac{5}{2}$ **25.** $y = -\dfrac{4}{3}x + 5$ **27.** $y = \dfrac{1}{2}x - \dfrac{3}{2}$

29. $y = \dfrac{5}{4}x - 5$ **31.** $x = \dfrac{b' - b}{m - m'}$ **33.** $y = -\dfrac{3}{2}x + 3$ **35.** $y' = -y/x$

37. $y' = -x/y$ **39.** $y' = \dfrac{2y - x^2}{y^2 - 2x}$

Exercise Set 4.5

1. 6 **3.** 6 **5.** 12 **7.** π **9.** 5 **11.** 5 **13.** 5 **15.** 5 **17.** $|x - 5| < 2$

19. $|x - 2.5| < 1.2$ **21.** $|x - 1| < 4$ **23.** $|x + 1.55| \geq 3.05$ **25.** $|x + 3.5| \geq 2.5$

27. $|x - 2| \geq 1.4$ **29.** The set of numbers x such that $1 \leq x \leq 9$

31. The set of numbers x such that $-1 < x < 5$ **33.** The set of numbers x such that $-6 \leq x \leq 0$

35. The set of numbers x such that $-5/4 < x < 3/4$

37. The set of numbers x such that $x < -17$ or $x > 7$

39. The set of numbers x such that $x \leq -9$ or $x \geq -3$

41. The set of numbers x such that $3.555 < x < 3.585$

43. The set of numbers x such that $2.05 < x < 2.07$

Exercise Set 4.6

1. $14\frac{1}{2}$ meters by $22\frac{1}{2}$ meters **3.** 15/4 meters by 75/4 meters **5.** 25 feet

7. 4 inches by 7 inches **9.** Rate of boat is 12 mph; rate of airplane is 412 mph.

11. Rate of boat is 15 mph; rate of plane is 300 mph. **13.** 4 hours 20 minutes **15.** 15 liters

17. At least 3 liters **19.** At most 160 liters **21.** At least 7 units

23. At least 7.5 fluid ounces **25.** At most 5 units **27.** At most 10 fluid ounces

29. 30/11 hours **31.** 6/5 hours **33.** $22,000

35. a. $|P - 8.24| < 0.04$ b. $|x - 2.06| < 0.01$

37. a. $|P - 14.28| < 0.06$ b. $|x - 3.57| < 0.015$

39. a. $|C - 28.15| < 0.02$ b. $|r - 4.48| < 0.003$

41. 9 **43.** 120 apples **45.** 12/61 of a day **47.** 30 pearls **49.** 20 cubits

51. $29 **53.** 43/125 scudi per pound **55.** $10\frac{1}{2}$ denarii

Review Exercises for Chapter 4

1. -8 **3.** The set of numbers x such that $x \geq 5/3$ **5.** 1/2

7. The set of numbers x such that $x < 3$ **9.** 3

11. The set of numbers x such that $x > 4$

13. The set of numbers x such that $x \leq 21/2$ **15.** 1/2 **17.** 2 **19.** No solution

21. 1 **23.** No solution **25.** The set of numbers x such that

27. The set of numbers x such that $-1 \leq x \leq 2$

29. The set of numbers x such that $x \leq 2$ **31.** -30 **33.**

35. $s = \dfrac{S - \pi r^2}{\pi r}$ **37.** $y = -\dfrac{4}{3}x - 8$ **39.** $y' = \dfrac{y - 2x}{2y - x}$ **4**

47. 2 **49.** $|x - 5| < 3$ **51.** $|x - 1| < 4$ **53.** $|x + 2.5| < 1.5$

55. The set of numbers x such that $-5 < x < 7$

57. The set of numbers x such that $x \leq 1$ or $x \geq 7/3$

59. The set of numbers x such that $-15 \leq x \leq 9$ **61.** 4 meters by 10 meters

63. 33.3 liters **65.** At most 8 ounces

67. a. $|P - 64.16| < 0.04$ b. $|x - 16.04| < 0.01$

Chapter 5 • Quadratic Equations and Inequalities

Exercise Set 5.1

1. 3 and 8 **3.** -5 and -6 **5.** 1/2 and 2/3 **7.** $-4/5$ and $-3/2$ **9.** -4 and 1

11. 3 and 4 **13.** -15 and 2 **15.** 3 and $-1/2$ **17.** -4 and $-2/3$ **19.** 2/5

21. -12 and 4 **23.** 9 and -3 **25.** -8 and 6 **27.** 9 and -3 **29.** -3 and 7/2

31. 5 and $-9/2$ **33.** 3 and 4 **35.** -1 and $-1/2$ **37.** 0 and $-3/2$

39. 1 and -3

Exercise Set 5.2

1. $\sqrt{2}$ and $-\sqrt{2}$ **3.** $2\sqrt{3}$ and $-2\sqrt{3}$ **5.** $2i$ and $-2i$ **7.** $2\sqrt{2}i$ and $-2\sqrt{2}i$

9. $x^2 + 2x + 1 = (x + 1)^2$ **11.** $x^2 - 2x + 1 = (x - 1)^2$

13. $y^2 - 8y + 16 = (y - 4)^2$ **15.** $x^2 + 6x + 9 = (x + 3)^2$

17. $x^2 - x + \dfrac{1}{4} = \left(x - \dfrac{1}{2}\right)^2$ **19.** $x^2 - \dfrac{5}{3}x + \dfrac{25}{36} = \left(x - \dfrac{5}{6}\right)^2$

21. $x^2 + \dfrac{4}{3}x + \dfrac{4}{9} = \left(x + \dfrac{2}{3}\right)^2$ **23.** $-1 + \sqrt{3}$ and $-1 - \sqrt{3}$ **25.** 6 and -4

27. $-1 + i$ and $-1 - i$ **29.** 3 and 5 **31.** $-3 + \sqrt{6}$ and $-3 - \sqrt{6}$

33. $4 + \sqrt{2}$ and $4 - \sqrt{2}$ **35.** $(-1 + \sqrt{5})/2$ and $(-1 - \sqrt{5})/2$

37. $(1 + \sqrt{3}i)/2$ and $(1 - \sqrt{3}i)/2$ **39.** $(-2 + i)/2$ and $(-2 - i)/2$ **41.** 2 and $-1/3$

43. $(1 + \sqrt{2})/2$ and $(1 - \sqrt{2})/2$ **45.** $-3/2$ **47.** $(1 + \sqrt{10})/3$ and $(1 - \sqrt{10})/3$

Exercise Set 5.3

1. $D = 8$ **3.** $D = -4$ **5.** $D = 5$ **7.** $D = -3$ **9.** $D = 25$ **11.** $D = -16$

13. $D = 41$ **15.** $D = -4$ **17.** $D = -3$ **19.** $D = -44$

Now we give the solutions for exercises 1–20.

1. $-2 + \sqrt{2}$ and $-2 - \sqrt{2}$ **3.** $1 + i$ and $1 - i$ **5.** $(-1 + \sqrt{5})/2$ and $(-1 - \sqrt{5})/2$

$(1 + \sqrt{3}i)/2$ and $(1 - \sqrt{3}i)/2$ **9.** -2 and $1/2$ **11.** $(1 + 2i)/5$ and $(1 - 2i)/5$
$(-5 + \sqrt{41})/2$ and $(-5 - \sqrt{41})/2$ **15.** $(3 + i)/5$ and $(3 - i)/5$
$(-3 + \sqrt{3}i)/2$ and $(-3 - \sqrt{3}i)/2$ **19.** $(1 + \sqrt{11}i)/4$ and $(1 - \sqrt{11}i)/4$
21. $3 + 2\sqrt{3}$ and $3 - 2\sqrt{3}$ **23.** $1 + \sqrt{6}$ and $1 - \sqrt{6}$ **25.** $-3 + \sqrt{7}$ and $-3 - \sqrt{7}$
27. $1 + i$ and $1 - i$ **29.** $1 + \sqrt{5}$ and $1 - \sqrt{5}$ **31.** $7 + 3\sqrt{6}$ and $7 - 3\sqrt{6}$
33. $(1 + \sqrt{17})/4$ and $(1 - \sqrt{17})/4$ **35.** $(8 + \sqrt{34})/6$ and $(8 - \sqrt{34})/6$ **37.** No solution
39. $\sqrt{5}$ and $-\sqrt{5}$

Exercise Set 5.4

1. Sum is 7, product is 12; solution is two real numbers.
3. Sum is -4, product is -12; solution is two real numbers.
5. Sum is $-1/3$, product is $-1/3$; solution is two real numbers.
7. Sum is $-5/4$, product is $5/4$; solution is two complex numbers. **9.** $3/4$ and $-5/6$
11. $5/2$ and -15 **13.** 10 (repeated solution) **15.** No real number solution
17. No real number solution **19.** $-75 + 5\sqrt{465}$ and $-75 - 5\sqrt{465}$
21. 70 and 80 **23.** No real number solution **25.** No real number solution
27. 1 and -5 **29.** 3 and $-7/2$ **31.** No real number solution **33.** 1.0 and -0.3
35. 119.0 and 3.4 **37.** 5.5 and -0.9 **39.** -0.80 and -6.82 **41.** 1.95 and -38.47
43. 2.19 and -0.62

Exercise Set 5.5

1. The set of numbers x such that $x \leq 1$ or $x \geq 4$
3. The set of numbers x such that $-2 < x < 1$
5. The set of numbers x such that $x \leq -4$ or $x \geq -2$
7. The set of numbers x such that $-3 < x < -1/2$
9. The set of numbers x such that $x \neq 1$
11. The set of numbers x such that $-1/4 < x < 1$
13. The set of numbers x such that $x < -1 - \sqrt{2}$ or $x > -1 + \sqrt{2}$
15. The set of numbers y such that $-3 - \sqrt{6} \leq y \leq -3 + \sqrt{6}$
17. The set of all real numbers **19.** No solution
21. The set of numbers x such that $x < 0$ or $x > 1/2$ **23.** No solution
25. The set of all real numbers
27. The set of numbers x such that $x \leq -2/5$ or $x \geq 1/2$
29. The set of numbers x such that $x \leq -2/3$ or $x \geq 3/2$
31. The set of numbers x such that $x \leq -6$ or $x \geq 6$

Exercise Set 5.6

1. $2, -2, \sqrt{3}$, and $-\sqrt{3}$ **3.** $1/2, -1/2, \sqrt{2}$, and $-\sqrt{2}$
5. $\sqrt{2}, -\sqrt{2}, 2\sqrt{2}$, and $-2\sqrt{2}$ **7.** $6, -6, 8$, and -8 **9.** 9 and 16 **11.** 16
13. $1/3$ and $1/4$ **15.** -3 **17.** 64 and 729 **19.** 2 and 17

21. $x = \pm\sqrt{1-y^2}$ for $-1 \le y \le 1$ 23. $y = \pm\sqrt{a^2 - x^2}$ for $-a \le x \le a$
25. $y = \pm\sqrt{x^2 - 1}$ for $x \le -1$ or $x \ge 1$ 27. $y = \pm\frac{1}{2}\sqrt{4 - x^2}$ for $-2 \le x \le 2$
29. $y = \pm\frac{3}{2}\sqrt{4 - x^2}$ for $-2 \le x \le 2$ 31. $y = \pm\frac{a}{b}\sqrt{b^2 - x^2}$ for $-b \le x \le b$
33. $y = \pm\frac{2}{3}\sqrt{x^2 - 9}$ for $x \le -3$ or $x \ge 3$
35. $y = -3 \pm \sqrt{8 - x^2 - 3x}$ for $8 - x^2 - 3x \ge 0$ 37. $-a \pm a\sqrt{2}$
39. $-c$ and $a + b + c$

Exercise Set 5.7

1. 1 3. 7 5. 8 7. 9 9. 1/2 11. 14 13. No solution 15. 2 and 3
17. 6 19. 8 21. 9 23. 2 25. 100 27. 72 29. 21 31. No solution
33. 3 35. 2 37. 1 39. 1

Exercise Set 5.8

1. 4 meters by 12 meters 3. 3 centimeters by 9 centimeters 5. 6 inches and 8 inches
7. 3 mph 9. 45 mph 11. At 2:40 13. 9 inches 15. 13.5 centimeters
17. 2 hours and 3 hours
19. It is possible to have two such fields; one is 70 meters by 160 meters and the other is 80 meters by 140 meters.
21. Since the equation has no real number solution, it is impossible to have such a field.
23. After 3.4 seconds (going up) and 119.0 seconds (coming down)
25. Since the equation has no real number solution, the rocket never reached a height of 15 meters.
27. After 13.8 seconds and 29.5 seconds 29. A price of at least $1.50 per gallon
31. A price of at least $6 per toy 33. $0.75
35. $2.50 before the tax and $2.57 after the tax 37. 35 pesos 39. 100 arrows

Review Exercises for Chapter 5

1. 3 and -2 3. $-3/4$ and $1/2$ 5. $2\sqrt{3}$ and $-2\sqrt{3}$ 7. -2 and -5
9. 3 and -6 11. $3i$ and $-3i$ 13. $1 + 3i$ and $1 - 3i$ 15. $-1 + \sqrt{2}$ and $-1 - \sqrt{2}$
17. -1 19. 0 and $-2/3$ 21. 6 and $-9/2$ 23. 6 and -4 25. 3 and -1
27. $-1 + \sqrt{17}$ and $-1 - \sqrt{17}$ 29. -5 (repeated solution)
31. $(-1 + \sqrt{5})/2$ and $(-1 - \sqrt{5})/2$ 33. 1 35. 1 and $-1/2$ 37. 2 and -1
39. No solution 41. The set of numbers x such that $-4 < x < 1$
43. The set of numbers x such that $x < -2$ or $x > 2$
45. The set of numbers x such that $x < -1/2$ or $x > 3$
47. The set of all real numbers 49. $\sqrt{5}$ and $-\sqrt{5}$ 51. 4
53. $y = \pm\sqrt{9 - x^2}$ for $-3 \le x \le 3$ 55. $y = \pm\frac{2}{3}\sqrt{9 - x^2}$ for $-3 \le x \le 3$ 57. 5
59. 8 61. 0 and 2 63. No solution 65. 5 meters by 12 meters
67. 3 kilometers per hour
69. There are two such fields: one is 100 meters by 300 meters and the other is 150 meters by 200 meters.

Chapter 6 • Polynomial Equations

Exercise Set 6.1

	Degree	Leading Coefficient	Constant Term
1.	2	3	7
3.	17	1	1
5.	87	12	0
7.	8	−7	0
9.	6	−1	0

11. $3x^2 - 1$ **13.** $8x^2 - 2x - 13$ **15.** $3x^4 - x^3 + 3x^2 - 2x - 3$

17. $x^4 + 4x^3 - 3x^2 - x$ **19.** $2x^5 + x^3 + 2x + 1$

21. $P(-2) = 6$ (not a root); $P(-3) = 8$ (not a root); $P(-4) = 0$ (root); $P(4) = 120$ (not a root)

23. $Q(0) = -1$ (not a root); $Q(2) = 15$ (not a root); $Q(-2) = 15$ (not a root); $Q(-i) = 0$ (root)

25. $P(1) = 0$ (root); $P(-1) = 0$ (root); $P(2) = 0$ (root); $P(-2) = -12$ (not a root)

27. $P(1) = 15$ (not a root); $P(-1) = 15$ (not a root); $P(3) = 63$ (not a root); $P(-3) = 15$ (not a root)

29. $Q(1) = -2$ (not a root); $Q(-1) = -2$ (not a root); $Q(\sqrt{2}) = 0$ (root); $Q(-\sqrt{2}) = 0$ (root)

31. $P(0) = -1$ (not a root); $P(1) = 0$ (root); $P(-1) = 0$ (root); $P(i) = 0$ (root)

Exercise Set 6.2

1. $P(x) + Q(x) = 3x^3 + 2x^2 + x - 2$; $P(x) - Q(x) = x^3 + 4x^2 - 5x + 4$

3. $P(x) + Q(x) = 4x^3 - 3x^2 + 3x + 2$; $P(x) - Q(x) = 2x^3 - 3x^2 - 3x + 6$

5. $P(x) + Q(x) = 5x^3 - x^2 - 2x + 2$; $P(x) - Q(x) = x^3 + x^2 - 4x + 6$

7. $P(x) + Q(x) = x^4 + x^3 + 2x^2 + x - 1$; $P(x) - Q(x) = x^4 - x^3 - x - 3$

9. $P(x) + Q(x) = 2x^4 + 2x^2 + 2$; $P(x) - Q(x) = 2x^3 + 2x$

11. $P(x) + Q(x) = 2x^5$; $P(x) - Q(x) = 2$ **13.** $x^3 + 6x^2 + 5x - 6$

15. $3x^3 - 9x^2 - 7x + 21$ **17.** $x^4 + 3x^3 + 2x^2 + x - 1$ **19.** $2x^4 - 9x^3 + 8x^2 - 16x$

21. $x^6 - 1$ **23.** $x^5 + 8x^4 + 12x^3 + 9x^2 - 2x + 2$

25. Quotient is $3x^2 + 2x - 1$ and remainder is 0

27. Quotient is $3x^2 - 7x + 2$ and remainder is 0

29. Quotient is $3x^2 + 2x + 2$ and remainder is $11x + 5$

31. Quotient is $x^4 + x^2 + 1$ and remainder is 0

33. Quotient is $x^3 + 2x^2 - 3x + 4$ and remainder is -10

35. Quotient is $x^3 + x^2 + x + 1$ and remainder is 0

Exercise Set 6.3

1. Quotient is $4x^2 - 5$; remainder is -2 **3.** Quotient is $x^2 - 8x - 33$; remainder is 0

5. Quotient is $3x^3 + 3x^2 + 6x + 6$; remainder is 10 **7.** Quotient is $x^2 - 3x + 2$; remainder is 0

9. Quotient is $x^2 + x + 9$; remainder is -29

11. Quotient is $2x^4 - 2x^3 + 5x^2 - 21x + 21$; remainder is -45

13. Quotient is $x^2 - 1$; remainder is 0 **15.** Quotient is $x^2 + 2x + 5$; remainder is 8

17. Quotient is $x^3 - 7x - 6$; remainder is -24 **19.** Quotient is $x^2 - 2x + 5$; remainder is -12

21. Quotient is $3x^3 - 6x^2 + 15x - 30$; remainder is 64

23. Quotient is $2x^4 - 4x^3 + 11x^2 - 38x + 76$; remainder is -176

25. Quotient is $2x^2 - 4x - 6$; remainder is 0 **27.** Quotient is $4x^2 - 2x - 6$; remainder is 0

29. Quotient is $2x^2 - 6x - 1$; remainder is 7/2 **31.** Quotient is $8x^2 - 4x + 2$; remainder is 0

Exercise Set 6.4

1. $P(1) = 0$ and $P(-1) = 0$ **3.** $P(1) = -4$ and $P(-1) = 0$

5. $P(1) = 10$ and $P(-1) = 10$ **7.** $P(2) = 0$ and $P(-2) = 0$

9. $P(2) = 56$ and $P(-2) = 0$ **11.** $P(2) = 64$ and $P(-2) = 64$

13. $P(3) = 63$ and $P(-3) = 15$ **15.** $P(3) = 0$ and $P(-3) = -84$

17. $P(0.5) = 0$ and $P(-0.5) = 3.5$ **19.** $P(0.5) = 25.625$ and $P(-0.5) = 15.375$

21. Yes; the other factor is $x^2 + x + 1$ **23.** No **25.** No **27.** Yes **29.** Yes

31. Yes **33.** No **35.** No **37.** Yes **39.** Yes; $P(x) = (x - 1)(x^2 - 4)$

41. No **43.** Yes; $P(x) = (x - 2)(x^2 + x - 2)$ **45.** No

47. If $P(x) = x^n - 1$, then $P(1) = 1^n - 1 = 0$; so $x - 1$ is a factor of $P(x)$.

49. If $P(x) = x^n + 1$, then $P(-1) = (-1)^n + 1 = -1 + 1 = 0$ since n is odd; so $x + 1$ is a factor of $P(x)$.

Exercise Set 6.5

1. $\pm 1, \pm 2, \pm 3$, and ± 6 **3.** $\pm 1, \pm 2, \pm 4$, and ± 8 **5.** ± 1 and ± 19

7. $\pm 1, \pm 3, \pm 5, \pm 15, \pm 25$, and ± 75 **9.** $\pm 1, \pm 5, \pm 7$, and ± 35

11. $\pm 1, \pm 7, \pm 13$, and ± 91 **13.** $1, -1$, and 2 **15.** -4 **17.** No integer roots

19. $1, -3$, and 11 **21.** -1 and 3 **23.** -1 **25.** No integer roots

27. No integer roots **29.** 2 **31.** 1 and -1 **33.** -1 **35.** 1, 2, and -2

37. $-2, \sqrt{10}i$, and $-\sqrt{10}i$ **39.** 5, $(1 + \sqrt{3}i)/2$ and $(1 - \sqrt{3}i)/2$ **41.** $1, -1$, and 2/3

Review Exercises for Chapter 6

1. Sum is $x^3 + x + 2$; difference is $x^3 - x$; product is $x^4 + x^3 + x + 1$; quotient is $x^2 - x + 1$ with remainder 0

3. Sum is $x^3 + x$; difference is $x^3 - 2x^2 + x - 2$; product is $x^5 - x^4 + 2x^3 - 2x^2 + x - 1$; quotient is $x - 1$ with remainder 0

5. Sum is $x^3 - x$; difference is $x^3 - 2x^2 - x + 2$; product is $x^5 - x^4 - 2x^3 + 2x^2 + x - 1$; quotient is $x - 1$ with remainder 0

7. Quotient is $x + 4$; remainder is -14 **9.** Quotient is $x^2 - 2x + 1$; remainder is 0

11. Quotient is $3x^2 + 6x + 24$; remainder is 44 **13.** Quotient is $x^2 + 10$; remainder is -40

15. Quotient is $3x^3 + 4x^2 + 3x + 6$; remainder is 4

17. Quotient is $x^4 - x^3 - x^2 + 2x - 2$; remainder is 1 **19.** $P(1) = 1$ **21.** $P(-3) = -7$

23. $P(-1) = 0$ **25.** $P(-3) = 109$ **27.** $P(2) = 13$ **29.** $(x - 2)(x^2 + 1)$

31. $(x - 1)(x^2 - 2x + 1)$ **33.** $(x + 1)(x^4 - x^3 + x^2 - x + 1)$ **35.** -1

37. $-1, 3$, and -3 **39.** 1 and -2 **41.** No integer roots **43.** 1, 3, and -2

45. $-3, i$, and $-i$ **47.** 1 and 1/2

Chapter 7 • Functions and Graphs

Exercise Set 7.1

1. $y = f(x) = 3x$ **3.** $y = f(x) = x + 3$ **5.** $y = f(x) = x - 2$ **7.** $y = g(x) = x^3$
9. $y = g(x) = 4 - x$ **11.** $y = f(x) = x(x + 3)$
13. $f(2) = 6; f(1) = 3; f(0) = 0; f(-1) = -3; f(-2) = -6$
15. $f(2) = -2; f(1) = -1; f(0) = 0; f(-1) = 1; f(-2) = 2$
17. $g(2) = 4; g(1) = 1; g(0) = 0; g(-1) = 1; g(-2) = 4$
19. $f(2) = 3; f(1) = 4; f(0) = 5; f(-1) = 6; f(-2) = 7$
21. $g(2) = 8; g(1) = 1; g(0) = 0; g(-1) = -1; g(-2) = -8$
23. $f(2) = 1/2; f(1) = 1; f(0)$ is not defined; $f(-1) = -1; f(-2) = -1/2$
25. $g(2) = 6; g(1) = 2; g(0) = 0; g(-1) = 0; g(-2) = 2$
27. $f(2) = 14; f(1) = 10; f(0) = 6; f(-1) = 2; f(-2) = -2$
29. $f(2) = 10; f(1) = 4; f(0) = 0; f(-1) = -2; f(-2) = -2$
31. $f(2) = 4; f(1) = 3; f(0) = 0; f(-1) = -5; f(-2) = -12$
33. $-29{,}500; -27{,}000; -22{,}000$ **35.** $-34{,}000; -22{,}000; 2000$

Exercise Set 7.2

1. (1, 2), (2, 4), and (0, 0) **3.** (3, 0) and (0, −3) **5.** (0, 0), (2, 2), (−1, −1), and (−2, 2)
7. (0, 0), (2, 1), (1, −2), and (0, −1) **9.** (−1, 1), (0, 0), (1, −1), (2, −2), (3, −3), and (4, −4)
11. (−1, −3), (0, −4), (1, −3), (2, 0), (3, 5), and (4, 12)
13. (−1, 0), (0, −1), (1, 0), (2, 3), (3, 8), and (4, 15)
15. (−1, 9/2), (0, 3), (1, 3/2), (2, 0), (3, −3/2), and (4, −3)
17. (−1, 3), (0, 0), (1, −1), (2, 0), (3, 3), and (4, 8)
19. (−1, −4), no ordered pair for $x = 0$, (1, 4), (2, 2), (3, 4/3), and (4, 1)

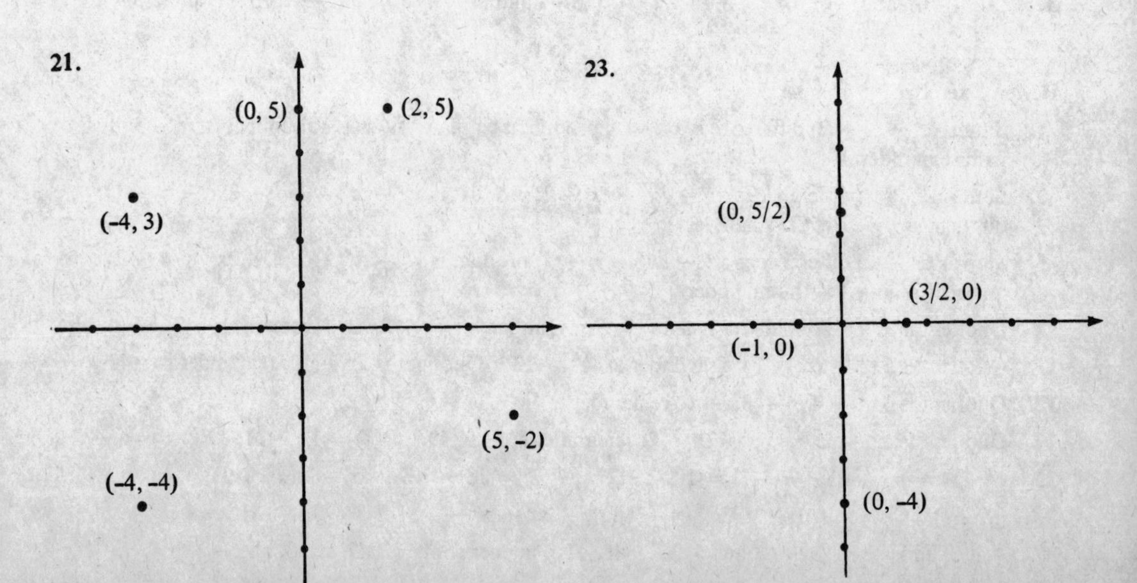

25.

(2, -1)
(1, -2)
(0, -3)
(-1, -4)
(-2, -5)

27.

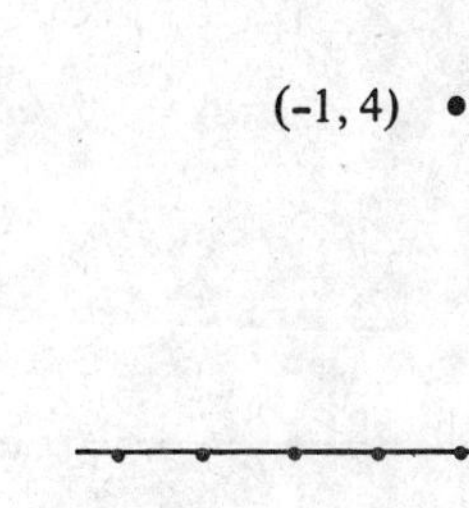

(-1, 4)
(0, 2)
(1, 0)

29.

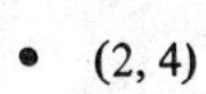

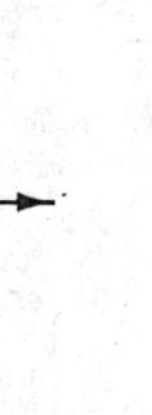

(2, 4)
(1.5, 2.25)
(1, 1)
(0, 0)

31.

(1, 1)
(2, 1/2)
(-2, -1/2)
(-1, -1)

33.

35.

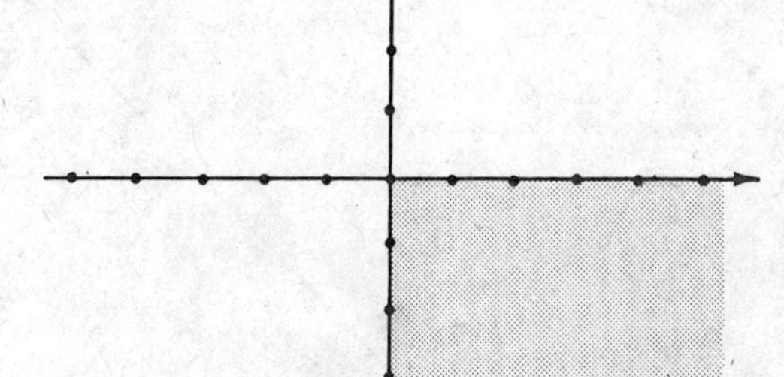

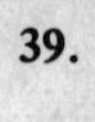

37.

39.

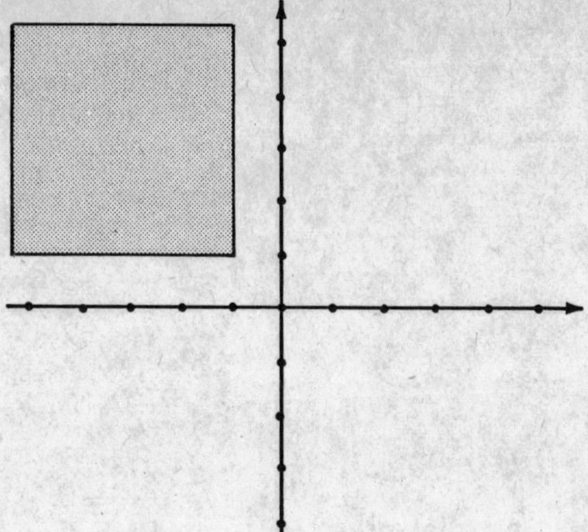

Exercise Set 7.3

1.

$f(x) = x$

3.

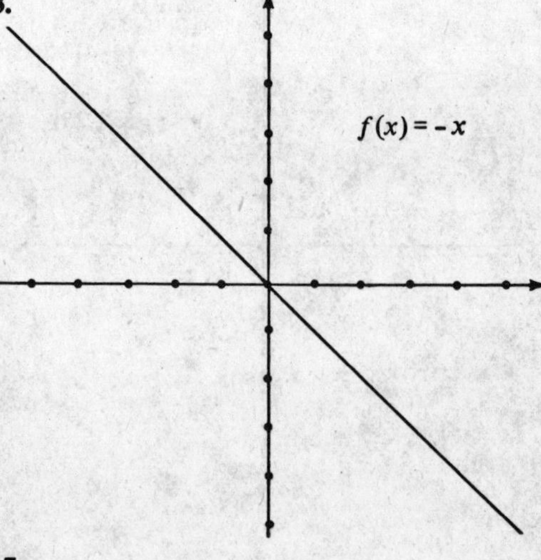

5.

$g(x) = x - 3$

7.

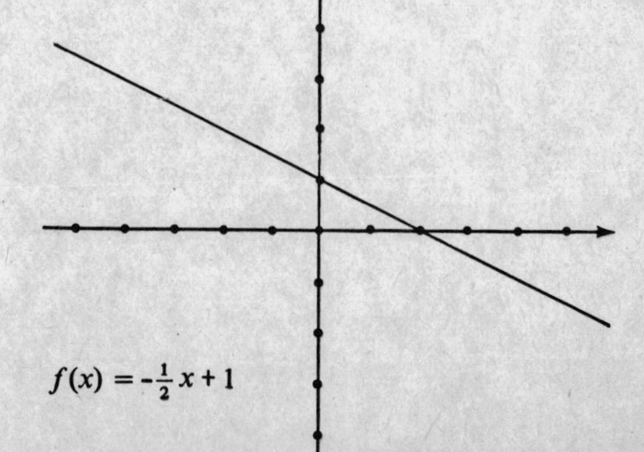

9.

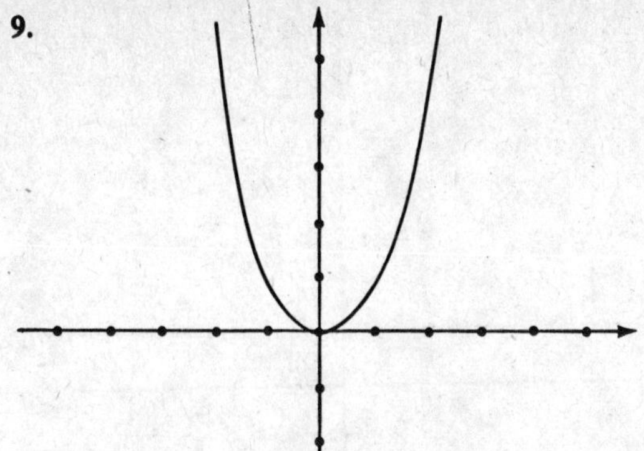

11.

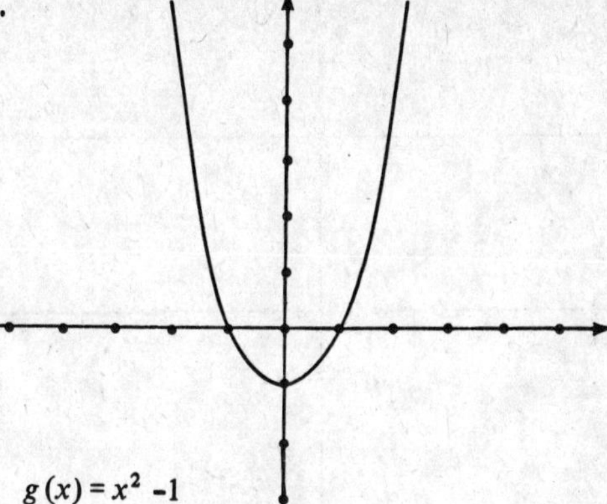

13.

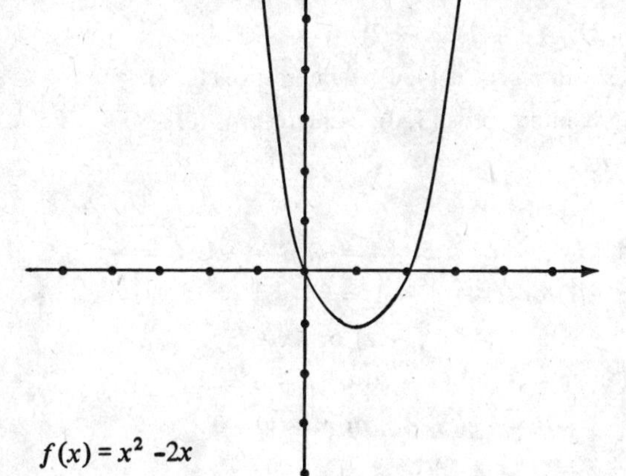

15.

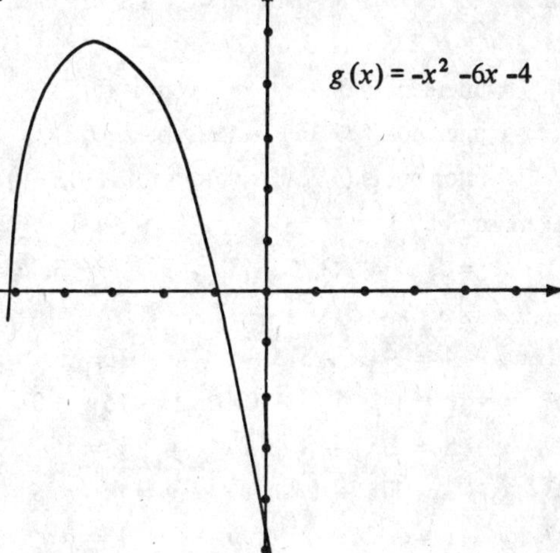

17.

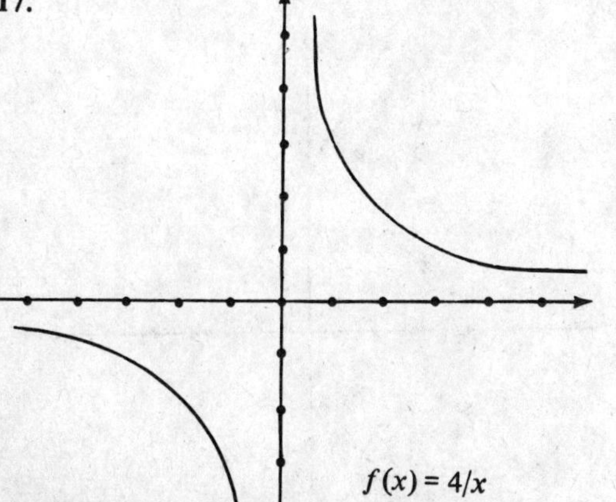

19.

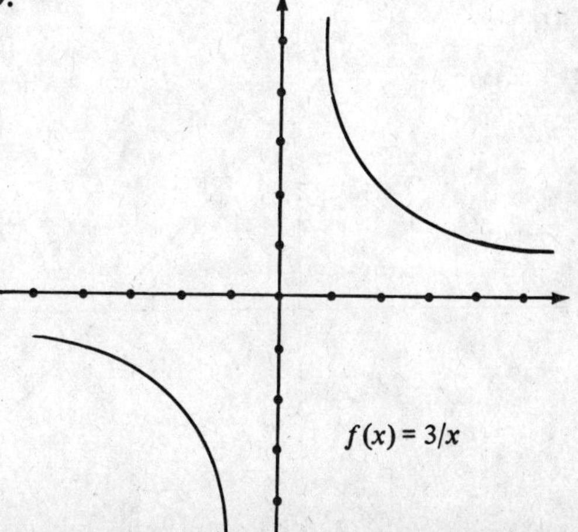

21.

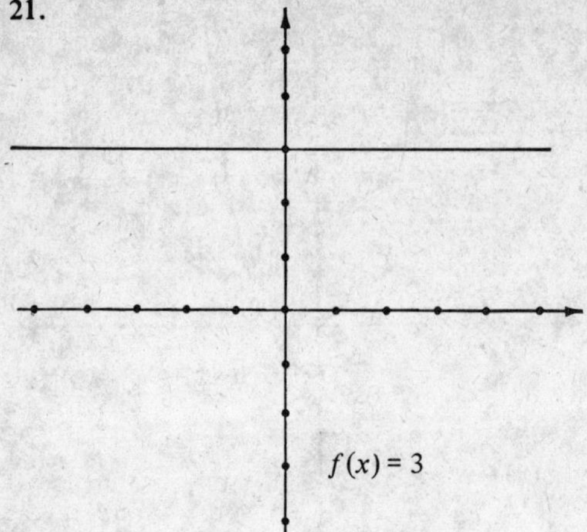

23.

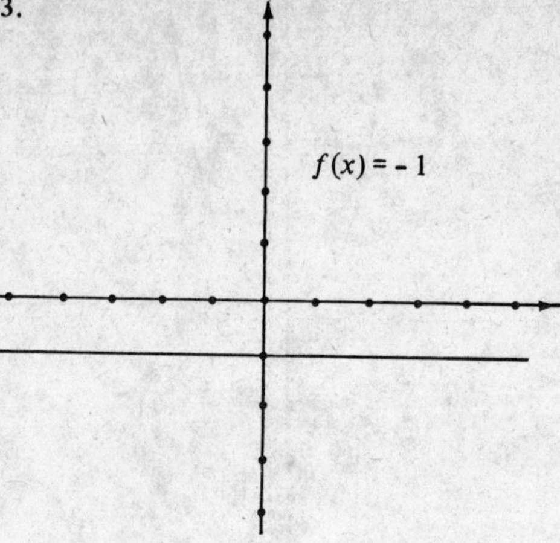

Exercise Set 7.4

1. 5 **3.** $2\sqrt{2}$ **5.** $5\sqrt{5}$ **7.** $\sqrt{74}$ **9.** $\sqrt{x^2 + y^2}$ **11.** $\sqrt{(x + 1)^2 + (y - 2)^2}$
13. isosceles **15.** isosceles **17.** In each case the slope is 1/2. **19.** 1 **21.** -2
23. 3/4 **25.** $-4/3$ **27.** $(y - 1)/(x - 1)$ **29.** $(y + 1)/(x + 2)$
31. x-intercept is (3, 0); y-intercept is (0, 2) **33.** x-intercept is (4, 0); y-intercept is (0, 4)
35. x-intercept is (2, 0); y-intercept is (0, 3) **37.** x-intercept is (3, 0); y-intercept is (0, -4)
39. x-intercept is (4/3, 0); y-intercept is (0, -4)

Exercise Set 7.5

1. $A = 2, B = -2, C = 3$ **3.** $A = 3, B = 4, C = -12$ **5.** $A = 2, B = -1, C = -5$
7. $A = 3, B = 5, C = -32$ **9.** $y - 1 = 2(x - 1)$ or $2x - y - 1 = 0$
11. $y - 4 = -(x + 3)$ or $x + y - 1 = 0$ **13.** $y + 5 = (4/3)(x - 2)$ or $4x - 3y - 23 = 0$
15. $y - 0 = (-3/5)(x - 0)$ or $3x + 5y = 0$ **17.** $y - 3 = 0(x - 2)$ or $y - 3 = 0$
19. $y - 8 = 2(x - 3)$ or $2x - y + 2 = 0$ **21.** $y - 1 = -(x + 1)$ or $x + y = 0$
23. $y - 3 = 3(x - 1)$ or $3x - y = 0$ **25.** $y - 2 = -(x + 2)$ or $x + y = 0$
27. $y - 1 = -(x + 2)$ or $x + y + 1 = 0$ **29.** $y = x + 3$ **31.** $y = -2x - 3$
33. $y = \frac{1}{2}x - 4$ **35.** slope is 4; y-intercept is (0, 6) **37.** slope is -1; y-intercept is (0, 6)
39. slope is $-3/2$; y-intercept is (0, 3)

41.

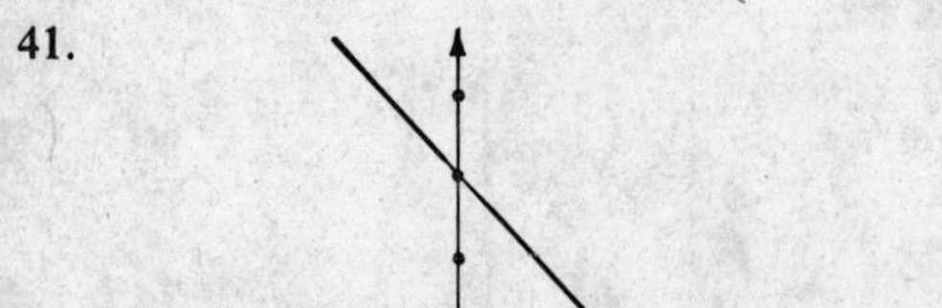

43.

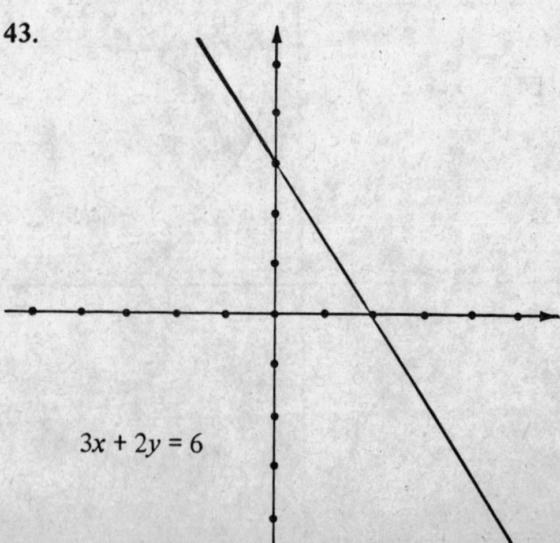

45.

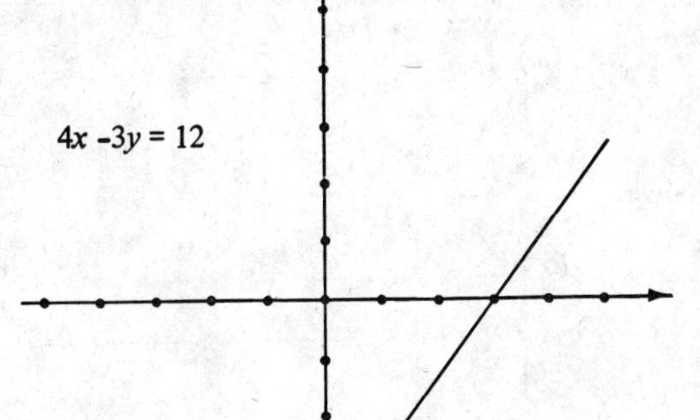

47.

$3x - y - 4 = 0$

Exercise Set 7.6

1.

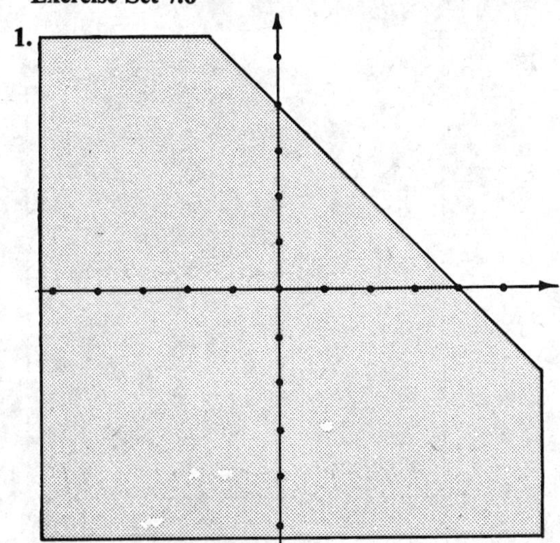

3.

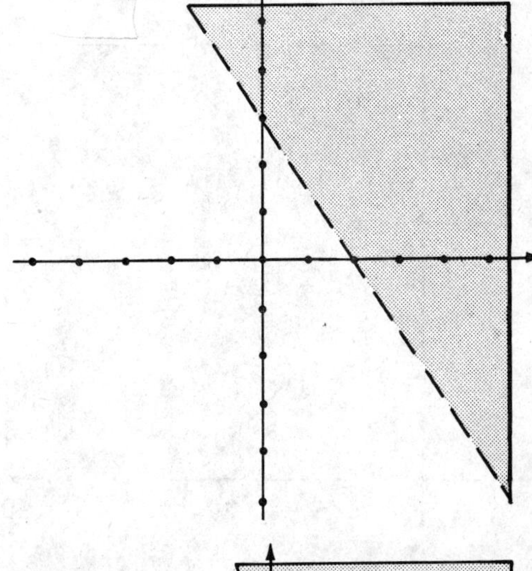

5.

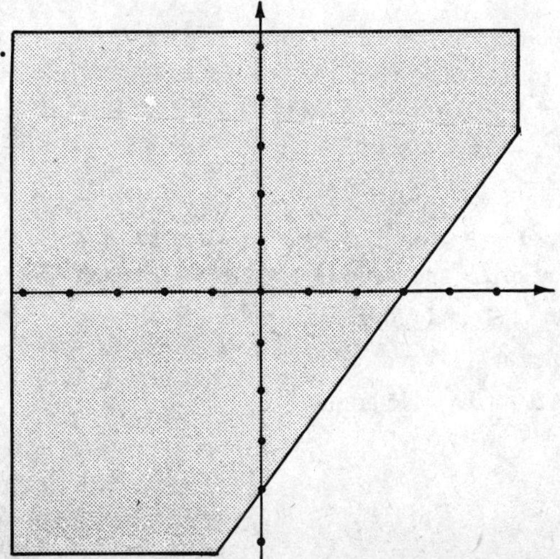

7.

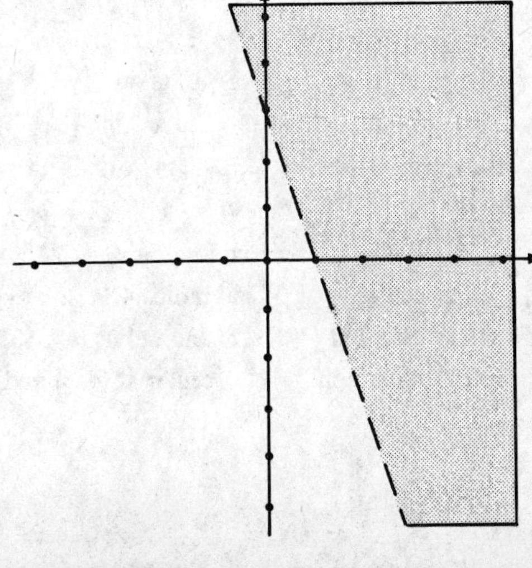

9.

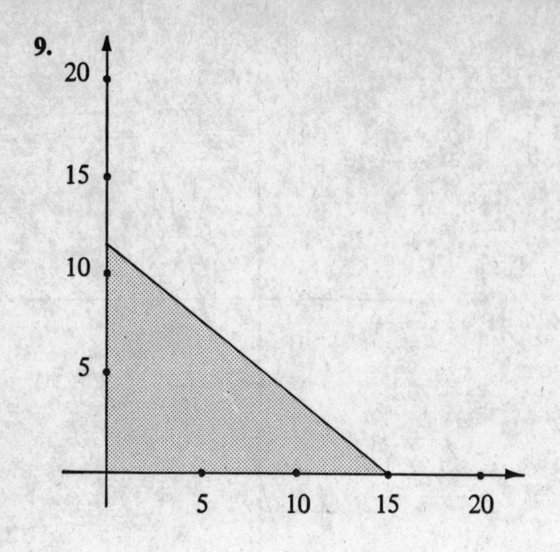

11.

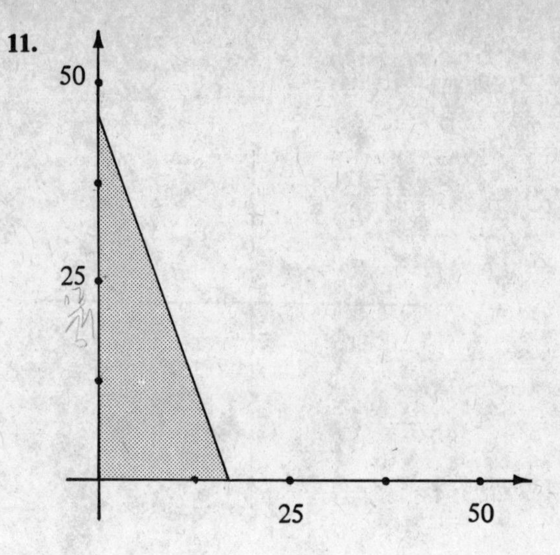

13.

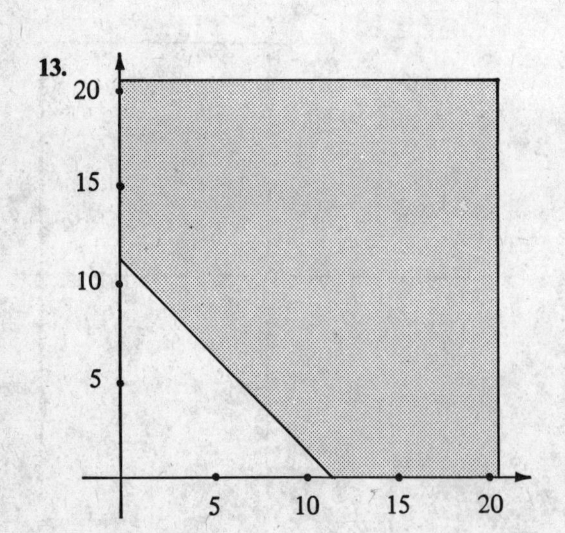

15.

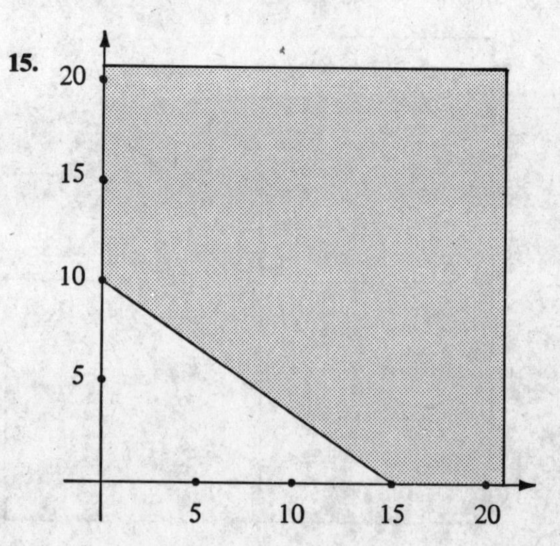

Exercise Set 7.7

1. $x^2 + y^2 - 10x - 10y + 49 = 0$ **3.** $x^2 + y^2 + 2x + 4y - 20 = 0$ **5.** $x^2 + y^2 - 4 = 0$

7. $x^2 + y^2 + 4x - 2y + 3 = 0$ **9.** $x^2 + y^2 - 4x - 4y = 0$

11. $x^2 + y^2 - 8x - 6y + 9 = 0$ **13.** $x^2 + y^2 - 6x + 6y + 9 = 0$

15. $x^2 + y^2 - 25 = 0$ **17.** $x^2 + y^2 - 2x - 2y = 0$ **19.** $x^2 + y^2 + 2x - 6y = 0$

21. center at origin and radius 5 **23.** center at origin and radius 5/2 **25.** No graph

27. center at (0, 3) and radius 4 **29.** center at (2, 2) and radius 3

31. center at $(-3, -5)$ and radius 6 **33.** center at (1, 1) and radius $\sqrt{2}$

35. No graph **37.** center at (4, 3) and radius 5 **39.** No graph

Exercise Set 7.8

1. $y = f(x) = 6 - x$; $f(1) = 5$; $f(2) = 4$; $f(3) = 3$; $f(5.5) = 0.5$

3. a. $y = 4 - x$ b. $A = x(4 - x)$ **5.** $y = g(x) = 4/x$; $g(1) = 4$; $g(2) = 2$; $g(3) = 4/3$; $g(0.5) = 8$ **7.** a. $P = 4x + 6$ b. $A = x(x + 3)$ **9.** a. $P = 4x$ b. $A = x^2$

11. a. $x = P/4$ b. $x = \sqrt{A}$ **13.** $P = f(n) = 0.70n - 28{,}000$; $f(50{,}000) = 7000$; $P(100{,}000) = 42{,}000$; break-even point is 40,000 **15.** $P = f(n) = 2.40n - 46{,}000$; $f(50{,}000) = 74{,}000$; $f(100{,}000) = 194{,}000$; break-even point is 19,167

17. $P = f(n) = 1.10n - 143{,}000$; $f(100{,}000) = -33{,}000$; $f(200{,}000) = 77{,}000$; break-even point is 130,000

19.

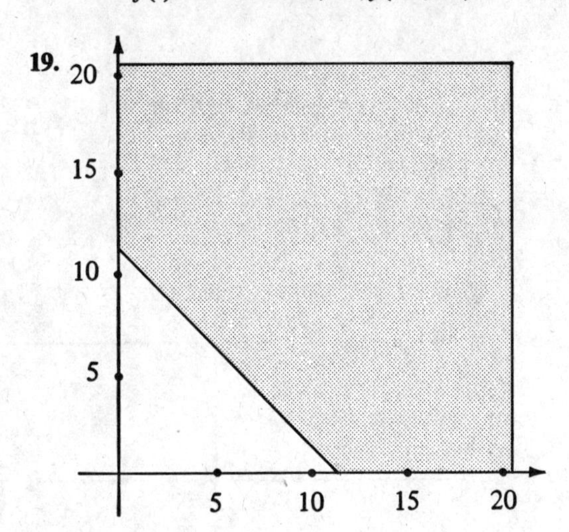

21.

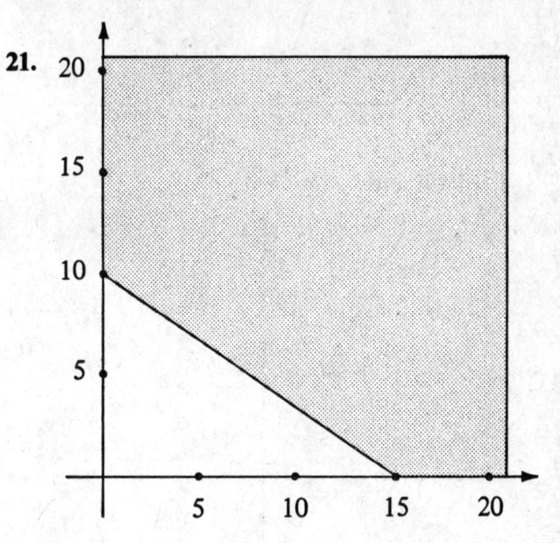

23.

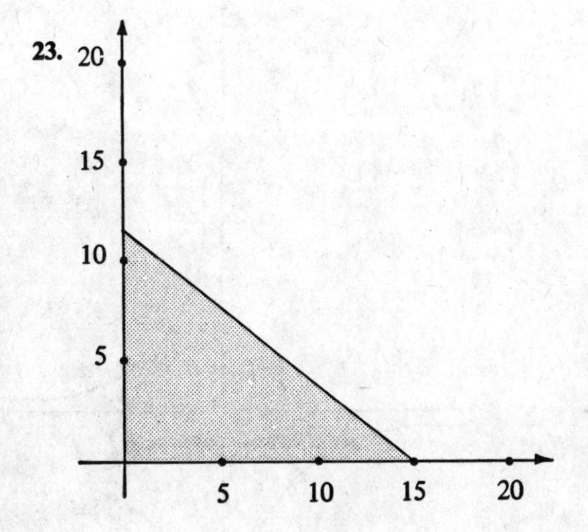

25.

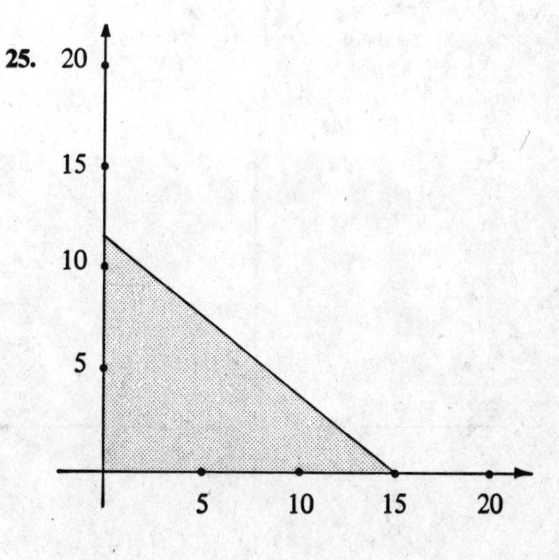

27.

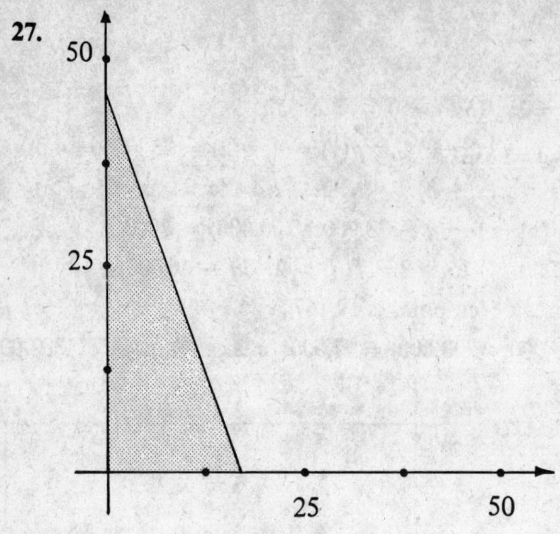

29. 21.84 meters **31.** 62.64 meters

33. 0.417 **35.** $5x - 2y + 15 = 0$

37. $x - 2y + 10 = 0$

Review Exercises for Chapter 7

1.

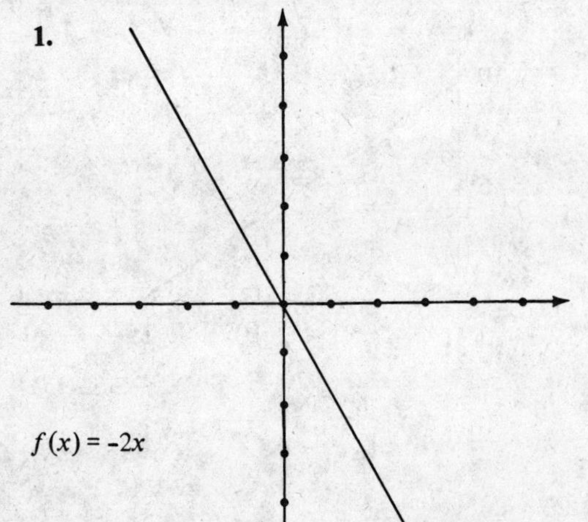

3.

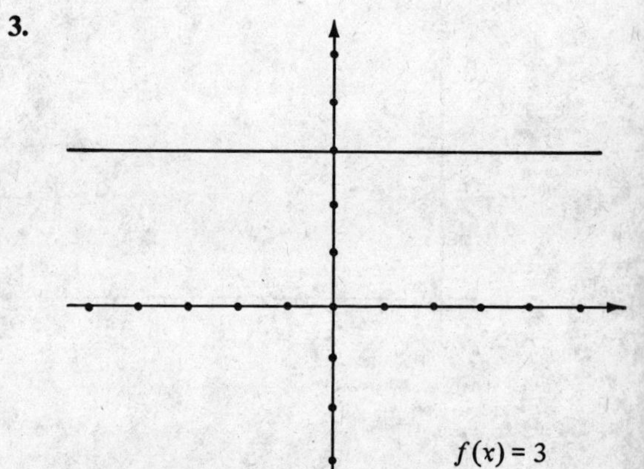

5.

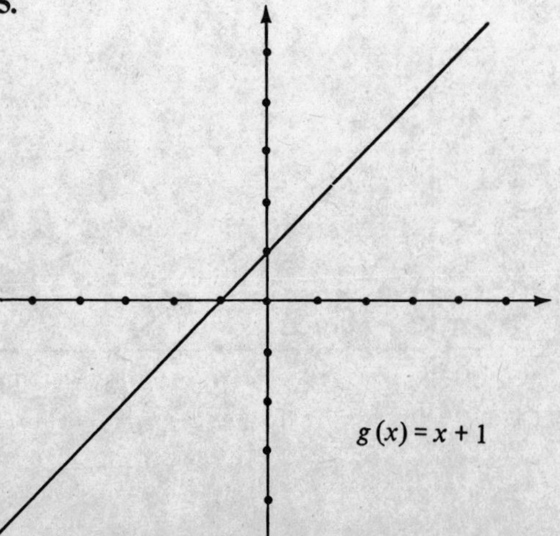

7.

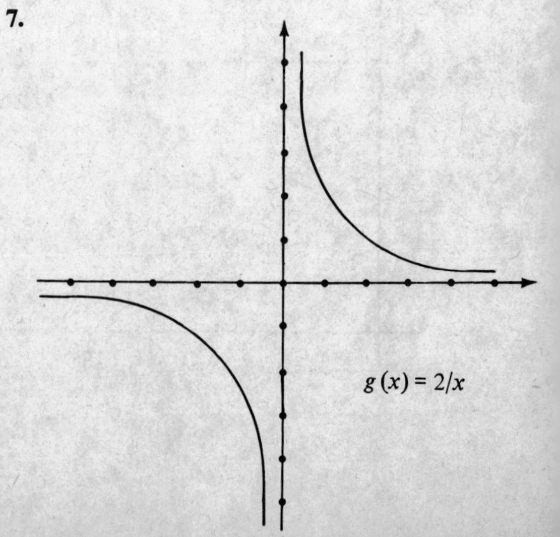

9.

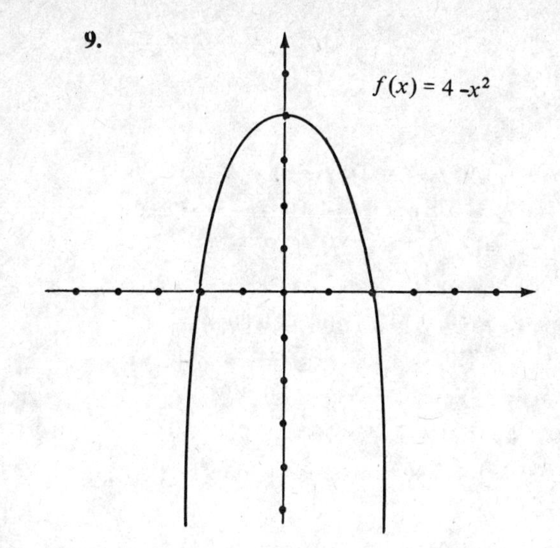

11.

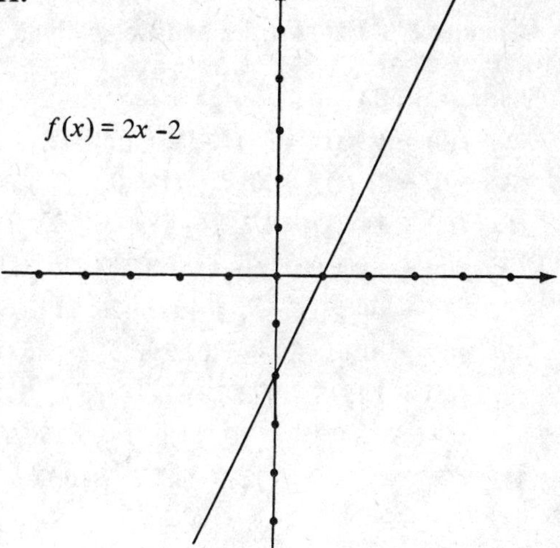

13.

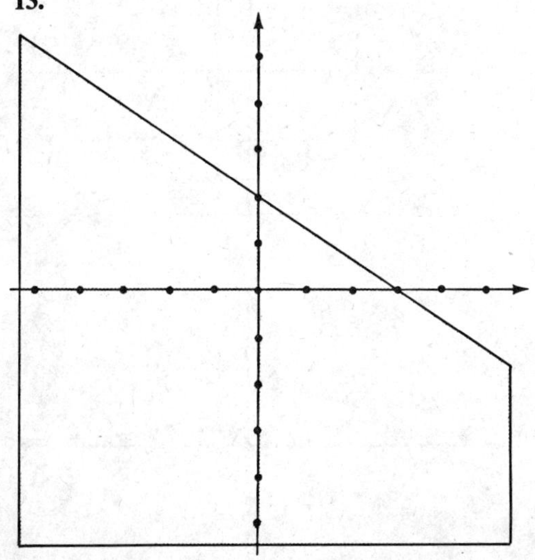

15.

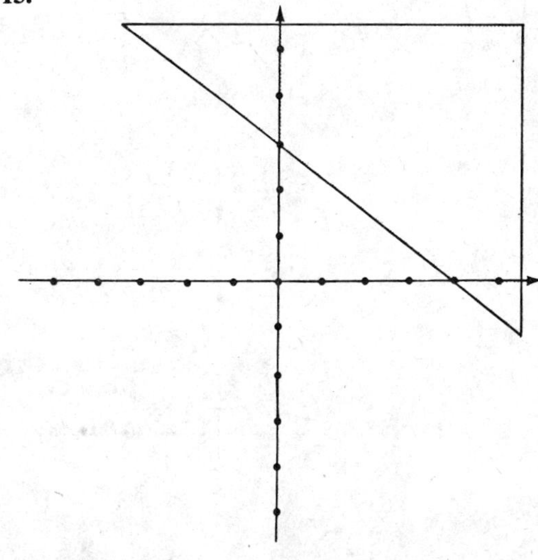

17. a. $2\sqrt{2}$ b. 1 c. $y = x$ **19.** a. 5 b. 3/4 c. $4y = 3x$ **21.** a. $5\sqrt{2}$ b. 1/7 c. $x - 7y - 11 = 0$ **23.** a. $4\sqrt{2}$ b. -1 c. $x + y + 2 = 0$
25. x-intercept is (5, 0); y-intercept is $(0, -3)$; slope is 3/5 **27.** x-intercept is (3/2, 0); y-intercept is (0, 9/4); slope is $-3/2$ **29.** $x^2 + y^2 + 2x - 4y - 4 = 0$ **31.** $x^2 + y^2 = 3$
33. $x^2 + y^2 + 6x - 10y + 25 = 0$ **35.** center is origin; radius is 5
37. center is (1, 1); radius is 2 **39.** No graph **41.** $f(2) = 3$; $f(1) = 1$; $f(0) = -1$; $f(-1) = -3$; $f(-2) = -5$ **43.** $f(2) = 6$; $f(1) = 2$; $f(0) = 0$; $f(-1) = 0$; $f(-2) = 2$
45. $f(2) = 0$; $f(1) = 1$; $f(0) = 0$; $f(-1) = -3$; $f(-2) = -8$ **47.** a. $y = 1 - x$ b. $A = x(1 - x)$ **49.** $P = f(n) = 0.40n - 24{,}000$; $P(50{,}000) = -4000$; $P(100{,}000) = 16{,}000$; break-even point is 60,000.

Chapter 8 • Exponential and Logarithmic Functions

Exercise Set 8.1

1. $f(0) = 1$; $f(1) = 3$; $f(-1) = 1/3$; $f(2) = 9$; $f(-2) = 1/9$; $f(3) = 27$; $f(-3) = 1/27$
3. $f(0) = 1$; $f(1) = 10$; $f(-1) = 0.1$; $f(2) = 100$; $f(-2) = 0.01$; $f(3) = 1000$; $f(-3) = 0.001$
5. $f(0) = 1$; $f(1) = 1/3$; $f(-1) = 3$; $f(2) = 1/9$; $f(-2) = 9$; $f(3) = 1/27$; $f(-3) = 27$
7. $f(6) = 1{,}000{,}000$; $f(-6) = 0.000001$; $f(9) = 1{,}000{,}000{,}000$; $f(-9) = 0.000000001$
9. $g(0) = 10$; $g(1) = 5$; $g(-1) = 20$; $g(2) = 2.5$; $g(-2) = 40$ **11.** $g(0) = 1000$; $g(1) = 1050$; $g(2) = 1102.5$; $g(3) = 1157.625$ **13.** $f(10) = 1.708144$; $f(25) = 3.813392$
15. $f(8) = 1.137639$; $f(20) = 1.380420$ **17.** $f(1.5) = 2.8$; $f(2.5) = 5.6$; $f(-0.5) = 0.7$; $f(-1.5) = 0.35$ **19.** $f(0.4) = 1.96$; $f(0.6) = 2.74$; $f(1.2) = 7$; $f(-0.2) = 0.71$
21. $f(0.3) = 2.197$; $f(0.5) = 3.71$; $f(-0.1) = 0.77$; $f(-0.3) = 0.46$
23. $f(x) = 5^{-x}$ **25.** $f(x) = 2^{2x}$ **27.** $f(x) = 10^{2x}$

29. **31.**

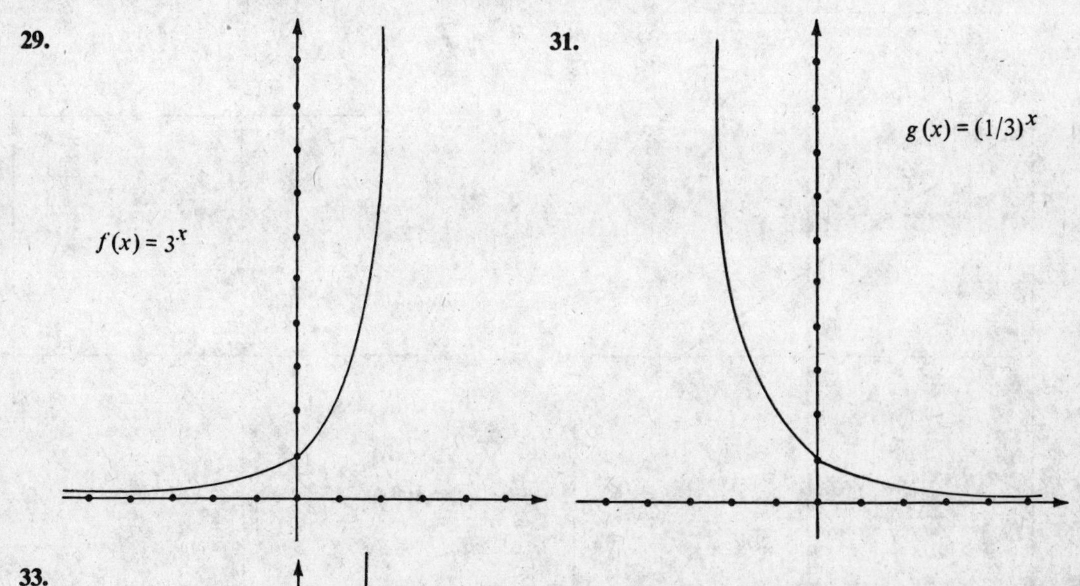

33.

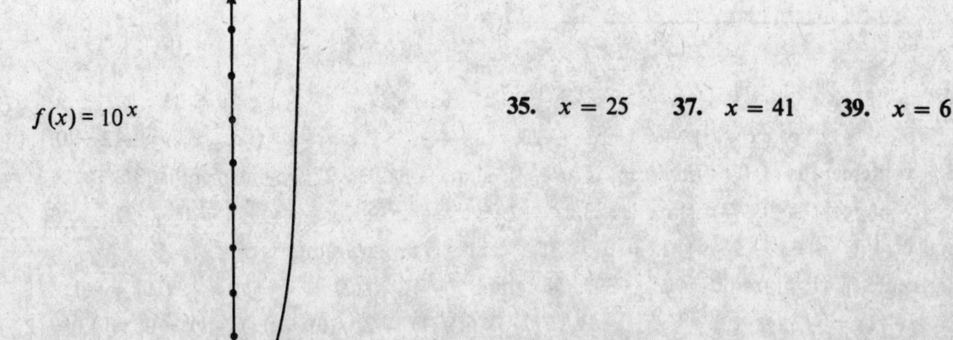

35. $x = 25$ **37.** $x = 41$ **39.** $x = 6$

Exercise Set 8.2

1. 2; 3; 1; 0; -1 **3.** 1; 2; 3; -1; -2; -3 **5.** 1; 2; 1/2; 0; $-1/2$

7. $g(1) = 0$; $g(5) = 1$; $g(\sqrt{5}) = 1/2$; $g(0.2) = -1$

9. $g(1) = 0$; $g(10) = 1$; $g(\sqrt{10}) = 1/2$; $g(\sqrt{1000}) = 3/2$; $g(1/\sqrt{10}) = -1/2$

11. $g(9) = 1$; $g(81) = 2$; $g(3) = 1/2$; $g(1) = 0$; $g(1/3) = -1/2$

13. 0.20; 0.35; 0.70; 0.45; 0.85

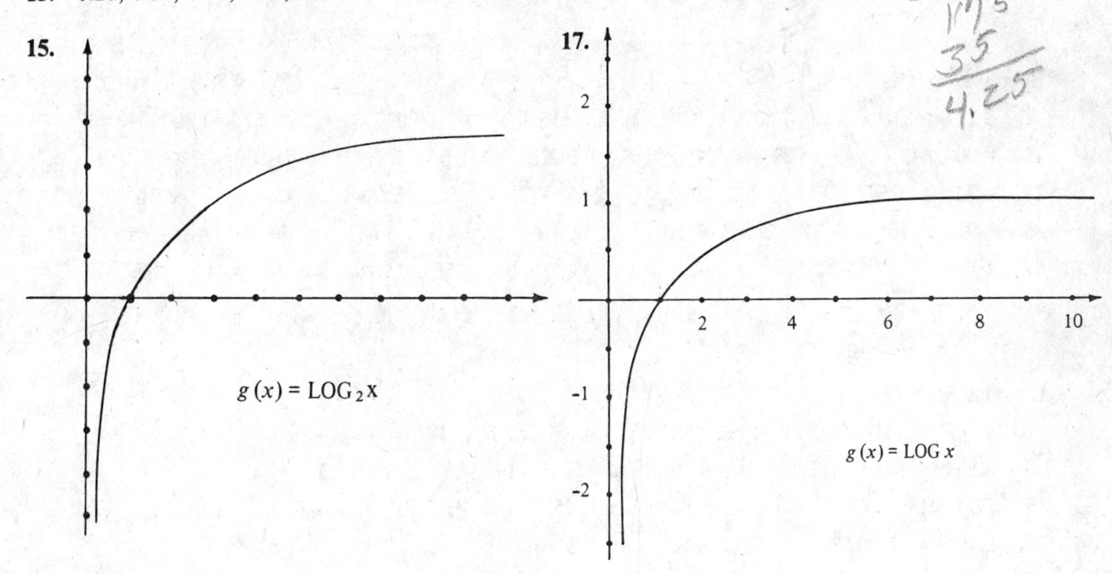

19. $x = \log_2 100$ **21.** $x = \log_3 10$ **23.** $x = 2 = \log_5 5^2$ **25.** $x = 1 + \log 4$

27. $x = -1 + \log 6$ **29.** $x = \frac{1}{2} \log 6.28$

Exercise Set 8.3

1. $\log 10 = 1$, $\log 100 = 2$, and $\log 1000 = 3$ **3.** $\log 100 = 2$, $\log 1000 = 3$, and $\log 0.1 = -1$

5. $\log 10 = 1$ and $\log 1000 = 3$

7. No; $\log(2 + 1) = 0.4771$ and $\log 2 + \log 1 = 0.3010$

9. No; $(\log 3)(\log 1) = 0$ and $\log(3 \cdot 1) = 0.4771$ **11.** $\log_b y + \log_b z$ **13.** $-\log_b x$

15. $-3 \log_b x$ **17.** $3[\log_b x - \log_b y]$ **19.** $\frac{1}{2}[\log_b x + \log_b y]$

21. $\frac{1}{2}[\log_b x + \log_b y - 3 \log_b z]$ **23.** $\log_b x + \log_b y - \log_b w - \log_b z$

25. $\frac{1}{2}[\log_b x + \log_b y] - 2 \log_b w - 3 \log_b z$ **27.** $\log_b w + \frac{1}{2}[\log_b x - \log_b y]$

29. $\frac{1}{2}[1 + \log_b x - \log_b y]$

31. This follows from $\log_b (x_1/x_2) = \log_b (t_1/t_2)$ and the logarithm of a quotient.

33. $\log 36.52 = 1.5625$; $\log 3652 = 3.5625$ **35.** $\log 314 = 2.4969$; $\log (3.14)^2 = 0.9938$

37. $\log 25 = 1.3980$; $\log 125 = 2.0970$; $\log 0.2 = -0.3010$ **39.** 0.6020 **41.** 1.5050

43. 1.2552 **45.** 3.0791 **47.** -0.6990 **49.** -1.6990 **51.** 0.3522 **53.** 0.8293

55. 0.1505 **57.** 0.3891 **59.** 0.4515

Exercise Set 8.4

1. 4.9894 **3.** 2.9238 **5.** 8.6274 – 10 **7.** 3.8235 **9.** 8.3874 – 10 **11.** 2.9165
13. 0.3010 **15.** 8.8876 – 10 **17.** 3.9952 **19.** 9.4346 – 10 **21.** 43.7 **23.** 789
25. 56,000 **27.** 0.467 **29.** 0.0661 **31.** 0.000273 **33.** 1.3298 **35.** 8.8239 – 10
37. 2.5698 **39.** 4.7869 **41.** 9.7225 – 10 **43.** 0.0233 **45.** 54.55 **47.** 7.523
49. 0.6187 **51.** 696.3 **53.** 1.585 **55.** 0.03678

Exercise Set 8.5

1. 3.211×10^9 **3.** 4.607×10^{-6} **5.** 116.3 **7.** 0.003724 **9.** 22,850,000
11. 0.0002847 **13.** 922,800,000 **15.** 1792 **17.** 0.7820 **19.** 8050 **21.** 1.710
23. 1.738 **25.** 1.136 **27.** 992.8 **29.** 259.7 **31.** 0.6805 **33.** 1.259×10^{30}
35. 2.923×10^{19} **37.** 3.568×10^{13} **39.** 1.513 **41.** 5.081 **43.** 146.4 **45.** 3.070
47. 10.83 **49.** 88.53 **51.** 41.23 **53.** 10.91 **55.** 11.44 **57.** 44.41 **59.** 49.83
61. 2000 **63.** –14.41

Exercise Set 8.6

1. a. $P = 211.9\,(1.008)^t$ b. 261 million c. In 2017 d. In 2060
3. a. $P = 825\,(1.017)^t$ b. 993 million c. In 1985 d. In 2015
5. Population of Pakistan will be double in 1995. **7.** In 2010 **9.** 68.0%
11. a. $G = G_0\,(1/2)^{t/15}$ b. 33% c. 49.8 hours **13.** 2000 years old
15. a. Compounded yearly: \$1055; compounded quarterly: \$1056.14
b. Compounded yearly: \$1708.14; compounded quarterly: \$1726.77
c. Compounded yearly: \$3813.39; compounded quarterly: \$3918.20
17. a. $A = 1000\,(1.01625)^{4t}$ b. \$1137.63 c. \$3631.15 d. 10.75 years
19. \$11,016

Review Exercises for Chapter 8

1. a. $f(2) = 36; f(1) = 6; f(0) = 1; f(-1) = 1/6; f(-2) = 1/36$
b. $f(0.2) = 1.44; f(0.3) = 1.728; f(0.9) = 5$
c. $f(20) = 3.664 \times 10^{15}; f(50) = 8.128 \times 10^{38}$
3. $f(4) = 25; f(6) = 125; f(-2) = 1/5; f(1) = \sqrt{5}; f(3) = 5\sqrt{5}$ **5.** 1; 2; 0; –2; 1/2
7. $2 \log x + \log y$ **9.** $-2 \log x$ **11.** $\log x - 2 \log y$ **13.** $3 \log x - 2 \log y$
15. $(3/2) \log x$ **17.** $(1/2) \log x$
19. a. 1.6021 b. 9.6021 – 10 c. 2.6021 d. 7.6021 – 10 e. 1.2042 f. 0.3010
g. –0.6021 h. –0.6021
21. a. 1.4772 b. 0.4772 c. 1.1762 d. 0.1762 e. 0.3495 f. 0.3891
g. –0.6990 or 9.3010 – 10 h. 1.0792
23. 48.22 **25.** 75.32 **27.** 7.140×10^{-7} **29.** 9.824×10^{47} **31.** 35.55
33. 0.01421 **35.** 7.27 **37.** 1.71 **39.** –6.21
41. a. $P = 52.5\,(1.008)^t$ b. 64.6 million c. In 2063
43. a. $A = 1000\,(1.05)^t$ b. \$11,467.39 c. 14.2 years

Chapter 9 • Systems of Equations and Inequalities

Exercise Set 9.1

1. (10, 2) **3.** (1, 2) **5.** (−2, −6) **7.** (−2, −1) **9.** (1/2, −1/2) **11.** (2, 1)
13. (5, 4) **15.** (2, −3) **17.** (2, −5) **19.** (1/2, 1/3) **21.** Inconsistent
23. Simultaneous, with solution (4, 1) **25.** Dependent **27.** Inconsistent
29. Dependent **31.** (1.2, 0.8) **33.** (10, 4) **35.** (2.5, 7.5) **37.** (1.30, 0.06)
39. (1.53, 1.52) **41.** (1.11, 0.55)

Exercise Set 9.2

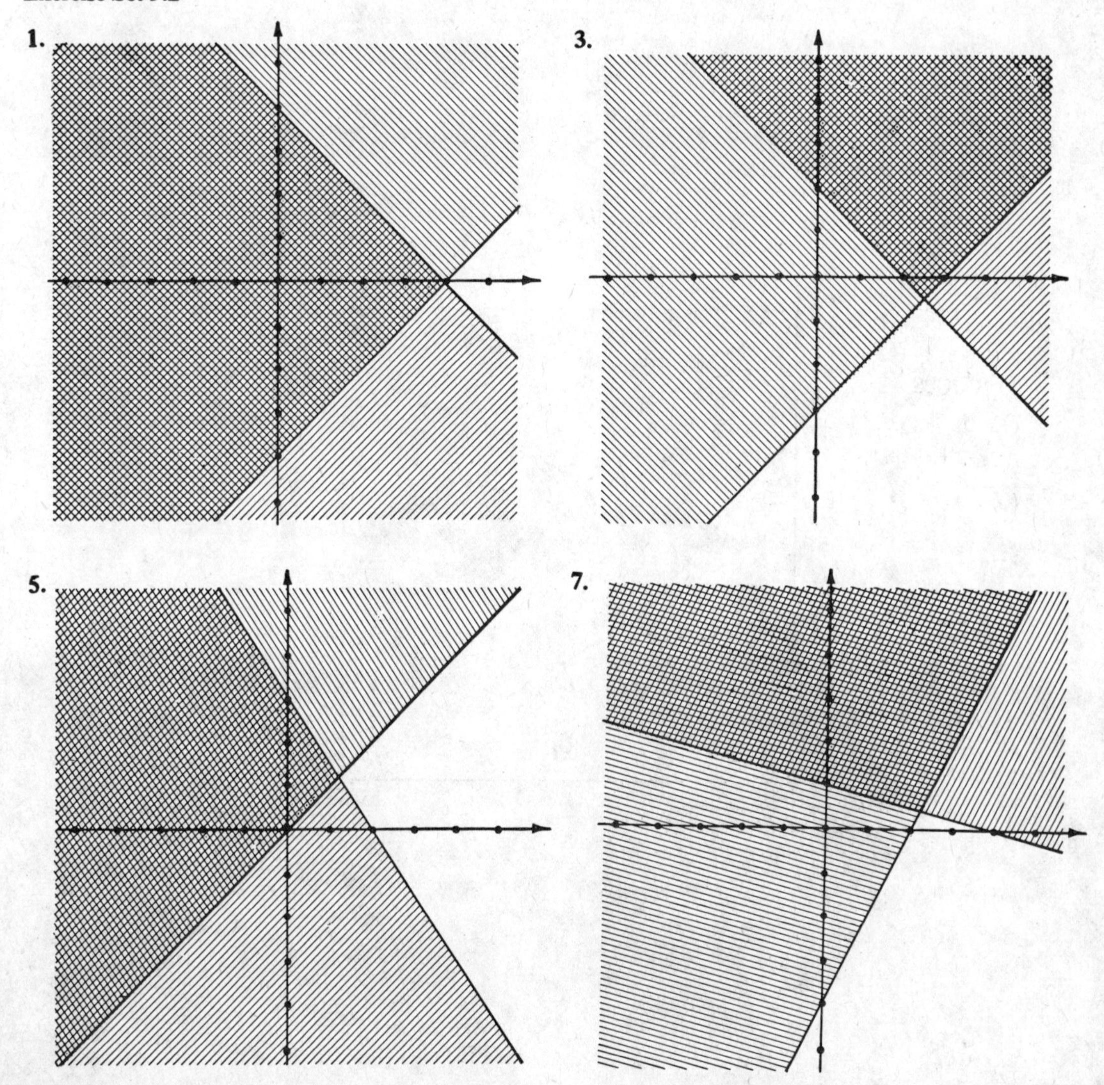

9.

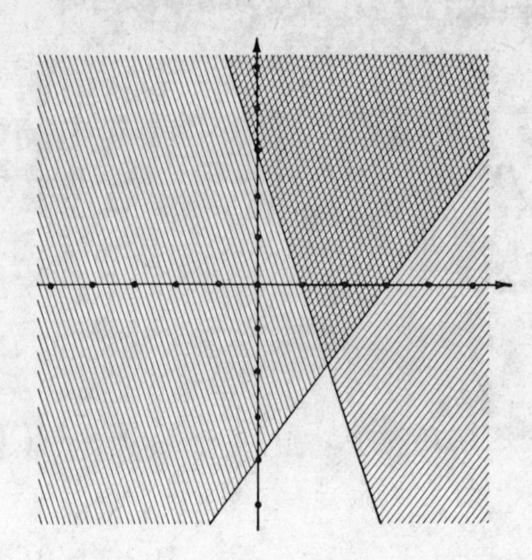

11.

VERTICES AT
(15, 0)
(6, 6)
(0, 12)

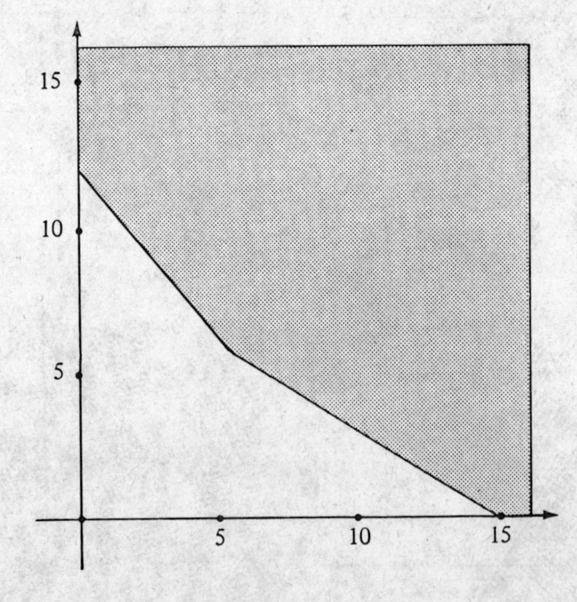

13.

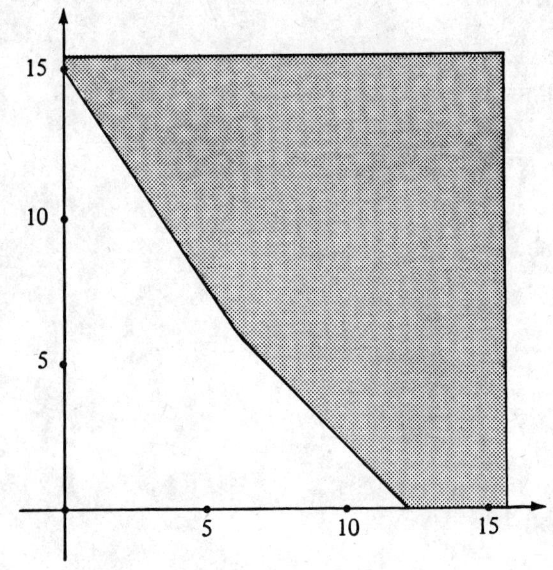

VERTICES AT
(12, 0)
(6, 6)
(0, 15)

15.

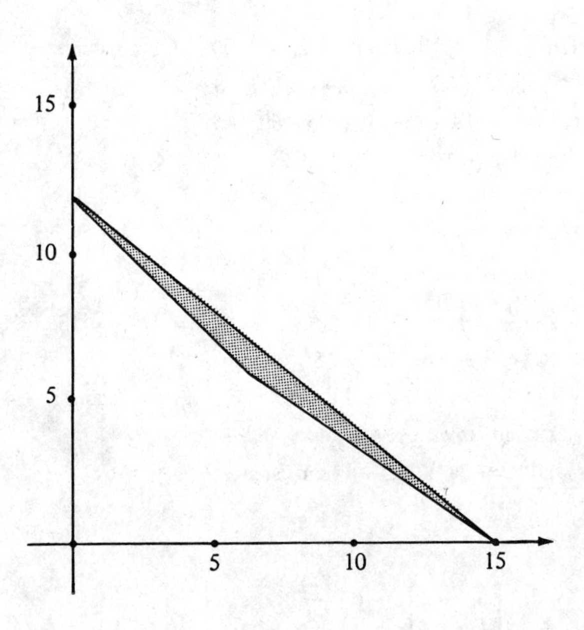

VERTICES AT
(15, 0)
(6, 6)
(0, 12)

17.

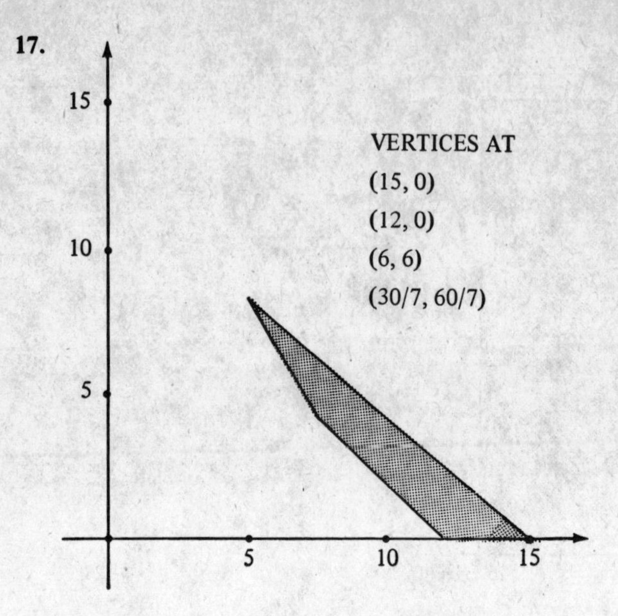

Exercise Set 9.3

1. (−1, −7) and (2, 2) **3.** Inconsistent **5.** (2, 3/2) and (3/2, 2) **7.** Inconsistent
9. (10, 8) **11.** (0, 2) and (4, 2) **13.** (8, 6), (−8, −6), (6, 8), and (−6, −8)
15. (4, 7) and (7, 4) **17.** (4, 12) and (12, 4) **19.** (50, 6) and (−60, −5)
21. (8, −6) and (−6, 8) **23.** (1, 1/2) and (1/6, −1/3) **25.** (2, 2) **27.** (4, 4)
29. (9, 12) and (−12, −9)

Exercise Set 9.4

1. $x = 1, y = -1, z = 3$ **3.** $x = -1, y = -1, z = -1$ **5.** $x = 9, y = 16, z = 13$
7. $x = -2, y = -8, z = 3$ **9.** $w = 12, x = 9, y = 6, z = 16$
11. $x = 1/2, y = -1/2, z = 3/2$ **13.** $x = 1, y = -1, z = 1$ **15.** $x = 4, y = 4$, and $z = 4$
17. $w = 2, x = 0, y = 0, z = 0$ **19.** Dependent with $y = 0$ and $z = 1 - x$
21. Simultaneous with $x = 3/2, y = -1$, and $z = 1/4$ **23.** Inconsistent

Exercise Set 9.5

1. a. 3 b. $\begin{bmatrix} -2 & -3 \\ 4 & 2 \end{bmatrix}$ c. 1 d. $\begin{bmatrix} 2 & -2 \\ 1 & 4 \end{bmatrix}$ e. −2 f. $\begin{bmatrix} 3 & 1 \\ 1 & 2 \end{bmatrix}$

3. a. 1 b. $\begin{bmatrix} 5 & -4 \\ 4 & 1 \end{bmatrix}$ c. 5 d. $\begin{bmatrix} 2 & 5 \\ 3 & 4 \end{bmatrix}$ e. 5 f. $\begin{bmatrix} 1 & 5 \\ 3 & 1 \end{bmatrix}$

5. a. 2 b. $\begin{bmatrix} 2 & -3 \\ 0 & 2 \end{bmatrix}$ c 0 d. $\begin{bmatrix} 0 & 2 \\ -3 & 0 \end{bmatrix}$ e. 2 f. $\begin{bmatrix} 2 & 0 \\ -3 & 2 \end{bmatrix}$

7. −1 **9.** −2 **11.** 0 **13.** 29 **15.** −145 **17.** 0 **19.** 27 **21** 0
23. −19 **25.** 48 **27.** 0 **29.** 0 **31.** 3 **33.** 8

Exercise Set 9.6

1. (3/25, −4/25) **3.** (3, −1) **5.** (2, 2) **7.** (−2/3, 1/3)

9. $x = 0, y = 1, z = 2$ **11.** $x = 1, y = 1, z = 1$ **13.** $x = 3, y = -3, z = 1$

15. $x = -1, y = 1, z = 0$ **17.** $4abc$ **19.** $ab(a + b - c)$

21. $(a - b)(b - c)(c - a)$ **23.** $x = 1, y = 1$ **25.** $x = 1, y = 1$

27. $x = 1/a, y = 1/b, z = 1/c$

29. $x = (d - b)(c - d)/(a - b)(c - a), y = (a - d)(d - c)/(a - b)(b - c), z = (b - d)(d - a)/(b - c)(c - a)$

31. $x = 1, y = 0, z = 0$

Exercise Set 9.7

1. The speed of the boat in still water is 4 mph; the speed of the water is 1 mph.

3. The speed of the plane in still air is 115 kilometers per hour; the speed of the wind is 15 kilometers per hour.

5. The original speed was 50 mph; the original time was 6 hours.

7. The larger pipe fills the tank in 30 minutes; the smaller pipe fills it in 60 minutes.

9. The sides are 4 meters and 12 meters.

11. The sides are 4 kilometers and 2.5 kilometers.

13. The sides are 6 meters and 8 meters. **15.** The length is 2 and the width is 1.5.

17. The length is 12 and the width is 5.

19.

VERTICES AT
(12, 0)
(6, 6)
(0, 15)

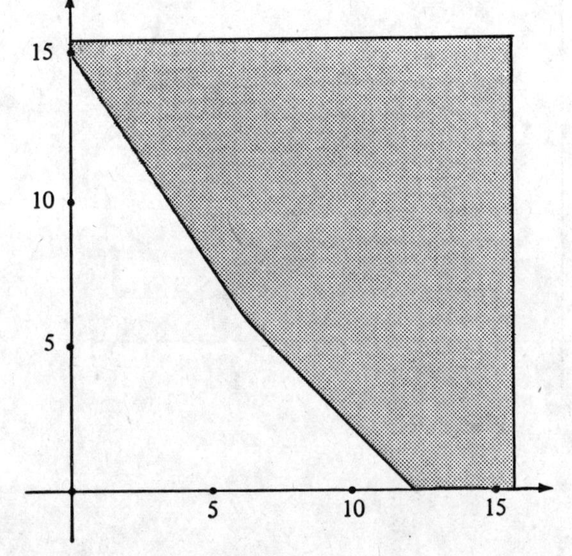

21.

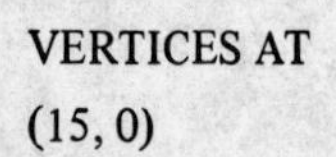

VERTICES AT
(15, 0)
(6, 6)
(0, 12)

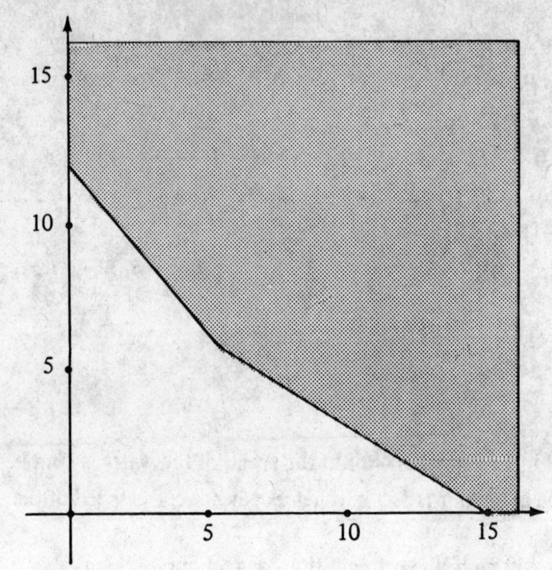

23.

VERTICES AT
(15, 0)
(12, 0)
(6, 6)
(30/7, 60/7)

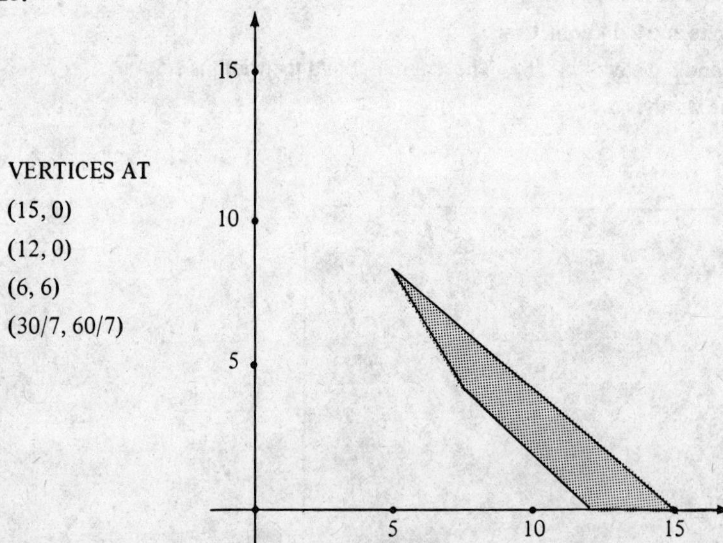

25.

VERTICES AT
(15, 0)
(6, 6)
(0, 12)

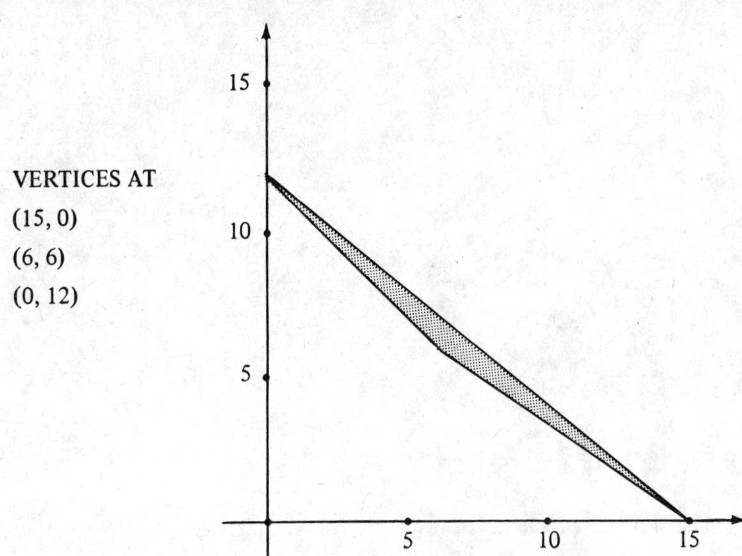

27. We mix 10 liters of the 40% solution with 4 liters of the 75% solution.

29. We mix 2.5 liters of the 12% solution with 7.5 liters of the 28% solution.

31. 5 liters of each **33.** 10 liters of each **35.** The conditions specified are inconsistent.

37. The first man has 12 denarii, the second man has 9 denarii, the third man has 6 denarii, and the fourth man has 16 denarii.

39. 8 cubits and 10 cubits

Review Exercises for Chapter 9

1. (3, −2) **3.** (−2, 2) **5.** (−2, −4) **7.** Inconsistent **9.** (11/12, 1/4)

11.

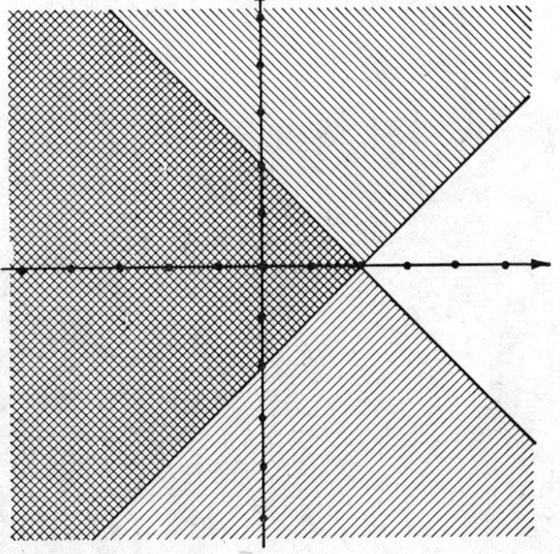

13.

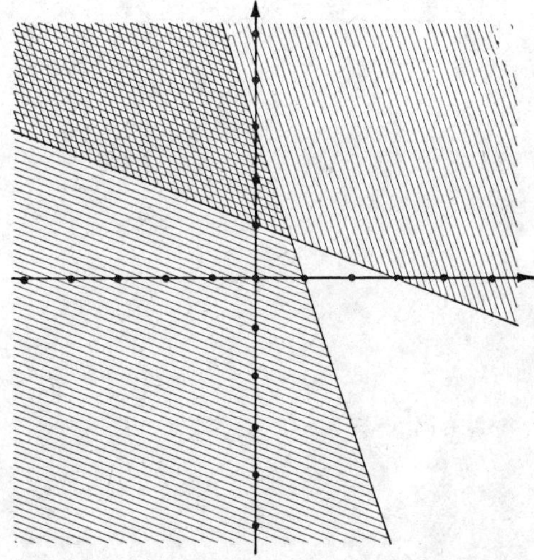

15.

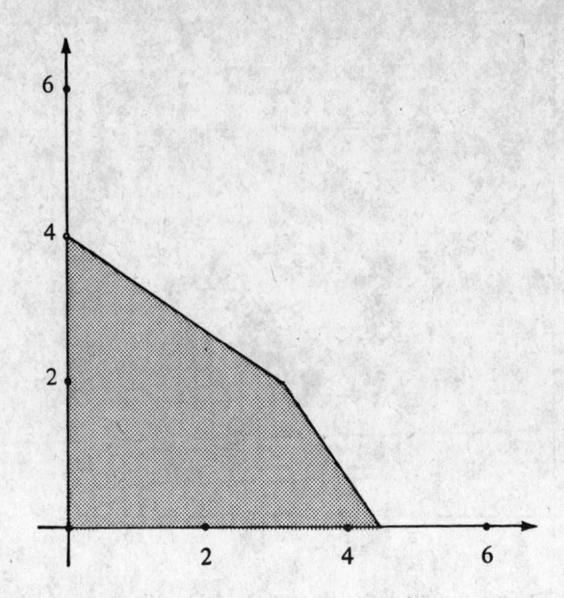

17. (2, 8) and (−5, 15) **19.** (1, 1) **21.** (1, 2), (−1, −2), (2, 1), and (−2, −1)

23. (2, 3) and (3, 2) **25.** (0, −2) and (2, 0) **27.** $x = 3, y = 2, z = 1$

29. $x = 1, y = -1, z = 1$ **31.** Dependent; $y = -x$ and $z = 1$ **33.** −2 **35.** 0

37. −20 **39.** 0

41. The speed of the boat is 6.5 kilometers per hour; the speed of the water is 1.5 kilometers per hour

43. 8 liters of the 10% solution with 12 liters of the 60% solution

Index